Cognitive Sensors, Volume 1

Intelligent sensing, sensor data analysis and applications

Online at: https://doi.org/10.1088/978-0-7503-5326-7

IOP Series in Sensors and Sensor Systems

The IOP Series in Sensors and Sensor Systems includes books on all aspects of the science and technology of sensors and sensor systems. Spanning fundamentals, fabrication, applications, and processing, the series aims to provide a library for instrument and measurement scientists, engineers, and technologists in universities and industry.

The series seeks (but is not restricted to) publications on the following topics:

- Advanced materials for sensing
- Biosensors
- Chemical sensors
- Industrial applications
- The Internet of Things (IoT)
- Lab-on-a-chip
- Localization and object tracking
- Manufacturing and packaging
- Mechanisms, modelling and simulations
- Microelectromechanical systems/nanoelectromechanical systems
- Micro- and nanosensors
- Non-destructive testing
- Optoelectronic and photonic sensors
- Optomechanical sensors
- Physical sensors
- Remote sensors
- Sensing for health, safety, and security
- Sensing principles
- Sensing systems
- Sensor arrays
- Sensor devices
- Sensor networks
- Sensor technology and applications
- Signal processing and data analysis
- Smart sensors and monitoring
- Telemetry

Authors are encouraged to take advantage of electronic publication through the use of colour, animations, video, and interactive elements to enhance the reader experience.

A full list of titles published in this series can be found here: https://iopscience.iop.org/bookListInfo/iop-series-in-sensors-and-sensor-systems.

Cognitive Sensors, Volume 1

Intelligent sensing, sensor data analysis and applications

Edited by
G R Sinha
International Institute of Information Technology Bangalore, Bangalore, India

Varun Bajaj
Indian Institute of Information Technology Design and Manufacturing, Jabalpur, India

IOP Publishing, Bristol, UK

ISBN 978-0-7503-5326-7 (ebook)
ISBN 978-0-7503-5324-3 (print)
ISBN 978-0-7503-5327-4 (myPrint)
ISBN 978-0-7503-5325-0 (mobi)

DOI 10.1088/978-0-7503-5326-7

Version: 20221201

IOP ebooks

British Library Cataloguing-in-Publication Data: A catalogue record for this book is available from the British Library.

Published by IOP Publishing, wholly owned by The Institute of Physics, London

IOP Publishing, No.2 The Distillery, Glassfields, Avon Street, Bristol, BS2 0GR, UK

US Office: IOP Publishing, Inc., 190 North Independence Mall West, Suite 601, Philadelphia, PA 19106, USA

Dedicated to my late grandparents, my parents, my teachers and Revered Swami Vivekananda

—G R Sinha

Dedicated to the late Mahendra Bajaj (my father) and family members

—Varun Bajaj

Contents

Preface xiii

Acknowledgements xv

Editor biographies xvi

List of contributors xviii

1 Introduction to the cognitive Internet of Things 1-1
Chandramouleeswaran Sankaran (Mouli)

1.1 Introduction 1-1
1.2 From the IoT to the CIoT 1-4
1.2.1 An overview 1-5
1.2.2 Conglomerates of technologies 1-5
1.2.3 The principles behind the CIoT 1-10
1.2.4 The architecture and layers of the CIoT 1-11
1.2.5 The need for massive data analytics in the CIoT 1-14
1.2.6 Knowledge discovery in the CIoT 1-14
1.2.7 Intelligent decision-making in the CIoT 1-16
1.2.8 Protocols in the CIoT 1-16
1.2.9 The framework of intelligent decision-making in the CIoT 1-17
1.3 The changing landscape due to the CIoT 1-18
1.4 Smart city applications using CIoT 1-19
1.5 Challenges in the CIoT 1-21
1.6 Advantages of the CIoT 1-21
1.7 Conclusions 1-22
Acknowledgments 1-23
References 1-23

2 Internet of Things-based cognitive wireless sensor networks: applications, merits, and demerits 2-1
Tanu Rizvi, Ravi Mishra, Priyanka Nandkishor Chopkar, Anupama Gomkale, Prajakta Upadhye and Devanand Bhonsle

2.1 Introduction 2-1
2.1.1 The Internet of Things 2-1
2.1.2 Wireless sensor networks 2-2
2.1.3 WSNs versus the IoT 2-3
2.1.4 CWSNs 2-5

2.2	Literature survey	2-9
2.3	Methodology	2-11
	2.3.1 Applications of WSN technologies	2-11
	2.3.2 CWSN countermeasures	2-13
2.4	Results and discussion	2-14
2.5	Conclusions and future scope	2-17
	References	2-17
3	**Psychiatric disorders and cognitive impairment following COVID-19: a comprehensive review and its implications for smart healthcare design** *Tayebe Azimi, Amirhossein Koochekian, Hamid Reza Marateb, Mehdi Shirzadi, Mónica Rojas-Martínez, Joan F Alonso, Alejandro Bachiller Matarranz, Marjan Mansourian, Manuel Rubio Rivas, Sergio Romero Lafuente and Miquel Angel Mañanas*	**3-1**
3.1	Introduction	3-1
3.2	Psychiatric disorders and post-COVID symptoms	3-3
	3.2.1 Potential tools for analyzing the psychiatric outcomes of post-COVID patients	3-3
3.3	Cognitive impairment and post-COVID symptoms	3-5
	3.3.1 Potential tools for analyzing the cognitive outcomes of post-COVID patients	3-5
3.4	Comprehensive literature review	3-7
3.5	Smart cities and post-COVID symptoms	3-17
3.6	Conclusions and future scope	3-18
	Acknowledgments	3-18
	References	3-18
4	**The use of the cognitive Internet of Things for smart sensing applications** *Dayarnab Baidya, Ruchee Bhagwat and Mitradip Bhattacharjee*	**4-1**
4.1	Introduction	4-1
4.2	The IoT and the advent of the CIoT	4-2
	4.2.1 The Internet of Things	4-2
	4.2.2 Limitations of the IoT	4-3
	4.2.3 The cognitive Internet of Things	4-4
	4.2.4 The CIoT architecture	4-5
4.3	The roles of big data and cognitive computing in the CIoT	4-7

4.4 Cognitive radio and its applications in IoT 4-10
4.5 Applications of the CIoT 4-12
4.5.1 Smart transport 4-12
4.5.2 Industry 4-13
4.5.3 The military 4-14
4.5.4 Smart cities 4-15
4.5.5 Smart healthcare 4-18
4.5.6 Smart homes 4-21
4.6 Conclusions 4-22
References 4-23

5 Design challenges and issues in cognitive sensor networks: a mitigation and deployment perspective **5-1**

Chandramouleeswaran Sankaran (Mouli)

5.1 Introduction 5-1
5.2 Wireless sensor networks 5-2
5.2.1 Sensor network architecture 5-4
5.2.2 Congestion management with common planes 5-7
5.2.3 The need for cognitive abilities 5-7
5.2.4 Summary 5-8
5.3 The knowledge plane in cognitive networks 5-9
5.4 Cognitive networks 5-10
5.5 Cognitive radio in wireless sensor networks 5-11
5.6 Areas of application of cognitive radio wireless sensor networks 5-13
5.6.1 Military and public security applications 5-13
5.6.2 Security threats in wireless sensor networks 5-14
5.6.3 Healthcare 5-16
5.6.4 Home appliances and indoor applications 5-17
5.6.5 Real-time surveillance applications 5-17
5.6.6 Transportation and vehicular networks 5-18
5.6.7 Multipurpose sensing 5-18
5.7 Challenges in cognitive radio wireless sensor networks 5-18
5.7.1 False alarms and misdetection in cognitive radio wireless sensor networks 5-18
5.7.2 Complex hardware design 5-19
Acknowledgments 5-19
References 5-20

6 Cognitive wireless sensor networks **6-1**
P Saranya, N Jeevitha and V Senthil Kumar

6.1 Introduction to wireless sensor networks 6-1
6.1.1 Defining wireless sensor networks 6-2
6.1.2 Utilizing networks of wirelessly connected sensors 6-3
6.1.3 Constraints of networks in wireless sensor systems 6-4
6.2 An introduction to cognitive radio networks 6-6
6.2.1 Purposes of cognitive radio 6-8
6.3 Wi-Fi sensor network integration with cognitive radio 6-9
6.4 The structure of a cognitive wireless sensor network 6-10
6.5 Spectrum-sensing device approaches in cognitive wireless sensor networks 6-15
6.5.1 Non-cooperative system sensing 6-16
6.5.2 System-wide cooperative sensing 6-19
6.5.3 The interference-based sensor technique 6-19
6.6 Implementing cognitive wireless sensor networks 6-20
6.7 Spectral optimization and new technology spaces 6-21
6.8 Applications and issues in cognitive wireless sensor networks 6-22
6.8.1 Public safety and military applications 6-22
6.8.2 Healthcare 6-23
6.8.3 Bandwidth-intensive applications 6-24
6.8.4 Transport and automobile networks 6-24
6.8.5 Virtual surveillance applications 6-24
6.9 Summary 6-24
Acknowledgments 6-25
References 6-25

7 Applications and challenges of IoT-based smart healthcare systems that use cognitive sensors: an overview **7-1**
Devanand Bhonsle, Yogiraj Bhale, Anu G Pillai, Shruti Tiwari, Vishal Moyal and Chih-Peng Fan

7.1 Introduction 7-2
7.2 Types of sensor 7-4
7.2.1 Sensors classified according to type 7-4
7.2.2 Sensors classified according to application 7-4
7.2.3 Sensors classified according to sensor placement 7-4
7.3 Smart healthcare using cognitive sensors 7-5
7.4 Services 7-6

7.5 Applications of the IoT in healthcare 7-7
7.6 Technologies used in IoT-based healthcare systems 7-10
7.7 Challenges in IoT-based healthcare systems 7-14
7.8 Security measures 7-15
7.8.1 Technical security measures 7-15
7.8.2 Non-technical security measures 7-17
7.9 Experimental prototype of a smart healthcare system using the IoT 7-18
7.10 Conclusions 7-19
References 7-20

8 Redundancy issues in wireless sensor networks 8-1
Veena I Puranikmath, Sridhar Iyer and Rahul Pandya

8.1 Introduction 8-1
8.1.1 Spatial redundancy 8-5
8.1.2 Temporal redundancy 8-6
8.1.3 Information redundancy 8-7
8.2 Related work 8-7
8.3 Proposed algorithm 8-9
8.4 Quality factors 8-9
8.5 Research avenues 8-17
8.6 Conclusions 8-17
Acknowledgments 8-18
References 8-18

9 Sensor-based devices and their applications in smart healthcare systems 9-1
Sapna Singh Kshatri, Devanand Bhonsle, Sourabh Tiwari, Ravi Mishra, Tanu Rizvi and Rimjhim Pandey

9.1 Introduction 9-1
9.2 Sensors used in medicine 9-3
9.2.1 Invasive sensors 9-3
9.2.2 Non-invasive sensors 9-4
9.3 Cognitive smart healthcare systems that use the Internet of Medical Things 9-9
9.4 The Internet of Medical Things and its applications 9-11
9.5 Conclusions 9-13
References 9-13

10 Electromagnetic fields and the effects of Internet of Things infrastructure on human health **10-1**

Hitesh Singh, Vivek Kumar, Kumud Saxena, K V Arya and Ramjee Prasad

10.1 Introduction 10-1
10.2 EMF exposure studies 10-2
10.3 EMF guidelines 10-5
10.4 The effects of EMF exposure on human health 10-6
10.5 Experimental setup and results 10-10
10.6 Conclusions 10-12
References 10-13

11 Cognitive sensing in the brain–computer interface: a comprehensive study **11-1**

Dhairya Shah, Haard Shah and Manan Shah

11.1 Introduction 11-1
11.2 Classification of advanced brain–computer interface technologies 11-3
11.2.1 Microelectrodes 11-4
11.2.2 Semiconductors 11-4
11.2.3 Polymer probes 11-5
11.3 The role of cognition in brain–computer interfaces 11-5
11.4 General architecture 11-6
11.4.1 Data acquisition 11-7
11.4.2 Preprocessing 11-7
11.4.3 Feature extraction 11-7
11.4.4 Feature selection 11-7
11.4.5 Motor imagery algorithms 11-8
11.5 Comparative analysis 11-9
11.5.1 Related work 11-10
11.6 Exciting research in the brain–computer interface field 11-12
11.7 Challenges and future scope 11-14
11.8 Conclusions 11-15
References 11-16

Preface

Cognitive sensors have become essential parts of a number of Internet of Things (IoT) devices and modern gadgets. In fact, the sensors which behave and act smartly in all modern applications are cognitive sensors. Sensors that have the ability to understand and interpret the data they sense, with the help of signal processing and machine learning techniques, are known as cognitive sensors. Cognitive sensors are heavily used in healthcare applications, smart city smart transport, smart manufacturing, smart automobiles, and a number of other such applications in which automation is achieved using sensor-based data and the analysis of sensor-based data using artificial intelligence (AI)-based approaches.

In the modern world, computer vision is dominating all advances and their applications in science, engineering, and technology; for example, computer-aided diagnosis, industrial robotics, driverless cars, the brain–computer interface (BCI), human-computer interaction (HCI), telemedicine, etc. are emerging areas in which computer vision plays a significant role in implementing, simulating, and improving their performance. Cognitive Science is actually a multidisciplinary area of research that mainly covers artificial intelligence, neuroscience, mathematics, philosophy, psychology, linguistics, which are the building blocks of humanoid robots, automated diagnosis, self-driving cars, etc. Cognitive sensors are most important components of cognitive science research and studies. For example, brain images such as those produced using electroencephalography (EEG) dominate among a number of other brain signal modalities; they are used in the analysis and study of various types of brain disorder and disability. In the computing part of cognitive science related to physics and images, a number of signal processing and soft computing techniques are used to make images for suitable applications. Cognitive sensors are basically smart and intelligent sensors that have the ability to exploit cognitive aspects of science and engineering, namely understanding, decision-making, improving, adapting to requirements, computing, etc.

In this volume of the book, we focus on smart sensors, the data captured by these sensors and its analysis, an overview of the different types of cognitive sensor, their characteristics, uses, and their potential applications. The readers of this book will gain an understanding of cognitive sensors, their fundamentals, their scope, and the possible applications in which cognitive sensors can play a significant role in smart sensing and decision-making. This volume takes 11 chapters to cover the complete scope of the book; a chapter-by-chapter description of this book as follows:

Chapter 1 presents and insight into various types of application areas in which cognitive IoT devices provide enhanced user experiences and smart features that are critical and essential to making human life safe and secure. Chapter 2 presents various applications and advantages of various sensors used in today's scenarios; for example, cognitive wireless sensor networks (CWSNs) are widely used for various military purposes, public security, healthcare, home appliances, real-time surveillance, transportation networks, etc. Chapter 3 discusses the role of sensors in cases of moderate COVID-19 infection that may be linked to cognitive impairments; such prediction systems are promising for smart healthcare design. Chapter 4 presents various facets of cognitive IoT devices used in smart sensing, including their architectures and applications in different fields such as healthcare, the military, intelligent home applications, and cognitive radio. Chapter 5 provides an overview of wireless sensor network (WSN) design challenges and additional design issues that need to be addressed in making them cognitive WSNs. Chapter 6 highlights CWSNs and discusses enhancing the communication flow range, diminishing the sensor node as required to cover a definite neighborhood, and working toward reduced power utilization. Chapter 7 presents the use of cognitive IoT devices for healthcare that can take intelligent decisions based on the physical state of the patient. Chapter 8 discusses various types of redundancy and presents a survey of the most recent studies which have aimed to minimize the effect of redundancy on WSN performance. Chapter 9 discusses the roles and capabilities of sensors, communication networks, and cloud storage in healthcare systems. Chapter 10 focuses on the important issue of interference and elucidates the different EMF radiation limit guidelines applicable worldwide. It also describes an experimental study that has been conducted in order to discover the effects of EMF in a college premises. Chapter 11 analyses the advances in the field of cognitive sensing using the BCI and provides the reader with a very insightful take on the topic to foster more research in this direction.

The structure of the book is arranged in such a way that the reader will easily understand the different topics covered in different chapters. Thus, the basic understanding of the fundamentals is discussed in the early chapters, which provide an overview, a necessary discussion, and a description of the possible applications. This is followed by the major challenges facing the implementation strategies used for WSNs; finally, cognitive sensing and its utilization in smart applications are discussed. We believe that the reader will enjoy learning about cognitive sensors using a different, case-study-based approach.

Acknowledgements

Dr Sinha expresses his sincere thanks to his family members, his wife Shubhra and daughter Samprati, parents, teachers, and colleagues for their unconditional love and support for his continuing literary work on his areas of interest.

Dr Bajaj also expresses his heartfelt appreciation to his mother Prabha, his wife Anuja, and his daughter Avadhi for their wonderful support and encouragement throughout the completion of this important book, Cognitive Sensors, Volume 1 with a focus on Intelligent Sensing, Sensor Data Analysis and Applications. His deepest gratitude goes to his mother-in-law and father-in-law for their constant motivation. This book is an outcome of sincere efforts that could only have been devoted to it with the great support of the family. He also thanks Prof. P N Kondekar, Director of Pandit Dwarka Prasad Mishra (PDPM) Indian Institute of Information Technology, Design and Manufacturing (IIITDM), Jabalpur for his support and encouragement. Dr Bajaj especially thanks his family, who encouraged him throughout the time spent editing this book. This book is wholeheartedly dedicated to his late father who has been his great source of motivation.

We would like to thank all our friends, well-wishers, and all those who keep us motivated to do more and more, better and better. We sincerely thank all our contributors for writing about the relevant theoretical backgrounds and real-time applications of cognitive sensors. We are also deeply thankful to many whose names are not mentioned here, but we wish to acknowledge and appreciate their help during this work.

We express our humble thanks to Dr John Navas, Senior Commissioning Editor of IOP Publications for his great support, necessary help, appreciation, and quick responses. We also wish to thank IOP Publishing for giving us this opportunity to contribute to a relevant topic via a reputed publisher. Finally, we want to thank everyone who helped us in one way or another while we were editing this book.

Last but not least, we would also like to thank God for showering his blessings and strength on us so that we can do this type of novel and quality work.

G R Sinha and Varun Bajaj

Editor biographies

G R Sinha

G R Sinha (PhD, FIETE, FIE) is Adjunct Professor at the International Institute of Information Technology Bangalore (IIITB), India. Prior to his role at the IIITB, he worked as a professor at the IIITB-mentored Myanmar Institute of Information Technology (MIIT), Mandalay, Myanmar. He has been Visiting Professor (Online) at the National Chung Hsing University, Taiwan and the University of Sannio, Italy and Visiting Professor (Honorary) at the Sri Lanka Technological Campus, Colombo. He has published 310 research papers, book chapters, and books at international and national levels and edited 20 books in the fields of cognitive science, biomedical signal processing, biometrics, optimization techniques, sensors, outcome-based education (OBE), and data deduplication with internationally reputed publishers such as Elsevier, Institute of Physics Publishing (IOPP), Springer, Taylor and Francis, and IGI Global. He owns two Australian patents. He is an associate editor of five Science Citation Index (SCI)/Scopus indexed journals and has been a guest editor at various SCI journals. Dr Sinha has been an Association for Computing Machinery (ACM) distinguished speaker in the field of digital signal processing (DSP) (2017–21). He has been an expert member of the Vocational Training Program of the Tata Institute of Social Sciences (TISS) for two years. He has been a contributing Cancer Science Institute (CSI) distinguished speaker in the field of image processing since 2015. He has also served as a distinguished IEEE lecturer on the IEEE India's Council for Bombay section. He has received more than 12 national and international level awards and recognitions. He has delivered more than 60 keynote/invited talks and chaired many technical sessions in international conferences across the world. He has been vice president of the Bhilai chapter of the Computer Society of India for two years. He is a regular referee of project grants under the Department of Science and Technology Extramural Research (DST-EMR) scheme of the Government of India. He has been an expert member of the Professor Promotion Committee of the German Jordanian University, Jordan and the Project Proposal Evaluation Committee of UK–Israel Research Grants. Dr Sinha has supervised eight PhD scholars, 15 M Tech scholars, and 100 undergraduate-level students and is supervising one more PhD scholar. His research interests include biometrics, medical/biomedical image processing and cognitive science applications, computer vision, OBE, and the assessment of student learning outcomes.

Varun Bajaj

Varun Bajaj (PhD, SMIEEE20) has been working as an associate professor in the discipline of electronics and communication engineering, at the IIITDM Jabalpur, India since July 2021. He worked as an assistant professor at IIITDM Jabalpur from March 2014 to July 2021. He also worked as visiting faculty at IIITDM Jabalpur from September 2013 to March 2014. He worked as an assistant professor at the Department of Electronics and Instrumentation, Shri Vaishnav Institute of Technology and Science, Indore, India from 2009 to 2010. He received his PhD degree in electrical engineering at the Indian Institute of Technology Indore, India in 2014. He received an MTech degree with honors in microelectronics and very large-scale integration (VLSI) design from the Shri Govindram Seksaria Institute of Technology and Science, Indore, India in 2009 and a BE degree in electronics and communication engineering from the Rajiv Gandhi Technological University, Bhopal, India in 2006. He is an associate editor of the IEEE Sensor Journal and a subject editor-in-chief of the Institution of Engineering and Technology (IET) Electronics Letters. He served as a subject editor of IET Electronics Letters from November 2018 to June 2020. He has been a senior member of the Institute of Electrical and Electronics Engineers (IEEE) since June 2020 and was a member of the IEEE (MIEEE) from 2016 to 2020; he has also contributed as an active technical reviewer of leading international journals published by the IEEE, IET, Elsevier, etc. He has 145 publications, which include 93 journal papers, 31 conference papers, 10 books, and 11 book chapters. The citation impact of his publications is around 4277 citations; they have an *h* index of 36, and an i10 index of 87 (Google Scholar, December 2021). He has guided seven (five completed and three in process) PhD scholars and eight MTech scholars. He has been listed in the world's top 2% of researchers/scientists by Stanford University, USA (October 2020 and October 2021). He has worked on research projects funded by the Department of Science and Technology (DST) and the Council of Scientific and Industrial Research (CSIR). He is the recipient of various reputed national and international awards. His research interests include biomedical signal processing, AI in healthcare, the BCI, pattern recognition, and electrocardiogram (ECG) signal processing.

List of contributors

Joan F Alonso

Biomedical Engineering Research Centre (CREB), Automatic Control Department (ESAII), Universitat Politècnica de Catalunya-Barcelona Tech (UPC), Spain

Joan Francesc Alonso, PhD belongs to the Biosignal Analysis for Rehabilitation and Therapy research group (BIOART, bioart.upc.edu) of the Universitat Politècnica de Catalunya (UPC, upc.edu). He defended his doctoral thesis in 2011, achieving an Excellent qualification and a Cum Laude distinction and was awarded the Extraordinary Prize by the UPC in the field of Industrial Engineering. He has worked on different projects as a postdoctoral researcher, while at the same time teaching as a part-time associate professor. He currently holds a full-time tenure-eligible lecturer position (Serra Húnter) and teaches subjects related to programming, control, and biomedical signal and image processing at the Barcelona East School of Engineering (EEBE, eebe.upc.edu). His research activities include non-invasive signal acquisition and processing, especially that of muscular (electromyography (EMG), mechanomyography (MMG)) and cerebral origin (EEG, magnetoencephalography (MEG)), for the assessment of different pathologies, drugs, and therapies, with a focus on the analysis of interactions between classical and other nonlinear and nonparametric measures derived from information theory. He has been an active member of more than 20 research projects (and the principal investigator two of them) in the field of biomedical engineering, funded by competitive public calls and private companies, and he is the co-author of more than 50 scientific publications, more than half of them in recognized international indexed journals.

K V Arya

ABV-Indian Institute of Information Technology & Management, India

Dr Karm Veer Arya has received his BSc from Rohilkhand University, Bareilly, India, in 1986. He completed his ME and PhD at the Indian Institute of Science (IISC) Bangalore and the Indian Institute of Technology (IIT), Kanpur in 1991 and 2007, respectively. He is currently working as professor and head of the Department of Computer Science and Engineering at Atal Bihari Vajpayee (ABV) Indian Institute of Information Technology and Management, Gwalior, India. He has also served as the dean of postgraduate studies and research at the Dr APJ Abdul Kalam Technical University, Lucknow, India. He has had more than 200 research papers published in reputed refereed international journals and conferences as well as five patents. He has supervised 11 PhD scholars and 98 master's students and handled many R&D projects. His research interests include image processing, biometrics, wireless networks, and reliability analysis. He is a senior member of the IEEE, a fellow of the Institution of Electronics and

Telecommunications Engineers (IETE), a fellow of the Institution of Engineers (India), a member of the ACM, and a life member of the Indian Society for Technical Education (ISTE).

Tayebe Azimi

Biomedical Engineering Department, Engineering Faculty, University of Isfahan, Iran

Tayebe Azimi received her BS degree in biomedical engineering from the Hamadan University of Technology, Hamadan, Iran, in 2015. From 2016 to 2017, she was a biomedical engineer at a hospital under the supervision of Shiraz University of Medical Sciences, Shiraz, Iran. In 2021, she completed her master's degree in Biomedical Engineering at the University of Isfahan, Iran. Her master's thesis addressed cardiovascular risk assessment and COVID-19 mortality risk assessment using machine learning methods. Her research interests include data mining, artificial intelligence, and pattern recognition.

Dayarnab Baidya

i Lab, Electrical Engineering and Computer Science, Indian Institute of Science Education and Research, Bhopal, India

Mr Dayarnab Baidya received his BTech degree in electronics and instrumentation engineering from the National Institute of Technology (NIT), Agartala, India, in 2017. He is currently pursuing his PhD at i Lab within the Electrical Engineering and Computer Science department at the Indian Institute of Science Education and Research (IISER), Bhopal, India. His research interests include decision-making, machine learning, flexible electronics, control systems, optimization, the IoT, sensors, and systems. Mr Baidya has authored four journal papers, three conference papers and one book chapter. He is a student member of the IEEE and the IEEE Control Systems Society (CSS) and a member of the IEEE Sensor Council, the IEEE System Council, the IEEE Nanotechnology Council, the IEEE Council on Electronic Design Automation, the IEEE Council on RFID, the IEEE Biometrics Council and the International Society on Multiple Criteria Decision Making.

Ruchee Bhagwat

Vellore Institute of Technology, India

Ms Ruchee Bhagwat, received her BSc degree in Physics from D G Ruparel College of Arts, Science, and Commerce, affiliated with Mumbai University, India, in 2021. She is currently pursuing her MSc in Physics at the Vellore Institute of Technology (VIT), India. Her research interests include sensors and systems, the IoT, flexible electronics, and nanomaterials and their applications.

Yogiraj Bhale

Chandigarh University, India

Yogiraj Bhale received his BE from Pt. Ravishankar Shukla University, Raipur and an ME from Chhattisgarh Swami Vivekanand Technical University (CSVTU), Bhilai in Computer Science and Engineering in 2007 and 2015, respectively. He has been working as a senior assistant professor at the Apex Institute of Technology, Computer Science and Engineering (AIT-CSE), Chandigarh University since February 2021. He has 13 years' experience.

Mitradip Bhattacharjee

i Lab, Electrical Engineering and Computer Science, Indian Institute of Science Education And Research, Bhopal, India

Dr Mitradip Bhattacharjee received his PhD degree from the Indian Institute of Technology, Guwahati, India in 2018. He subsequently joined the University of Glasgow, UK, as a postdoctoral fellow in January 2019. He is currently an assistant professor at the Electrical Engineering and Computer Science Department at the Indian Institute of Science Education and Research, Bhopal, India where he leads the i-Lab research group. His research interests include electronic sensors and systems, biomedical engineering, bioelectronics, flexible/printed and wearable electronics, wireless systems, and reconfigurable sensing antennae. Dr Bhattacharjee has authored more than 48 research articles in reputed journals/conferences and filed more than 15 national/international patents. He has also authored several book chapters/books. He is a member of the IEEE and is the YP chair of the IEEE Sensors Council. He is also an organizer of various reputed international conferences.

Devanand Bhonsle

Chhattisgarh Swami Vivekanand Technical University, Bhilai, India

Dr Devanand Bhonsle, PhD received his BE from Pt. RSSU, Raipur and ME and PhD degrees from CSVTU, Bhilai in electronics and telecommunication in 2004, 2008, and 2019, respectively. He has been working as a senior assistant professor at the Faculty of Engineering, Shri Shankaracharya Technical Campus, Bhilai since July 2005. He has 17 years' experience. His research interests include signal processing and image processing. He has authored or co-authored 45 publications in various peer-reviewed journals. He has written four book chapters and one book and presented 15 papers at national/international conferences. He is a professional member of the Institute for Engineering Research and Publication (IFERP).

Priyanka Nandkishor Chopkar

Rashtrasant Tukadoji Maharaj Nagpur University, Nagpur, India

Ms Priyanka Chopkar received her BE in electronics and telecommunication (ETC) from Dr Babasaheb Ambedkar Marathwada University (BAMU), Aurangabad in 2009 and an MTech in electronics from Rashtrasant Tukadoji Maharaj (RTM) Nagpur University in 2013. She has been working as an assistant professor at the G H Raisoni Institute of Engineering and Technology (GHRIET), Nagpur for four years and nine months. Her research interests are network on chip, Very High-Speed Integrated Circuit Hardware Description Language (VHDL), and communications. She has published six research papers in various journals. She is a professional member of the ISTE.

Chih-Peng Fan

National Chung Hsing University, Taiwan

Dr Chih-Peng Fan works as a professor at the Department of Electrical Engineering at National Chung Hsing University. His research area is very large-scale integration (VLSI) design and system simulations of digital baseband transceivers, VLSI designs for digital image/video signal processing, and the fast prototyping of digital signal processing (DSP) systems using field-programmable gate arrays (FPGAs) or embedded system on a chip (SOC) platforms.

Anupama Gomkale

Rashtrasant Tukadoji Maharaj Nagpur University, Nagpur, India

Ms Anupama Gomkale received a BE in Electronics from Ramdeobaba Kamla Neharu Engineering Collage Nagpur (RTMNU) in 2007 and an ME in wireless communications and computing from G H Raisoni Collage of Engineering for Women (RTMNU) in 2015. More recently, she has been working as an assistant professor at GHRIET and has seven years' teaching experience.

Sridhar Iyer

KLE Technological University, MSSCET, India

Sridhar Iyer (a member of the IEEE) received an MS degree in electrical engineering from New Mexico State University, USA in 2008 and a PhD degree from Delhi University, India in 2017. He received the young scientist award from the Department of Science and Technology (DST)/Science and Engineering Research Board (SERB), Government of India in 2013 and the Young Researcher Award from Institute of Scholars in 2021. He was a recipient of the

'Protsahan Award' from the IEEE ComSoc, Bangalore as recognition of his contributions towards papers published and tutorials offered in recognized conferences/journals during January 2020–September 2021. He has completed two funded research projects as the principal investigator and is currently involved in an ongoing funded research project as the principal investigator. He serves on the review panel of high-impact journals such as IEEE, Elsevier, Springer, etc. His current research focus includes semantic communications and spectrum enhancement techniques for 6G wireless networks and efficient design and resource optimization of the flexi-grid eventcentric organizing network (EONs) enabled by space-division multiplexing (SDM). He has published over 90 reviewed articles in the aforementioned areas. He currently serves as an associate professor in the Department of Electronics and Communication Engineering (ECE), KLE Dr. M.S. Sheshgiri College of Engineering and Technology, Belagavi, Karnataka, India.

N Jeevitha

JNTUA University, India

Ms N Jeevitha is a senior assistant professor at the Faculty of Electronics and Communication Engineering of the Vaagdevi Institute of technology and Science, Proddatur. She obtained her ME degree in VLSI Design at Bannari Amman Institute of Technology in 2010 and a BE in electronics and communication engineering (ECE) at K S Rangasamy College of Technology in 2008. She started her teaching career as a lecturer at the Vivekananda Institute of Engineering and Technology for Women, Tiruchengode. She has more than 12 years of experience in the teaching profession. Her specific research interests include signal processing, VLSI design, and networking. She has published more than seven refereed articles in a wide range of well known international journals and conferences.

Amirhossein Koochekian

Pediatrics Department, Child Growth and Development Research Center, Research Institute for Primordial Prevention of Non-Communicable Disease, Isfahan University of Medical Sciences, Iran

Amirhossein Koochekian is currently a BS student of Biomedical Engineering at the Islamic Azad University Central Tehran Branch. He is also affiliated with the Pediatrics Department of the Child Growth and Development Research Center at the Research Institute for Primordial Prevention of Non-Communicable Disease, Isfahan University of Medical Sciences. His research interests include medical data mining and biomedical signal processing.

Sapna Singh Kshatri

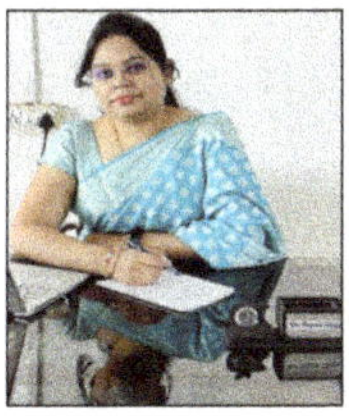

Chhattisgarh Swami Vivekanand Technical University, Bhilai, India

Dr Sapna Singh Kshatri has been awarded a PhD degree by MATS University, Raipur. She completed her MCA at Rungta College of Engineering and Technology, Bhilai, India. She has worked as an assistant professor at the Ashoka Institute of Technology and Management. She is the head of the Computer Science Department at Bharti University, Durg. She has recently been working as an assistant professor at Shri Shankaracharya Institute of Professional Management and Technology (SSIPMT). She has eight years of teaching and academic experience, has had 15 research articles, chapters, and national and international conference papers published together with an international book published by IGI Global, and has received Best Paper awards from international conferences. She is a reviewer for several journals. She has delivered five keynote/invited talks and chaired many technical sessions at international and national conferences. She has reviewed many papers for reputed publishers such as the IEEE, the IET, and Oxford University Press. Her research interests include biomedicine, machine learning, and data science.

Vivek Kumar

Noida Institute of Engineering and Technology, India

Dr Vivek Kumar is associated with the Noida Institute of Engineering Technology (NIET), Greater Noida, India as a professor in the IT department. He received his PhD degree in computer engineering and technology from Suresh Gyan Vihar University, Jaipur, Rajasthan, India in 2013 and has 20 years' experience in teaching and research. He has researched vendor selection through agents, fuzzy logic, and case-based reasoning. He has published papers in several peer-reviewed and indexed journals and presented papers at conferences of national and international repute. His area of research interest is machine learning. He has published patents and books in the fields of emerging technologies such as 6G. His current research interests focus on the role of machine learning (artificial neural networks and deep neural networks) in studying the rain attenuation of radio-wave propagation. His machine implementations are performed using the Google Colab—Python development environment.

Sergio Romero Lafuente

Biomedical Engineering Research Centre (CREB), Automatic Control Department (ESAII), Universitat Politècnica de Catalunya-Barcelona Tech (UPC), Building H, Floor 4, Av. Diagonal 647, 08028, Barcelona, Spain

Sergio Romero, PhD is an industrial engineer who has specialized in automatic control at the Universitat Politècnica de Catalunya

(UPC) since 2000. In 2010, he obtained his PhD in biomedical engineering from the UPC with a thesis entitled 'Artifact reduction in EEG signals by means of new automatic filtering techniques based on blind source separation.' In 2001, he joined UPC as a part-time assistant professor and as a research technician in the Center for Drug Research (CIM) of the Hospital de Sant Pau in Barcelona. In this role, he achieved experience in the management of clinical trials and multivariate statistical analysis, joining a large multidisciplinary team (medical doctors, pharmacists, nurses, and engineers). His current research focuses on the analysis of cerebral signals (EEG/MEG) to aid in the diagnosis of different neurodegenerative diseases and to improve and monitor treatments. He has published around 30 articles in Journal Citation Report (JCR) indexed journals and made many presentations at international and national conferences. Since 2000, he has participated in a large number of projects (around 25, two of them as principal investigator (PI)) grant funded by public and private entities. He currently belongs to the CREB of the UPC and the BIOART research group.

Miquel Angel Mañanas

Biomedical Engineering Research Centre (CREB), Automatic Control Department (ESAII), Universitat Politècnica de Catalunya-Barcelona Tech (UPC), Spain

Miquel Angel Mañanas, PhD received his telecommunications engineering and PhD in biomedical engineering (BME) degrees from the Universitat Politècnica de Catalunya (UPC) in 1993 and 1999, respectively. He is currently an associate professor at the ESAII at the same university, a member of the CREB at the UPC, and a member of the Bioengineering, Biomaterials and Nanomedicine Networking Biomedical Research Centre (CIBER-BBN). He is the founder and leader of the multidisciplinary BIOART, UPC, which is composed of engineers and physicians. His research is focused on applying engineering techniques to the field of health. His career has mainly focused on biomedical signal processing and biological system modeling in three fields: (1) neuromuscular: evaluating neuromuscular disorders and motor rehabilitation using high-density EMG; (2) neurological: evaluating brain activity by EEG and magnetoencephalography; (3) respiratory: modeling the respiratory control system and simulating pulmonary diseases and mechanical ventilation. He is the co-author of 65 JCR articles (31 in Q1 and 23 in Q2), has more than 100 peer-reviewed scientific communications, and co-invented two patents. He has co-supervised eight PhD theses. He has also been PI of 30 national/international projects. He received the Leonardo Award from the BBVA Foundation in 2016 in recognition of his career.

Marjan Mansourian

Biomedical Engineering Research Centre (CREB), Automatic Control Department (ESAII), Universitat Politècnica de Catalunya-Barcelona Tech (UPC), Spain

Marjan Mansourian, PhD is currently affiliated with the BIOART research group (UPC), the Marie Curie fellowship, the Biostatistics and Epidemiology Department, Faculty of Health, Isfahan University of Medical Sciences (Iran) as an associate professor, and also as the head of the Burden of Disease Department of Isfahan Cardiovascular research center, a WHO collaboration center. She received BS and MS degrees from the University of Isfahan and the Isfahan University of Medical Sciences, Isfahan, Iran, in 2003 and 2007, respectively. She received her PhD in biostatistics from Tarbiat Modares University, Tehran, Iran, in 2004. Her research interests include longitudinal data mining/modeling and survival analysis. She has authored numerous publications in indexed journals in the fields of data modeling and medical data mining.

Hamid Reza Marateb

Biomedical Engineering Research Centre (CREB), Automatic Control Department (ESAII), Universitat Politècnica de Catalunya-Barcelona Tech (UPC), Spain

Hamid R. Marateb, PhD received BS and MS degrees from the Shahid Beheshti University of Medical Science and Amirkabir University of Technology, Tehran, Iran, in 2000 and 2003, respectively. He received his PhD and a postdoctoral fellowship from the Laboratory of Engineering of Neuromuscular Systems, Politecnico di Torino, Turin, Italy, in 2011 and 2012, respectively. He was a visiting researcher at Stanford University in 2009 and at Aalborg University in 2010. He was a visiting professor at UPC, Barcelona, in 2012 and 2017. His research line is cognitive informatics in health and biomedicine, mainly focusing on clinical neurophysiology, computational neurosciences, and medical data mining. Dr Marateb is a reviewer for more than 30 International Scientific Indexing (ISI) journals and is now on the editorial boards of some international journals, including Frontiers in Integrative Physiology and Frontiers in Cognitive Neuroscience. He has also received European Union/Spanish grants (e.g. from the Marie Curie fellowship and MYOSLEEVE). He currently works for the Department of Biomedical Engineering, the Faculty of Engineering, the University of Isfahan, IRAN, and also for BIOART, ESAII, CREB, UPC, Barcelona, Spain.

Alejandro Bachiller Matarranz

Biomedical Engineering Research Centre (CREB), Automatic Control Department (ESAII), Universitat Politècnica de Catalunya-Barcelona Tech (UPC), Spain

Alejandro Bachiller Matarranz, PhD is a researcher at BIOART, UPC and a member of the CREB, UPC. He received his BSc and MSc degrees in telecommunications engineering and a PhD in biomedical engineering from the University of Valladolid (UVa) in 2011, 2012, and 2017, respectively. His doctoral thesis achieved a grade of Excellent and a Cum Laude distinction and was awarded an Extraordinary Prize by the UVa and declared to be the best thesis defended in the year 2017 by the Spanish Committee of Automatic Control (CEA), which belongs to the International Federation of Automatic Control (IFAC). Between 2011 and 2017, he belonged to Biomedical Research Group (GIB) of the University of Valladolid. He currently holds a full-time tenure-eligible lecturer position (Serra Húnter) at the UPC (School of Industrial, Aerospace, and Audiovisual Engineering of Terrassa (ESEIAT)). His main research interests include biomedical signal processing focused on pathologic and physiologic brain signals, the analysis of brain dynamics in terms of connectivity patterns, the localization of brain sources, and the evaluation of brain networks.

Ravi Mishra

Rashtrasant Tukadoji Maharaj Nagpur University, Nagpur, India

Dr Ravi Mishra received his BE in Electronics and Telecommunication (ETC) from Pt. RSSU, Raipur in 2002, his MTech (ETC) from the College of Engineering (COE), Pune in 2008, and his PhD (ETC) from Dr C V Raman University, Bilaspur in 2017. He has been working as an associate professor at GHRIET, Nagpur for three years and three months. His total experience is around 19 years. His research interests include video shot boundary detection, the design of IoT-based applications, and drone applications. He has authored/co-authored around 80 research papers in various reputed international publishers' journals/conferences, such as those published by the IEEE, Springer, and John Wiley & Sons, Inc. He has contributed two book chapters to reputed publications. He is a reviewer for IEEE Access, IET Image Processing, Electronics Letters, and IGI Global publications. He is a lifetime professional member of the ISTE.

Vishal Moyal

Dr. Babasaheb Ambedkar Technological University, Lonere, Raigad, India

Dr Vishal Moyal completed his PhD and ME in electronics and telecommunication engineering at CSVTU, Bhilai, Chhattisgarh and graduated in electronics and telecommunication engineering at Amravati University, Amravati. He completed a certificate course in VLSI design at the Department of Electronic Science, University of Pune. He is a senior member of the IEEE, a member of the Institution of Engineers (India) (IE(I)) and a life member of the Indian Society for Technical Education (MISTE). He has 18 years of teaching experience and currently works as an assistant professor at Shri Vile Parle Kelavani Mandal's (SVKM's) Institute of Technology, Dhule.

Rimjhim Pandey

Chhattisgarh Swami Vivekanand Technical University, Bhilai, India

Rimjhim Pandey is a final year undergraduate student pursuing a BTech in Electrical Engineering at Shri Shankaracharya Technical Campus, Bhilai. Her interest is reading books and learning about new technologies.

Rahul Pandya

Indian Institute of Technology, India

Rahul Jashvantbhai Pandya completed his MTech at the Electrical Engineering Department, Indian Institute of Technology, Delhi, New Delhi, in 2010. He completed his PhD at the Bharti School of Telecommunication, IIT, Delhi, in 2014. He worked as the senior network design engineer at the Optical Networking Industry, Infinera Pvt. Ltd, Bangalore, from 2014 to 2018. Later, from 2018 to 2020, he worked as an assistant professor at the ECE Department, National Institute of Technology, Warangal. He currently works at the Electrical Engineering Department of the Indian Institute of Technology, Dharwad. His research areas are wireless communication, optical communication, optical networks, computer networks, machine learning, and artificial intelligence. He is currently working on multiple projects, such as Science and Engineering Research Board (SERB), and Scheme for Promotion of Academic and Research Collaboration (SPARC), SGNF, and RSM.

Anu G Pillai

CSVTU, Bhilai, India

Anu G Pillai completed her Diploma in Electrical Engineering at the Government Polytechnic College, Durg (affliated to Rajiv Gandhi Prodhyogiki Vishwavidyalay, Bhopal, Madhya Pradesh) in 2004. She completed her BE in electrical engineering in 2007 at MPCCET Bhilai (Pt. RSSU, Raipur, Chhattisgarh) followed by an MTech (Power System Engineering) in 2014 at Shri Shankaracharya Technical Campus (SSTC), Bhilai (CSVTU, Bhilai). She is pursuing her PhD in Electrical Engineering at Kalinga University, Raipur. She has worked as an assistant professor in the Electrical Engineering Department, SSTC, Bhilai since June 2008. She has published more than ten papers in renowned international and national journals and conferences, including Scopus journals. Her areas of interest include renewable energy grid integration and power quality. She is a professional member of the Institute for Engineering Research and Publication (IFERP).

Ramjee Prasad

Aarhus University, Denmark

Dr Ramjee Prasad, a fellow of the IEEE, the IET, the IETE, and the Wireless World Research Forum (WWRF), is a professor of future technologies for business ecosystem innovation (FT4BI) at the Department of Business Development and Technology, Aarhus University, Herning, Denmark. He is the Founder President of the Core Tele Infrastructure (CTIF) Global Capsule (CGC). He is also the Founder Chairman of the Global Information and Communication Technologies (ICT) Standardization Forum for India (GISFI), established in 2009. The objective of the GISFI is to increase collaboration between European, Indian, Japanese, North American, and other worldwide standardization activities in the area of ICT and related application areas. He was honored by the University of Rome 'Tor Vergata,' Italy as a distinguished professor of the Department of Clinical Sciences and Translational Medicine on March 15, 2016. He is honorary professor of the University of Cape Town, South Africa, and the University of KwaZulu-Natal, South Africa. He received Ridderkorset af Dannebrogordenen (Knight of the Dannebrog) in 2010 from the Danish queen for the internationalization of top-class telecommunication research and education. He has published more than 50 books, more than 1000 journal and conference publications, and more than 15 patents and has supervised more than 145 PhD graduates and a larger number of masters students (over 250). Today, several of his students are worldwide telecommunication leaders in their own right.

Veena Puranikmath

S G Balekundri Institute of Technology, India

Mrs Veena I Puranikmath received a BE degree in electronics and communication engineering from S G Balekundri Institute of Technology, Belagavi in 2013, and a master's degree in digital communication and networking at Godutai College of Engineering for Women, Kalaburgi in 2015. She is currently working as an assistant professor at S G Balekundri Institute of Technology, Belagavi. She is pursuing a PhD degree at Viswerayya Technological University, Belagavi. She has authored a book titled 'Introduction to Information theory and Coding,' and has published 15 articles in various high-impact-factor, peer-reviewed journals. She is a member of the International Association of Engineers (IEANG), Teaching and Education Research Association (TERA) and the Institute for Engineering Research and Publication (IFERP). Her current research interests include wireless sensor networks and the IoT.

Manuel Rubio Rivas

Department of Internal Medicine, Bellvitge University Hospital, Hospitalet de Lobregat, Barcelona, Spain

Manuel Rubio Rivas, PhD is an internist who has specialized in systemic autoimmune diseases since 2005. He is currently working as a consultant at Bellvitge University Hospital (Barcelona, Spain), where he is in charge of the outpatient clinic for scleroderma and sarcoidosis. He has been in the front line of COVID-19 treatment since the beginning of the pandemic and has extensive clinical experience in the management of this disease. He is listed in the Spanish National Registry of COVID-19 practitioners and has extensive clinical research experience in this field.

Tanu Rizvi

Chhattisgarh Swami Vivekanand Technical University, Bhilai, India

Tanu Rizvi completed a BE in Electrical and Electronics Engineering in 2008 at SSTC Bhilai (Pt. RSSU, Raipur, C.G.) followed by an MTech (Power System Engineering) in 2013 at SSTC Bhilai (CSVTU, Bhilai). She reached second position (silver medalist) in the university rankings for her BE followed by honors in MTech. She is currently pursuing her PhD in Electrical Engineering at CSVTU, Bhilai. She has been working as an assistant professor in the Electrical and Electronics Engineering (EEE) Department, SSTC, Bhilai since 2008. She has published over 20 papers in international/national journals and conferences, including those of the IEEE, Scopus journals, and Taylor & Francis. Outside her own domain, she has published a book on machine learning and two book chapters on COVID-19 pandemics. Her areas of interest includes renewable energy grid integration, power quality, signal processing, and image processing She is the alumni relations officer of her department, the college-level coordinator of IIT Bombay's spoken tutorial, and a professional member of the IFERP.

Mónica Rojas-Martínez

Biomedical Engineering Research Centre (CREB), Automatic Control Department (ESAII), Universitat Politècnica de Catalunya-Barcelona Tech (UPC), Spain

Mónica Rojas-Martínez, PhD graduated from the Universidad de Los Andes in Bogotá, Colombia in 1999 as an electrical engineer and completed her PhD degree in biomedical engineering in 2012 at UPC (BarcelonaTech), Barcelona, Spain. She has worked as an associate professor and researcher at the Department of Bioengineering, Faculty of Engineering, Universidad El Bosque, Bogotá, Colombia. She is currently a member of BIOART, ESAII, CREB, BarcelonaTech (UPC), Barcelona, Spain. Her research interests include signal processing and machine learning techniques with applications in the study of motor rehabilitation, human–machine interfaces, prostheses, and assistive devices. During the last few years, she has been working on several research projects focused on analyzing EMG signals in order to diagnose muscular pathologies, monitoring rehabilitation therapies, and using high-definition electromyography (HD-EMG) to control external devices.

Chandramouleeswaran Sankaran (Mouli)

RV University, India

Prof. Chandramouleeswaran Sankaran (Mouli) received his BE Electronics and Communication Engineering (ECE) from the Thiagarajar College of Engineering, Madurai, India and an MTech in computer science engineering from the Indian Institute of Technology, Madras (IITM), India, in 1987 and 1997, respectively. He joined the Electronics and Radar Development Establishment (a Defence Research and Development Organization lab in Bangalore) as a scientist in 1988, where he was responsible for the Radar Data Processor (RDP) for the Rajendra Phased Array Radar, the heart of the Akash missile system. He has worked with semiconductor and product companies in various senior techno-managerial roles at Motorola, Lucent Microelectronics, Agere Systems, LSI, Sharp, and Honeywell. He headed the Networking Software Division as director of engineering at Agere Systems and LSI Bangalore, with a team of 100+ engineers, working on various networking products based on ARM SoCs. He headed the Automation Control Solutions team of 180+ engineers at Honeywell, Madurai, as Director of Engineering,

Mouli has immense interest and six years of experience in teaching. He has recorded 45 hours of a video course on 'ARM-based Development,' which is available via the National Programme on Technology Enhanced Learning (NPTEL) and at the National Digital Library of India. While Mouli was working as adjunct faculty at IIIT Bangalore in 2015, he was sent on deputation for four years to the Myanmar Institute of Information Technology (MIIT), Mandalay (an Indo-Myanmar Government project), where he designed and delivered 25+ theory and lab courses for the BE degree (Computer Science and Engineering and ECE).

As adjunct faculty at upGrad, he has recorded video lectures on electrical and electronics circuits and modules on multicore performance metrics and recurrent neural networks (RNNs). He has also been visiting faculty at IIIT Lucknow and taught courses on Python and AI for IoT. He worked as adjunct faculty at IIIT Bangalore before joining RV University, Bangalore as a professor in the School of Computer Science and Engineering.

Mouli holds three patents granted by the United States Patent and Trademark Office (USPTO). He is a Six Sigma Green Belt and also a certified scrum master.

P Saranya

Anna University, India

Ms. P. Saranya received her B.E and M.E degrees from SNS College of Technology, Coimbatore in 2010 & 2012. Working as an Assistant Professor at Dr. N.G.P. Institute of Technology, Coimbatore since 2022 and earlier had gained her experience from SriGuru Institute of Technology, Coimbatore. She has a total of 8 years of teaching experience. Authored 6 research publications both international and national journals and presented 5 papers at International and national conferences. Her areas of interest include Signal Processing and VLSI. She is a member of the ISTE.

Kumud Saxena

Noida Institute of Engineering and Technology, India

Dr Kumud Saxena received a PhD degree in computer science from Dr B R Ambedkar University, Agra, UP, India in 2016; her thesis was titled 'Design of image enhancement techniques using soft computing.' She has a rich experience of 13 years in teaching. She is currently associated with NIET, Greater Noida as Head of Department in the IT department. She has published many papers in several peer-reviewed and indexed journals and presented papers at conferences of national and international repute. She is a reviewer for some reputed journals. She is a member of the Board of Studies, Maharishi University of Information Technology, and the Computer Society Of India. She is associated with Linked Globe Consultancy as a Senior Consultant. She has worked on government-aided projects and is currently working as the principal investigator for a project on the 'Implementation and optimization of illumination switching patterns in smart cities using IoT' under the Collaborative-Research-and-Innovation-Program through the Technical Education Quality Improvement Programme, APJ Abdul Kalam Technical University. Her research interests include image enhancement, soft computing, biometric-based recognition, the IoT, AI, etc.

V Senthilkumar

Anna University, India

Mr V Senthilkumar is currently working as an assistant professor II at the Department of Computer Science and Engineering, Kumaraguru College of Technology, Coimbatore. He is pursuing his PhD in computer science and engineering at PSG College of Technology, Coimbatore. He has completed a master's degree in computer science and engineering at Sri Krishna College of Engineering, Coimbatore (2012). He obtained his BE degree in computer science and engineering from Coimbatore Institute of Engineering and Technology (CIET) in 2010. He has more than ten years of teaching experience. He has published one patent, five Scopus indexed research articles, and seven international conference papers. His areas of interest include sentiment analysis, machine learning, data analytics, database management systems, digital systems and design, and web technology.

Dhairya Shah

Nirma University, India

Mr Dhairya Shah is a senior at Nirma University, Gujarat pursuing a BTech in instrumentation and control engineering. He has worked with Glintsolar as a machine learning intern, where he lead the data collection team, and he recently completed a summer internship at the Indian Space Research Organisation (ISRO), where he automated the task of email data entry and researched flood forecasting using ISRO satellites. He prefers to research areas such as machine learning and algorithms and loves reading books.

Haard Shah

Nirma University, India

Mr Haard Shah is currently pursuing his BTech in instrumentation and control engineering at the Institute of Technology, Nirma University, Ahmedabad, Gujarat along with his minor degree in robotics and automation. He aspires to pursue a master's degree in robotics and automation, through which he will get the right combination of elements that will best contribute to his understanding of the subject. He has experience in the field of automation and contributes to the International Society of Automation (ISA); he is currently appointed as the president of the ISA Nirma University chapter. He was a member of Team ABU Robocon, which has now been a National Champion team ten times and has represented India in a global benchmark. He completed his internship at the Space Applications Centre (SAC), ISRO, working in the high power testing division, where he developed an autonomous variable-power combining system that made it possible for the division to efficiently undertake ring resonator tests.

Manan Shah

PDEU University, India

Dr Manan Shah has a BE in chemical engineering from L. D. College of Engineering and an MTech in petroleum engineering from the School of Petroleum Technology, Pandit Deendayal Energy University (PDEU). He completed his PhD on the topic of the exploration and exploitation of Geothermal Energy in the state of Gujarat. He is currently working as an assistant professor in the Department of Chemical Engineering, School of Technology (SOT), PDEU and he has published several articles in reputed international journals in various research sectors. He serves as an active reviewer for several reputable international publishers such as Springer and Elsevier. Dr Shah is a young and dynamic academician and researcher in various engineering fields. Dr Shah has received the Best Paper Award for world's second-best journal.

Mehdi Shirzadi

Biomedical Engineering Research Centre (CREB), Automatic Control Department (ESAII), Universitat Politècnica de Catalunya-Barcelona Tech (UPC), Spain

Mehdi Shirzadi received BS and MS degrees in Biomedical Engineering from the Islamic Azad University of Najaf Abad, Najaf Abad, Isfahan, Iran and the University of Isfahan, Isfahan, Iran, in 2015 and 2018, respectively (and was ranked in the top two students). He is a PhD candidate in biomedical engineering at the UPC, Barcelona, Spain.

He won a four-year scholarship from the Ministry of Economy and Competitiveness (MINECO) in Spain in 2019. His research interests include electromyographic signal processing, optimization, medical data mining, and machine learning.

Hitesh Singh

Noida Institute of Engineering and Technology, India

Dr Hitesh Singh holds a PhD awarded by the Technical University of Sofia, Bulgaria. His areas of research are wireless communications, machine learning, propagation studies, and cyber security. He works as an assistant professor at the NIET, Greater Noida. He has completed an MTech and a BTech in computer science engineering. He has published books in the field of computer science and patents in the fields of propagation studies, 6G, and IOT. He has published papers in several peer-reviewed and indexed journals and presented papers at conferences of national and international repute. His current research interest is the applicability of machine learning to the field of radio wave propagation and in particular, attenuation due to cloud, fog, and dust particles. His implementation is based on Python.

Shruti Tiwari

CSVTU, Bhilai, India

Dr Shruti Tiwari, PhD received her BE from Pt. RSSU, Raipur and her ME and PhD degrees from CSVTU, Bhilai in EEE and Electrical Engineering in 2007, 2009, and 2021, respectively. She has been working as a senior assistant professor at the Faculty of Engineering, SSTC, Bhilai since July 2007. She has 15 years' experience. Her research interests include energy management systems, sustainable energy, and signal processing. She authored/

co-authored various publications in various peer-reviewed journals. She has has produced patentable work. She has written one book chapter, one book and papers presented in various national/international conferences.

Sourabh Tiwari

Data Scientist, Airbus, Bangluru, India

Sourabh Tiwari received his BE in mechanical engineering with distinction from Shri Shankaracharya College of Engineering and Technology, Bhilai, Chhattisgarh in 2010 and his MTech in Thermal and Fluids Engineering from the Indian Institute of Technology, Bombay in 2012. From July 2012 to May 2018, he was a deputy manager at Whirlpool of India Ltd. in Pune, Maharashtra, India. From May 2018 to June 2020, he worked as a lead engineer at Carrier HVAC, Hyderabad, Telangana, India. From May 2020 to the present, he has been employed as a data scientist at Airbus Bangalore, Karnataka, India. He has been selected to be the current Tech Champion in Health Engineering at Airbus, Banglore, Karnataka, India.

Prajakta Upadhye

Rashtrasant Tukadoji Maharaj Nagpur University, Nagpur, India

Prajakta Upadhye received her BE in ETC from Amaravati University in 2003 and an MTech in Electronics (Communication) from RTMNU, Nagpur in 2014. More recently, she has worked as an assistant professor at the Electronics and Telecommunication Department of GHRIET Nagpur, and has a total of 12 years' experience. She has published three research papers in various journals and conferences. She is a professional member of the ISTE and the IE(I).

Chapter 1

Introduction to the cognitive Internet of Things

Chandramouleeswaran Sankaran (Mouli)

The Internet of Things (IoT) is a network of futuristic sensors and actuators that are connected to each other; they exchange information among themselves and with the Internet. The term 'cognitive IoT' is applied to enhanced sensors in the IoT network which have the capability to analyze and understand the environment which they are part of and behave accordingly by seamlessly integrating themselves with humans and their surroundings. This chapter provides a deep insight into various types of application areas in which the cognitive IoT provides an enhanced user experience and smart features that are critical and essential to making human life safe and secure. This chapter also deals with the application of cognitive computing technologies to the data generated by the cognitive IoT devices and systems, which works by making them behave in a similar way to human cognition. This chapter provides a broad survey of various cognitive IoT paradigms and systems.

1.1 Introduction

The (IoT) is a network of physical objects with embedded technology that are capable of sensing the environment and exchanging information with other objects in the network as well as over the Internet. IoT elements can be controlled and activated to make changes to the environment through various actuators interfaced with the physical objects.

A *thing* in the IoT can be any physical object, which could, for example, be a heart monitor implanted within a human being to monitor his or her vital signs or a bridge health monitoring system that monitors the load on the bridge, makes visual inspection of cracks on the bridge, measures the water level under the bridge, etc. The IoT is evolving from supporting specific application deployments, such as healthcare monitoring, watering of plants, etc. to a platform that can support a variety of application scenarios with a broader set of sensors and actuators, such as

doi:10.1088/978-0-7503-5326-7ch1

smart city solutions that take care of traffic regulations, detect traffic rule violations, monitor weather, and even control and prevent criminal activities, etc.

In a network of IoT devices, each device in the network is uniquely identifiable and traceable and each of those devices are capable of communicating and exchanging information with other devices on the same network or any other objects or machines over the Internet.

The sensors embedded in IoT devices capture significant information from their physical environments and bring that information into the digital realm. They send it across the Internet to any other devices or machines across the world. Actuators such as motors connected to IoT devices can be driven or controlled based on the information which is processed and then exchanged, thus making changes to the environment.

This capability paves the way for various information-centric applications which were never thought of in the past, enabling a safer and human-centric society that is secure and comfortable to live in. For example, a smart home fitted with IoT devices can control the temperature of the interior of the home by tracking the movement of the owner and switching on the home air conditioning before the owner arrives home during a hot summer to bring the home temperature to the level the user has selected, based on his or her comfort.

Communication is the most important and necessary capability of an IoT device. Its optional capabilities include sensing, actuation, data capture, data analysis, data storage, and data processing.

IoT devices collect various data through their installed sensors; after the necessary processing, they share the data with other units and central control units through the data communication mechanisms provided.

The fundamental characteristics of IoT devices can be listed as follows:

- **Interconnectivity**: IoT devices need to be interconnected with the global information and communication infrastructure.
- **Heterogeneity**: IoT devices are normally heterogeneous in nature; they are based on different hardware platforms and networks. IoT devices can interact with other devices or service platforms through different networks.
- **Dynamic changes**: The state of IoT devices can change dynamically, e.g. sleeping and waking up, connected and/or disconnected. In addition, device contexts, including their location and speed, can change continuously, depending on whether the devices are stationary or mounted on a moving platform.

Applications such as smart homes, smart cities, or any other industrial applications have a variety of scenarios to be supported with the help of IoT devices and the networks they are part of. These applications are expected to gather information from a wide variety of IoT sensors spread across a large geographical area and take the necessary actions based on the data gathered after processing it to extract the information required for a particular application, users, or scenario under consideration (figure 1.1).

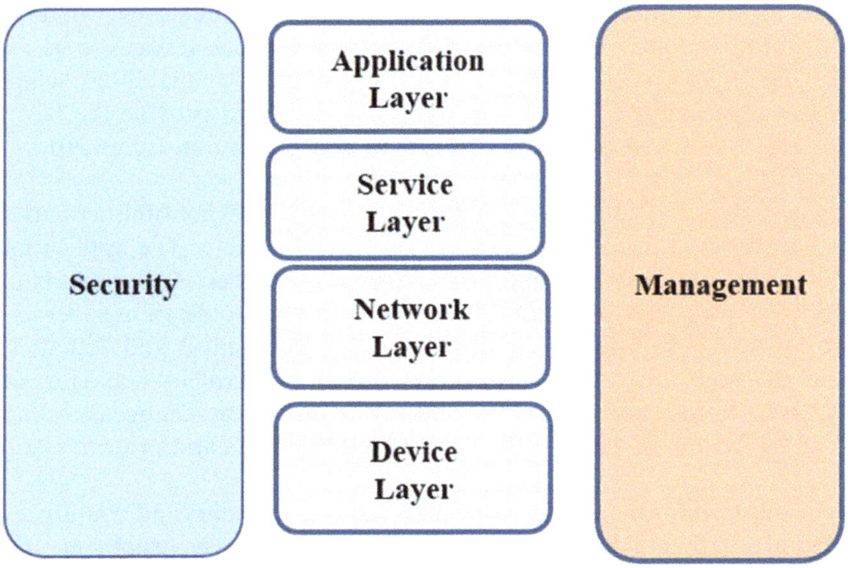

Figure 1.1. IoT reference model.

Figure 1.1 depicts the IoT reference model, which is similar to the Open System Interconnect (OSI) model for networking. The operation of an IoT system can be defined using the above reference model. Two of the vertical layers control and influence the functioning of all the horizontal layers.

The *device layer* is also called the *physical* or *perception layer*; it has all the sensing elements which gather information about the environment in which the IoT devices are installed. This layer also has actuators that modify or make necessary changes to the environment based on the actions taken by the system using the inputs provided by the sensors.

The *network layer* above the device layer takes care of providing a communication mechanism, either wireless or wired, enabling IoT systems to be connected to other devices in the neighborhood as well as to the Internet. This allow IoT devices to connect to the IoT cloud, where the data is stored and analyzed using big data and data analytics to extract hidden and useful information from the huge amounts of data collected by many IoT devices. This provides the ability for the entire IoT system to behave cohesively according to the system requirements.

Security features such as encryption and error correction (used if the data is corrupted due to noise in the environment) are also provided by the network layer. Many of the protocols which are covered in later chapters fall under this layer. A suitable protocol or access method is chosen based on the scenario and the system configuration, although many protocols and communication technologies are used by this layer in a typical system; some examples are 5G, WiFi 6, Bluetooth, low-power wide area networks (LPWANs), etc.

The *service layer* provides service management, which is a very important part of an IoT system, as it allows the system to be built, deployed, and managed by its vendors. It receives information from the network layer and stores it in a database after performing the necessary data processing based on the services to be provided by the IoT device or system. If the IoT system has cognitive capabilities, which is the case for the cognitive IoT, it can take local decisions and provide the best service possible under the given conditions sensed by the perception (physical) or device layers.

The *application layer* takes care of application-related responsibilities and uses the service layer below it to achieve the goals of the IoT system. The application layer takes care of the user interface and also receives and processes user inputs to allow the users to control the behavior of the system. The application layer is also a gateway into the IoT system, as it displays the current status and activities of the system to the users, thereby allowing the users to control its behavior and also allowing them to take any corrective actions or make any changes to the configuration of the system to cater to the varying requirements and priorities of the end users.

The current trend is that IoT platforms need to have smart and cognitive sensors that normally gather a huge amount of data from the environment. They are expected to performing initial processing and noise removal and then communicate the relevant data or a combination of them so that they can be used by multiple applications. Cognitive sensors are expected to be knowledgeable and to be able to segregate the data and share the relevant items, depending on the many applications that are being served by the sensors.

These smart cognitive sensors face many constraints, including operating as part of a noisy and unreliable wireless network. They also have the additional limitation that they need to operate in a power-constrained environment. They are expected to meet various requirements generated by the multiple applications they are expected to support; depending on the type of application, they also have other requirements to meet, such as higher reliability, minimum latency, reduced power or bandwidth consumption, etc.

These requirements mean that current IoT devices need to have additional intelligence and a higher level of awareness about different applications and users, coupled with the capability to learn and understand the physical and social world by themselves without any external help. IoT devices that have these capabilities and intelligence are collectively known as the cognitive IoT (CIoT).

This chapter explores in more detail how the CIoT has evolved and how it can be supported by resource-constrained IoT devices using networks which are noisy and unreliable.

1.2 From the IoT to the CIoT

We will now explore how the early IoT systems evolved over the decades since 1999, when Kevin Ashton introduced the concept of the IoT to the world. Over the decades, IoT devices have matured to have cognitive capabilities, thereby becoming CIoT systems. These are required to support the increased demands placed on IoT networks to cater to the varied range of requirements from different applications and

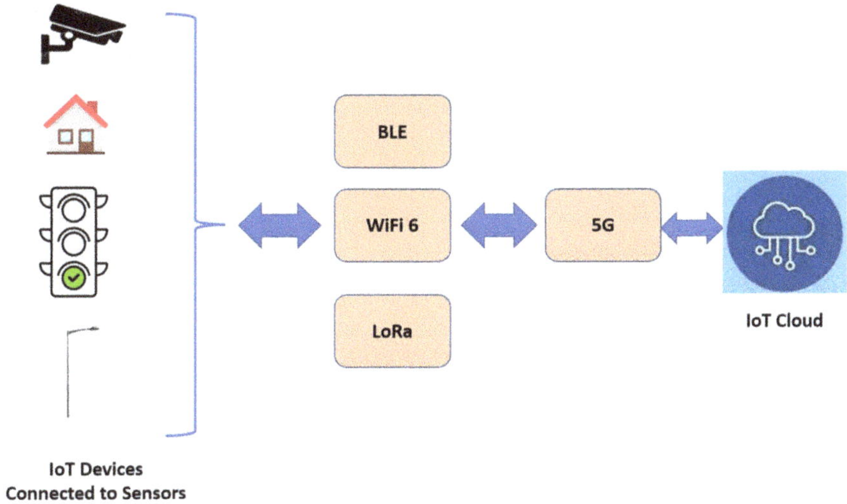

Figure 1.2. Generic IoT architecture.

scenarios that exploit the available IoT infrastructure by utilizing the massive amount of data generated by IoT devices (figure 1.2).

Advancements in data analytics and machine learning have made it possible to extract meaningful data from the huge amount of data produced by sensors, which is then used for different purposes and use-case scenarios.

1.2.1 An overview

IoT devices and systems are transformed from conventional data sources into smart devices by making use of advancements in converging technologies such as pervasive computing, sensor network over wireless technologies, embedded multicore microcontrollers with real-time operating systems (RTOSs) and network stacks running on them, and applications that exploit these capabilities.

The features supported by the CIoT are needed for applications that have broader sets of goals, which keep expanding due to new sets of user requirements that keep evolving as different technologies mature and benefit from improved capabilities. Examples of maturing technologies include wireless technologies with increased bandwidth, reduced latency, and reliable end-to-end connectivity (even if the lower physical layer is running over an inefficient and unreliable network infrastructure), improved computing abilities and low-power multicore microcontroller architectures, and RTOSs with deterministic schedulers and multitasking capabilities.

1.2.2 Conglomerates of technologies

Home, industrial, medical, and infrastructure applications as well as smart home and home automation systems are expected to provide safety and security features with intelligence that can address different scenarios and take appropriate decisions without requiring any human intervention.

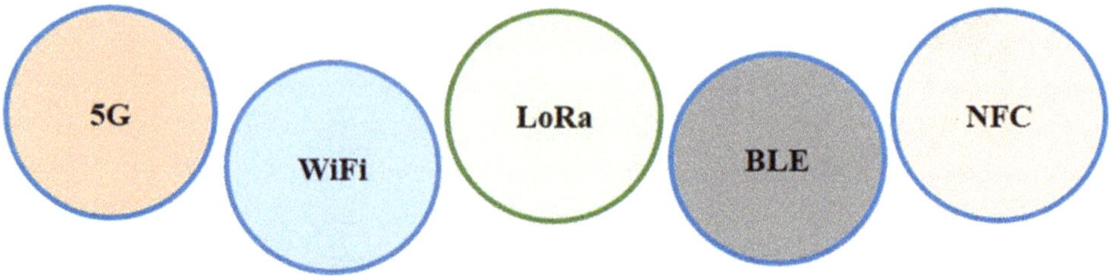

Figure 1.3. IoT technologies and protocols.

Smart city applications need intelligent traffic systems that can regulate traffic and take decisions to divert traffic based on traffic conditions such as congestion or unforeseen events such as accidents that need medical emergency teams to be guided to reach the scene of an emergency without much of a delay. This is achieved by diverting the traffic around the area without the need for police or city administrative authorities to get involved; this saves time and allows for quicker emergency actions to be taken at any time of the day to ensure that essential and critical medical support is provided to victims at the scene of an accident reliably, quickly, and efficiently.

This requires a conglomerate of technologies to be involved, because the above systems have numerous heterogeneous devices connected together as well as connectivity to the Internet, which allows information to be exchanged with central repositories using cloud technologies. These systems also need to receive commands from the control center in order to take the necessary actions based on decisions taken in the cloud or based on local intelligence provided by an individual device or a set of co-located CIoT devices (figure 1.3).

The promising technologies and architectures include low-power and low-cost connectivity technologies, such as radio-frequency identification (RFID), Bluetooth Low Energy (BLE), advanced machine-to-machine (M2M) communication, wireless sensor networks (WSNs), LPWANs, long range (LoRa), 5G, WiFi 6, cognitive radio (CR), and Internet Protocol version 6 (IPv6) over Low-Power Wireless Personal Area Networks (6LoWPAN). Some other protocols that are popularly used by IoT systems include Constrained Application Protocol (CoAP) and the Routing Protocol for Lower-Power and Lossy Networks (RPL).

We will now briefly review some of these diverse technologies to understand the gravity of the challenges and issues involved in integrating them into one system to enable them to work together seamlessly, though our intent here is not to go into the technical details of making them work, which is beyond the scope of this chapter.

1.2.2.1 Wireless sensor networks

WSNs are very popular and widely used in many IoT networks that need to interconnect a large number of devices for both civilian and military application scenarios. A WSN is a wireless network that can connect independent distributed sensor nodes that monitor physical or environmental conditions. WSN network nodes collect environmental data such as temperature, pressure, humidity, or the concentrations of pollutants or suspended particles (in parts per million (PPM)) using sensors and they pass this information across to a base station which is connected to the Internet to reach the central repository.

WSNs have four essential characteristics: redundancy exploitation, data-centric routing, data aggregation, and localized algorithms.

Sensor nodes are prone to failure due to their random placement in harsh environments; they are also expected to run out of energy because they run on limited battery power. Redundancy and increased device density are exploited to ensure full coverage of the intended regions.

WSNs are also expected to self-configure when communication links are disturbed or disrupted due to signal attenuation, noisy environments, or sudden device failures.

The sensor nodes in WSNs are not identified by unique IDs. The central unit which queries the sensors, known as the *sink*, identifies the sensors based on the data being received from multiple responding sensor nodes on the WSN, which is called data-centric routing. Traditional end-to-end network routing is not suitable here. Data-centric routing is employed for data aggregation and distribution operations in WSNs.

The data from the sensors are shared with all the neighboring nodes, which is called *network flooding* in WSN parlance. Due to flooding, multiple copies of a data set are sent over the network, which is called *data implosion*. Energy is wasted due to this phenomenon. To address this issue, *data suppression* is performed at the local level by eliminating duplicates. *Data fusion*, which uses sophisticated signal-processing techniques, is also performed to combine data from different sensors which are co-located in a WSN.

Localization is the process of computing the locations of different wireless devices on the WSN; this is particularly important because of the absence of unique IDs. In the *localization process*, the positions of unknown devices are extracted using input data from *reference nodes* whose locations are known.

WSNs follow a hierarchical system, which creates clusters for data aggregation to reduce data traffic over the power-constrained IoT network. The broad set of use cases for WSN-based IoT devices includes intrusion detection, traffic monitoring on highways, structural health monitoring of buildings and bridges, environmental monitoring systems, etc.

1.2.2.2 Low-power wide area networks

The narrowband IoT (NB-IoT) is one of the leading LPWAN technologies; it is used for large-scale deployments of the IoT. The NB-IoT enables IoT sensors to send their data directly to the cloud without the need for an intermediate gateway.

LPWANs reduce complexity, limit bandwidth requirements and also impose a limit on the maximum data rate supported by the network. This is acceptable, because IoT-based applications are tolerant of delays and operate in a constrained low-power environment with lower data rates. This reduces device costs, as it is only necessary to support frequency-division multiplexing (FDD), a single antenna implementation, and half-duplex mode, as opposed to a multiple input, multiple output (MIMO)-type antenna.

LPWAN is the last-mile communication technology and is used by the IoT nodes on the network to share data with the sink. Since sensor devices share the sensed data with the sink most of the time and only occasionally receive commands from the central controller, the NB-IoT has a higher uplink data rate than its downlink data rate.

1.2.2.3 Long-range WANs

LoRa is another LPWAN technology that is widely used for the deployment of IoT devices. LoRa WANs (IEEE 801.15.4g) use unlicensed spectrum in the industrial, scientific, and medical (ISM) bands. They operate in different frequency bands in different geographical regions, namely 915, 868, and 433 MHz. They operate with different ranges: 5 km in urban areas and 15 km of range in deployments in rural regions. The data rates of this technology are 290 bits per second (bps) to 50 kbps, and the battery life is more than 10 years for a wider maximum range of 20 km. It uses a 32-bit cyclic redundancy check (CRC) for error detection, which is a very important requirement in IoT systems design. LoRa has also been designed to be more reliable and to have very high interference immunity, because IoT systems normally have to be deployed in noisy environments.

1.2.2.4 5G

The ‘G’ in 5G refers to the fifth generation of mobile networks. It is a recent technology which is being deployed all over the world; it is a global wireless standard that follows its predecessors, 1G, 2G, 3G and 4G networks. 5G has advanced network capabilities to connect virtually everyone and everything, including machines, objects, and tiny IoT devices.

5G creates some new challenges for mobile network operators. 5G is designed to support faster mobile broadband speeds with lower latencies, making a whole lot of new applications possible on this network, such as on-demand video, autonomous vehicles, etc. This, in turn, requires the wireless operators to obtain access to larger amount of spectrum bandwidth to make these new services and applications a reality.

The lower bands of 5G provide greater ranges and lower data rates, whereas the higher bands of 5G provide complementary features, i.e. short ranges and higher data rates, thus enabling 5G networks to cater to a wide range of applications. 5G networks promise greater flexibility to support a wide range of devices using a small-cell architecture, covering very small areas by having many base stations in a given area, compared to the previous generations.

5G supports the *Tactile Internet*, which is an Internet that combines ultra-low latencies, very high availability nodes, and higher reliability and security. 5G also supports the convergence of different wireless technologies such as WiFi, Long Term Evolution (LTE), mmWave (millimetric wave), etc. It also uses software-defined networking (SDN) and network virtualization principles, enabling broader adaptability and configurability based on the traffic conditions and varying network traffic.

5G is the most suitable network infrastructure for IoT, as it facilities lower power consumption, more devices per base station (thus increasing device density and coverage), lower latency, higher reliability and, most importantly, a technology that supports the mobility of IoT devices, which opens up new avenues and applications for the IoT. This enables applications such as smart cities, agriculture-related applications, environmental control, healthcare, autonomous cars, and immersive real-time extended reality (XR) experiences which need high data rates. 5G can provide peak data rates of 20 gigabits per second (Gbps) and average data rates of 100+ megabits per second (Mbps). 5G is implemented using multiple isolated logical networks of varying sizes and structures that meet the needs of different types of service and application.

1.2.2.5 WiFi 6

Wireless Fidelity (WiFi) 6 is the sixth generation of WiFi technology. It provides advanced WiFi connectivity that has higher capacity, coverage, and performance in a hyper-dense environment than its earlier generations. Due to its ability to connect large numbers of nodes in a given area, WiFi 6 is well suited to the IoT. The seamless integration of WiFi with 5G networks enables lower battery consumption by making WiFi 6 devices perfectly suited for the new age of connected applications such as smart cities, smart homes, etc.

Multiuser MIMO (MU-MIMO) interfaces enable access points (APs) to handle larger number of devices simultaneously. It operates in the 2.4 GHz band (the free ISM band) and at 5 GHz simultaneously. *Beam forming* is the technology used to precisely focus the wireless signals from the devices toward the AP, thus improving the quality of data exchange and reducing interference in hyper-dense environments. This also enables *power efficiency*, which is most important for battery-operated IoT nodes.

1.2.2.6 Near-field communication

Near-field communication (NFC) allows users to make transactions which are secure and safe and to exchange digital content without contact or touch. It is used for communication between two devices which are very close to each other, over a distance of just 4 cm or less. NFC-based devices can be used as electronic identity documents or keys. NFC is also used for contactless payments. NFC protocols and standards are based on the existing radio-frequency identification (RFID) standards. NFC operates at 13.56 MHz in the unlicensed ISM band.

1.2.2.7 Bluetooth low energy

BLE, also referred to as 'Bluetooth Smart,' is a lightweight protocol best suited for the low-power environment of the IoT. It is based on the Bluetooth 4.0 core specification. Although BLE overlaps with the classic Bluetooth standards, it actually has a completely different protocol standard, which was started by Nokia as an in-house project called 'Wibree' and later adopted by the Bluetooth Special Interest Group (SIG), which is a conglomerate of hundreds of companies and industry leaders in the market.

Bluetooth is a short-range cable replacement technology. The name 'Bluetooth' was proposed in 1997 by Jim Kardach of Intel; it was then jointly developed by Intel, IBM, Toshiba, Ericsson, and Nokia. It is a short-range wireless communication technology standard. It uses the ISM band from 1.402 GHz to 2.480 GHz. Bluetooth has been growing exponentially over the last two decades. It is used in a wide range of applications, including audio and wearable devices, location services, and automation solutions.

Bluetooth is a personal area network (PAN) or *piconet* which can have a maximum of eight Bluetooth devices, usually a single master and and up to seven slaves. The master initiates communication with the slave devices. The master device also governs the communication link and the traffic on the piconet. Multiple piconets are called a *scatternet*. The same devices can also be members of multiple piconets in either the master or slave roles. BLE was introduced in the Bluetooth specification version 4.0 in 2010. The original Bluetooth defined in the previous versions is referred to as Bluetooth Classic. BLE is not an upgrade of the original Bluetooth, but a new technology that specifically caters to the requirements of IoT applications, in which small data units are sent over the network at lower speeds, thus consuming less energy; this is more suitable for the IoT.

The BLE specification deviates from Bluetooth Classic in terms of its technical specification and implementation as well as the types of application the BLE standard is suited for. Basically, BLE is used for sharing the sensor data collected by IoT devices in a low-bandwidth environment. BLE is capable of establishing connections with the master device much quicker than Bluetooth Classic devices.

A newer Bluetooth standard, Bluetooth 5, was released in 2016, which introduced many important upgrades that specifically target BLE. The most important ones are a doubling of its speed, a fourfold increase in its range, and an eightfold increase in advertising data capacity; these are most essential features needed by IoT devices and smart CIoT devices in particular.

1.2.3 The principles behind the CIoT

The word Cognitive Radio was introduced by Joseph Mitola in 1999. The most important concern in IoT communication was to explore ways to exploit the scarce radio frequencies that were used by IoT devices in a cooperative manner. The cognitive IoT was born because the cooperating components of an IoT network need to interact and exchange information in the most efficient manner by using the available bandwidth of the physical layer of the network.

CIoTs are designed to be aware of and having the capability to analyze and understand the status of the network behavior and its features and traits, so that based on the available information and knowledge about the network [2], CIoT devices can perform suitable operations and actions in response to the observations made, such that the overall performance metrics of the network are improved.

This intelligence and the ability to improve the performance of the network based on the network behavior is not a simple task. It requires the ability to understand the diverse nature of network technologies, as described in section 1.2.2, and to adopt the techniques used to leverage the available bandwidth in the network to improve the performance of the network. This capability of the CIoT network makes it possible to quickly integrate different network technologies that cooperatively work together, thus improving the performance.

The main advantage and the reason for the evolution of the CIoT in the recent years is the ease of designing IoT systems with diverse networking technologies that minimize human intervention to the maximum extent possible by designing a cognition process into the network. This enables the network to adapt itself to the changing conditions and traffic scenarios such that the available bandwidth is efficiently used by providing the best possible latencies to support the end applications running on the CIoT architecture.

The network performance enhancement is achieved by having a combination of human-level intelligence and a proper system design. Every device in the CIoT system is configured to be agile and most efficient in such a way that each sensor node deployed on the network uses a very minimal amount of power while at the same time providing the best possible performance from an overall system perspective.

Cognition is the technique which is used to enable uniform resource utilization of the network, such that all the available resources and bandwidth are homogeneously distributed and shared by all the devices in the network. The IoT sensors are arranged in a logical manner in a layered architecture so that the communications among them as well as with the external world are well coordinated with a minimum amount of latency and in the most power-efficient way in a resource-constrained environment.

The layers are categorized as the sensing layer, the network layer, and the application layer. The sensing layer is the lowest physical layer; it is related to the protocol and hardware features of the IoT devices, which are configured by the software versions running on the system [1]. The network layer takes care of information exchange through various network elements. The application layer that interacts with the user and the outside world is responsible for the different interactions between the devices and the underlying network.

The CIoT network entities that are part of the network are grouped into various domains. Each of the entities thus grouped is dependent on other devices in the network domain and coupled with them. Each of the nodes in the network has the ability to both transmit and received data over the wireless networks it is part of. As explained in the previous section, the wireless technologies used may be of different types and may have different properties and features, depending on the types of technology used.

1.2.4 The architecture and layers of the CIoT

Although the CIoT does not follow any standard layered architecture, it can be broadly classified as consisting of the following three layers, based on the responsibilities of each layer that a CIoT system is composed of (figure 1.4).

The CIoT consists of three broad layers, namely, the protocol layer, the cognitive layer, and the adjusting layer [3]. Each cognitive node interacts with its neighbors and configures itself using the information shared between them (figure 1.5).

We will briefly describe the roles and responsibilities of each of these layers in detail below.

1.2.4.1 The protocol layer

This layer is similar to the traditional seven layers of the OSI network standard and has multiple sublayers within it. The sublayers of this protocol layer of the CIoT are: the information perception layer (IPL), the near-field interconnection layer (NFIL), the network access layer (NAL), the network layer (NL) and also some other layers which are required to enable data exchange and information flow within the network.

The IPL is responsible for gathering information from different sources and sensors and the NFIL manages the transmission and reception of data from different sources and destinations. The information exchanged can be multimedia data, depending on the application and use-case scenarios of the CIoT network, including text, audio, and video data at times.

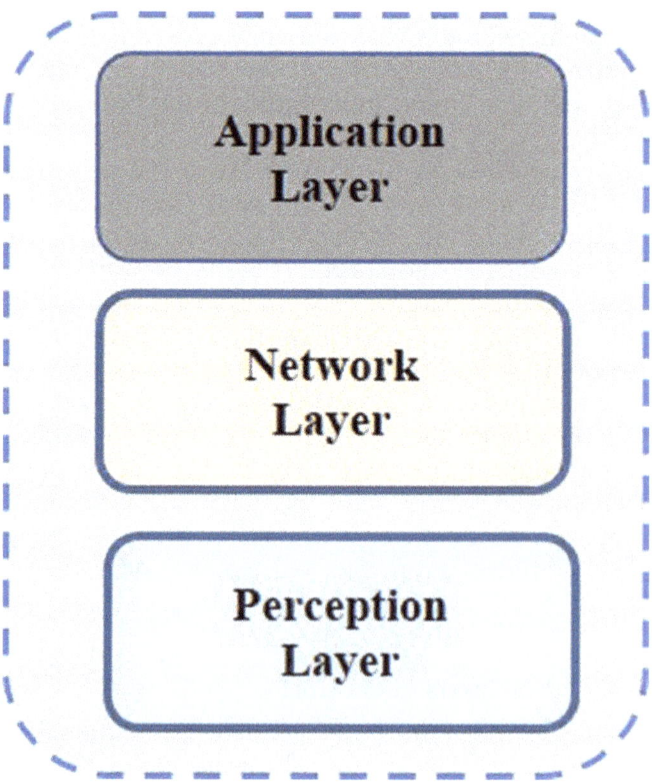

Figure 1.4. Generic IoT architecture.

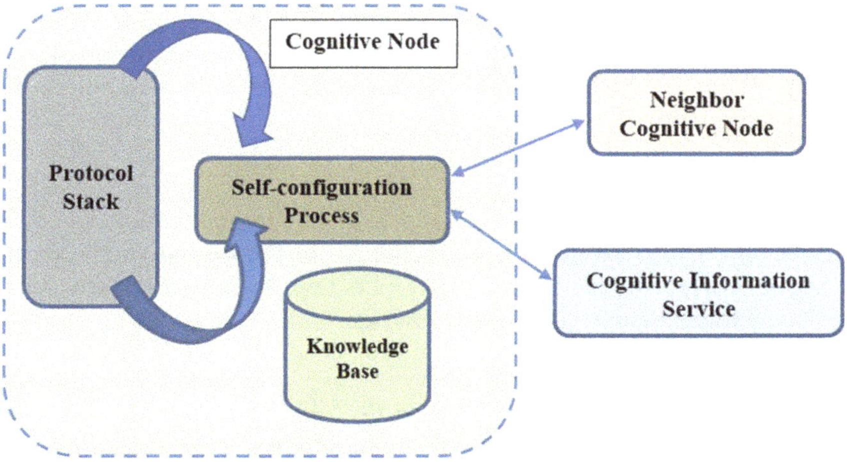

Figure 1.5. Architecture of the CIoT.

Although the transmission of video data over resource-constrained networks is a challenge, some security and monitoring applications need to be able to transmit video after the necessary compression and local processing are done by the transmitting node, depending on the application and capability of the nodes under consideration. The NAL provides generic and easy access to different kinds of data, including video and audio, in a similar way to that used for textual data. The main responsibility of the NAL layer is to provide uniform access to different kinds of data. The NAL provides access to the IoT device from the external world as well as from other devices, allowing it to share its status and sensing information with the outside world and the central repository or IoT cloud.

1.2.4.2 The cognitive layer

The main responsibility of the cognitive layer is to analyze the network's conditions, its current status, and the key requirements of the various applications running on the infrastructure with varying priorities in order to make decisions based on the analysis done by this layer. The cognitive processing is performed using a cycle which has observe, orient, decide, and act as its four key roles.

The nodes first observe the behavior of the network and the environment and sense the data which is not yet fully processed [11]. The sensed data is processed and data duplication is avoided by performing a data fusion operation to prevent duplicate data from being transmitted over the network and thus consuming precious network bandwidth. The status and content of the fused data decide the action to be taken by the node and the necessary action to be performed by the node with the data. This decides and defines the behavior of the network and its characteristics. Based on the decisions taken by the individual nodes, the collective knowledge that each of the nodes in the network has about the status of the network enables coordinated action to be taken by all of them.

The main aim of this observe–orient–decide–act (OOAD) cycle is to minimize and to the maximum possible extent avoid human intervention, while allowing the network to adapt to the continuously changing network conditions and the application requirements.

1.2.4.3 The control layer

The functionality of the control layer and that of the application layer are very similar. The aim of this layer is to make changes to the system with respect to the underlying network-protocol-related parameters and the network status by making changes to the logical connectivity of the system.

1.2.5 The need for massive data analytics in the CIoT

The CIoT will be heavily populated by a large number of heterogenous interconnected devices that generate massive amounts of data which will grow exponentially [4]. The data sensed by the nodes and devices on the network will be of no use if it is not analyzed, interpreted, and understood properly.

For example, traffic data collected and crowdsourced from various sources, such as cameras deployed over junctions and roads, data collected from the vehicles using the roads, the cell phones of the drivers in the vehicles, the Global Positioning System (GPS)-assisted data from the vehicles and also from the cell phones of the passengers are normally noisy, corrupted, erroneous, heterogeneous, high dimensional, and nonlinear in nature [7]. To exploit and efficiently extract value from this massive amount of data, there is a need for efficient data analytics algorithms to be developed for data reduction and data extraction which can be used by the applications to provide value to the end users.

1.2.6 Knowledge discovery in the CIoT

The data used in CIoT systems are obtained from the physical world and the social world; the data should be processed well in an organized manner. Since trillions of objects and devices are involved in generating the massive amounts of data, which are connected and expected to work cooperatively, it is still a difficult task to utilize the analyzed data effectively to take meaningful actions, due to the complexity of the system that is generating the data as well as the inefficiencies of the algorithms used for the data processing (figure 1.6).

Since trillions of objects need to work cooperatively, they need to have a better understanding of their individual behaviors and need to take collective and cooperative actions which maximize the overall efficiency of the system and its collective goal.

For example, traffic signals at a junction can take a decision by analyzing the number of vehicles waiting at each junction while also using information about whether an ambulance is present among them; they can take a collective decision that serves society [8]. A well thought-out decision can taken after data analysis has been performed.

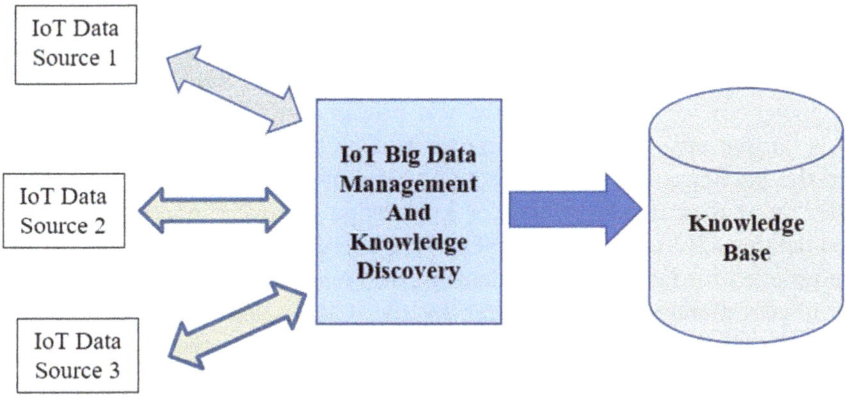

Figure 1.6. Knowledge discovery in the CIoT.

To make this kind of decision-making possible, the devices in the CIoT need to have a better understanding of the system they are in and their expected behavior in a given scenario by automatically deriving the semantic information from the analyzed data [6]. Based on the analyzed data and semantics, valuable rules and patterns of decision can be derived and stored as knowledge for use in future decisions by making the objects in the CIoT intelligent.

The derivation of semantics gives a context for the situation known to the objects in the CIoT which is relevant for the interaction between the user and the application being supported by the CIoT. For example, a real-time traffic situation map can be derived using crowdsourced data obtained from different sources such as drivers, vehicles, and different sensors on the road. Following data analysis and the application of semantic processing by incorporating the rules derived from the social world constraints and expected behavior of the system, a decision can be taken. The CIoT system is expected to be capable of collating and fusing the information from different contexts to arrive at a collective decision which maximizes the goal of the system.

In philosophy, an ontology is a theory about the nature of existence. An ontology is a formal, explicit specification of a shared conceptualization. It offers an expressive language that can represent the relationship between different contexts in the CIoT. If the fusion of different contexts is performed properly, the decisions derived from them are more appropriate for a given collective context of the system. This drives the need for standardization of semantics in the CIoT, which would enable effective semantic interoperability and extendibility [5]. The CIoT should support interactions between massive amounts of heterogenous data from different sources, including their interactions from various sources and contexts, which can be made possible by defining standard interfaces and models to make sure that a high degree of semantic interoperability is achieved between diverse systems.

This would enable the objects in diverse contexts to interact and exchange information unambiguously with higher efficiency and accuracy. To make this possible, the objects need to exchange semantic information at frequent intervals so that the data exchanged are reliable and reflect the real-time states of the different contexts that are to be considered in the collective decision-making process.

In the radio spectrum allocation scenario, standards related to spectrum allocation, the selection of nodes, transmission power control, and the communication protocols chosen ensure that the devices in the CIoT share the valuable spectrum between the participating nodes effectively and harmoniously.

Clustering analysis is performed for knowledge discovery by partitioning a set of analyzed data into subsets, in which each of the subsets is a cluster. Each of the data sets that are grouped into a cluster are identical, whereas the data that belong to different clusters have different characteristics. In the CIoT, depending on the clustering algorithms used, different clusters can be generated from the same set of diverse processed data. The clusters formed based on the number of vehicles on the road would be different from the clusters formed based on the number of passengers in them, those based on the types of vehicle involved, and those based on the number of passengers traveling in each of the vehicles. Thus, the clustering algorithms used for the types of data considered for the clustering result in different clusters being formed, which in turn changes the decisions derived from them that meet the collective goal of the system.

1.2.7 Intelligent decision-making in the CIoT

The general steps involved in intelligent decision-making in the CIoT are reasoning, planning, and selecting. Reasoning and planning involve analyzing the data collected from diverse sources after the necessary initial data preprocessing and data fusion (if required), and inferring useful information from them (figure 1.7).

The 'selecting' step refers to the decision-making, i.e. choosing the most appropriate outcomes from a possible set of different outcomes that the system could take. For example, some possible decisions could be choosing a particular path for a vehicle in a smart traffic system or choosing a channel in a wireless transmission system, etc.

1.2.8 Protocols in the CIoT

The types of protocol used between the devices of the CIoT have an impact on the types of decision made and the target applications that can be supported by such a CIoT

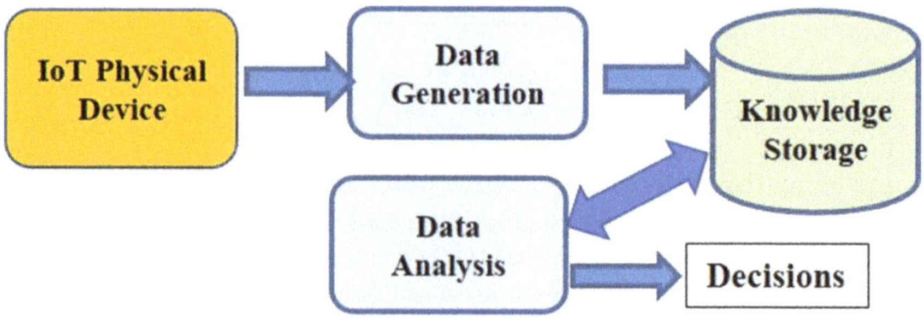

Figure 1.7. Decision-making in the CIoT.

system. A protocol can be real-time or non-real-time in nature, depending on whether the protocol supports the dynamic configuration of its networking parameters.

For example, in gaming applications with multiple users, the decisions taken by an individual player can affect and change the possible decisions the other players can take, which is a continuously changing scenario that impacts all the players in the game. In a similar way, since the CIoT has sensors which are diverse in nature and have different contexts from the perspective of the environment the CIoT devices are in, the networking and routing decisions that an CIoT system can take based on the continuously changing conditions are many. Using massive data analysis and knowledge-based decision-making, different strategies can be analyzed and the best one can be chosen by the system, thus improving the performance of the network. Even if the initial decisions move away from delivering the expected results, due to the knowledge-assisted learning techniques, the decisions converge toward the best possible outcomes.

1.2.9 The framework of intelligent decision-making in the CIoT

We understand that the steps involved in decision-making in the CIoT use the available information gathered from a diverse set of sensors, helped by data fusion and the removal of duplicates and any potential noise from the data [12]. The data are then processed before they are used for decision-making and stored in a knowledge base for use in future data analysis (figure 1.8).

The semantic information is also derived from the available information, using an understanding of the expected behavior of the CIoT system based on the user and the application under consideration to arrive at the most suitable decision.

The above diagram depicts the way in which individual nodes in the CIoT come up with the final decision using the derived knowledge extracted from the available

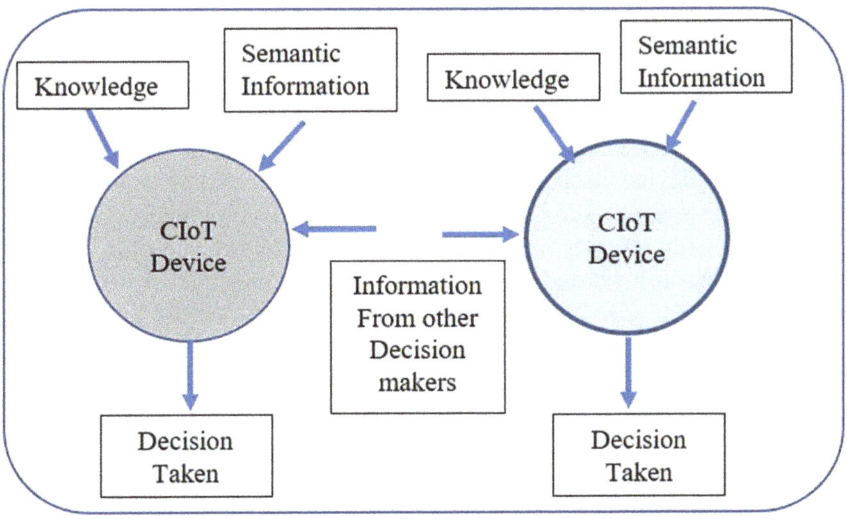

Figure 1.8. Framework of intelligent decision-making.

information of the system under consideration and the semantic information based on the application and the current state of the system.

In addition to using the semantic information and knowledge [9], the nodes also make use of information from other decision-makers, because in the case of the closely knit nodes of a CIoT system, the decisions taken by nearby nodes have a large impact on the decisions of individual nodes. *Swarm intelligence* is a good example, which is the collective behavior of decentralized, self-organizing systems in which a decision made by nearby nodes has an impact on the decisions taken by the closer nodes as well as impacting those further away, though in varying degrees.

1.3 The changing landscape due to the CIoT

We now turn to understanding how the IoT landscape, its use cases, and the end-user expectation of what IoT systems should deliver is changing drastically with the movement of AI and machine learning techniques into edge devices. This move is due to the proliferation of processors and systems with larger memories and computing capabilities built using low-power multicore microcontrollers that offer many more millions of instructions per second (MIPS) per watt than the systems available a decade ago.

The cognitive IoT has the power to resolve complicated issues and its devices are capable of taking decisions on their own without any human intervention while remaining within the minimal power budget that CIoT devices are provided with. Cognitive systems benefit significantly from learning, which is also true for humans. Standard IoT systems may find that a specific task cannot be accomplished, while a cognitive system finds a means and an efficient method of accomplishing it using its expanded intelligence and capability to adapt and take on challenging tasks which were never previously intended to be performed by these systems.

As per research by the firm Frost & Sullivan, the IoT will move to cognitive, predictive computing over the next 12–18 months. CIoTs are self-learning systems that utilize data mining, natural language processing, pattern recognition, and vision to imitate the way humans think.

As we learnt in this chapter, CIoT systems make use of machine learning algorithms and light versions of deep learning neural networks to run inferencing tasks on the IoT device itself without having to send massive amounts of data collected from the sensors across the Internet to the IoT cloud for decisions to be taken. This drastically reduces the traffic on the network and the time delay involved in arriving at a decision related to data from the diverse sources at the IoT device itself, thus making the CIoT more useful and responsive to the changing environment and conditions.

The ability for the CIoT to run inference engines using lighter versions of the machine learning models has been made possible both because of recent advances in processor design that can offer more MIPS per watt as well as lighter versions of the models defined by frameworks such as TensorFlow which are suitable for edge devices.

The machine learning framework TensorFlow provides a lighter version of models, defined using the TFlite format, which can run effectively on low-power and lower memory microcontroller-based systems without losing the ability to perform inference accurately at the edge on the IoT device itself. However, the weights and biases of the network are less accurate and represented by lower-resolution floating-point formats that only have 16 bits, which reduces memory and computational complexity on the edge computing devices.

This makes it possible to run applications and features which require decisions to be taken locally by the IoT devices themselves. Even if they are disconnected from the Internet and the central repository of the knowledge base maintained in the IoT cloud (due to any network issues or poor network infrastructure or perhaps natural calamities that disconnect the IoT devices from the central control repository) they can take decisions locally using the data they obtain from their connected sensors.

This makes CIoT systems more reliable and resilient in the face of connectivity disturbances and power outages. It is imperative that in the future, all IoT devices should be expected to have their own local intelligence and should be expected to be CIoT devices and systems.

1.4 Smart city applications using CIoT

The use of the IoT is common in smart city applications; CIoT systems are most suited for smart city applications because of the wide variety of sensors available and the different kinds of data generated by the innumerable sensors in the city.

Moreover, since these devices are deployed in the city, there is infrastructure available to power these IoT devices as well as access to very good Internet connectivity due to the widespread availability of WiFi-based connectivity in cities [10]. This allows smart city applications to make use of the advanced features and capabilities provided by CIoT systems with machine learning abilities in the edge, which are more useful and feasible to deploy and operate in city environments.

Since smart city solutions are expected to be closer to human activities, they need to work in conjunction with humans and understand their behavior to be more useful, effective, safe, and reliable solutions. The CIoT fits the bill perfectly, as it has the ability to learn from the massive data provided by reliable sensors and with the good connectivity available in cities, the CIoT systems can interact with other neighboring CIoT systems to understand their surroundings better and take collective action, so that the decisions taken by these systems are much closer to, or sometimes even better than, the way that humans react.

In a real-life scenario, a police officer at a traffic junction is only aware of the traffic conditions on the roads in front of him or her, whereas a CIoT system which is controlling the traffic in the city has a global view of the entire city traffic conditions, including the weather conditions and the forecast, which helps these systems to perform big data analytics and make more informed decisions in terms of either directing the traffic to ease congestion on the roads or to guide an ambulance ferrying a critical patient who needs urgent medical attention to the nearest hospital, etc.

A CIoT-based system monitoring a smart city can be more useful and efficient in providing a safer and more comfortable environment for the citizens of the city by continuously monitoring every person on the road and can possibly help security personnel to accomplish the difficult task of identifying criminals using advanced face recognition solutions running on the cameras installed on the roads, tracking their movements to guide the police team to arrest them with minimal delay.

CIoT systems installed in cities can potentially track any possible untoward incidents using gesture analysis and crowd behavior, thereby identifying and localizing issues and containing trouble before it spreads across the city. These systems can potentially catch these potential threats at the places where they originate and stop them before they spread to the rest of the city; thereby helping security personnel to contain and stop the damage.

Natural or man-made calamities such as floods or fire could also be tracked and damage could be contained using advancements in the CIoT systems' ability to perform video or image processing to identify natural calamities before they spread and endanger life and property.

Various smart city-based applications can be trained by applying cognitive data with machine learning techniques to make them more advanced. Models can be trained using available data collected from incidents that have happened in the past in different parts of the world.

These systems have ability to learn from systems that have gathered such data previously and use it to their advantage; reusing learnings on systems with pre-trained models can help these systems to be useful in a new environment as well.

Although such systems are more useful and their use cases are many, there are also associated issues in successfully designing, training, and implementing them and commissioning them to provide reliable solutions.

- We need to understand how a CIoT can be implemented successfully and devise a data model that allows them to cater to exponentially growing data that can be collected from the city. Scalability is a big concern in this situation, especially when one considers the requirement to provide a reliable solution.
- The existing cognitive systems which are capable of providing solutions for multiple smart city-based applications are still not flexible enough. The current solutions still lack scalability and flexibility, leading to the result that they are unable to take care of multiple solutions that have real-time constraints.
- Current research is limited, since it does not address the flexibility required to in provide cognitive data not only to a single cognitive computing AI-based solution but also to multiple solutions using the same data.
- The existing CIoT systems are not mature enough to be sufficiently generic that the models which are trained using data from the past events or data from another city or a region are directly usable.
- There are many different conditions and scenarios to be considered before these systems can be put in place, since any wrong decisions made by these

systems could be very costly and dangerous, because they would be directly interacting with humans and their safety and lives.

Lots of research is taking place to make CIoT-based solutions more effective and to reduce the lead time required to train and deploy them to manage real-life situations. The long lead time is due to the diverse nature of the sensors and systems they need to be interfaced with.

1.5 Challenges in the CIoT

The complexity of these systems is so large that performing system testing with all possible inputs and making sure that no erroneous decisions are produced by the system when a rare combination of input conditions is given to it is very hard and time-consuming.

The reliability of these systems will be questioned if they are not foolproof and well tested. As machine learning models run on these systems, an element of uncertainty is possible if they are not trained effectively, if all the corner cases have not been tested, or if the proper actions are not taken prior to deploying them in the field.

The massive amounts of data these systems process make them more complex in nature. Any potential disruption to the system or by a malicious entity trying to confuse the system with wrong or misleading data could thus make the system misbehave and perform an dangerous action that causes harm to humans. Protecting these systems from cyber criminals is the most important challenge, and needs to be addressed without any compromise.

1.6 Advantages of the CIoT

The CIoT has a much wider application scope. The same infrastructure is capable of supporting a multiuser environment and allowing multiple applications to run on it. This flexibility comes from the capabilities of performing data analytics using big data collected from multiple sources and extracting intelligence from the data collected from various sensors.

Adding intelligence or a 'brain' to the IoT devices in the CIoT enables us to exploit the information gathered and use it in various application scenarios. The CIoT gives us the benefit of being able to expand its usage and application domains even after the hardware is installed by updating its capabilities by changing the data analytics algorithms running on the system.

This enables users and service providers to make use of the money spent on the IoT infrastructure installation and maintenance by expanding the roles of IoT systems.

Scalability is the most important advantage of CIoT systems, which can accommodate the exponential growth of data generation without affecting the efficient operation of the system because of the use of big data and and the support of data analytics.

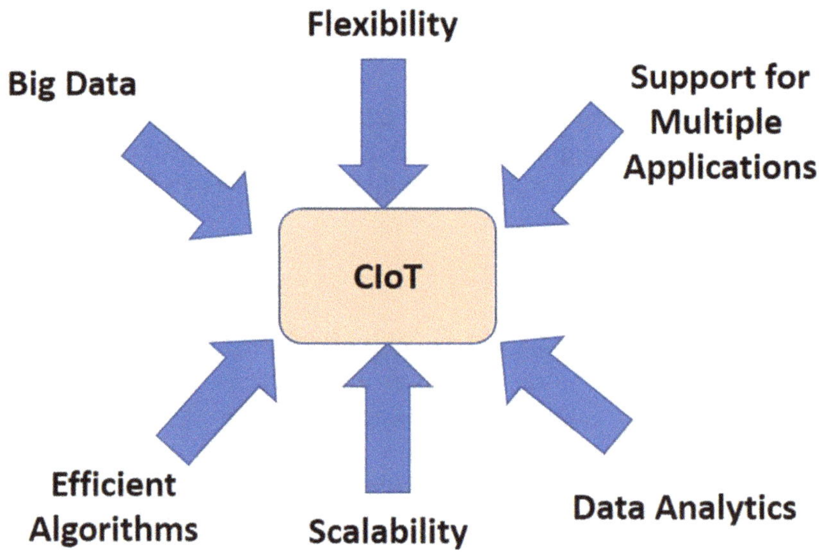

Figure 1.9. Advantages of the CIoT.

Secure communication and storage of data on the IoT cloud requires that the data in transit should be protected from adversaries. This means that the data needs to be encrypted in transit as well at rest, i.e. when the data is stored in the local system as well as when the data is sent across the Internet to repositories. Such repositories are used for big data analytics or for further processing of the data from different locations and time series; such data analysis allows a comparison of the changes happening to the data over time (figure 1.9).

The evolution and widespread presence of the CIoT is already happening in the marketplace due to its higher flexibility and higher return on investment for service providers. As a result of these advantages of the CIoT, its usage will become widespread across the world.

1.7 Conclusions

This chapter provided an introduction to the generic architecture of the IoT and its transition to the CIoT. We also discussed various connectivity or communication protocols and systems used in recent times which cater to the higher demands that IoT devices place on their communication technologies in terms of data bandwidth, reduced latency, and security features.

Such communication technologies enable CIoT devices to be connected to the Internet, with the result that that are finally able to store the huge amounts of data they generate in the IoT cloud. Prior to storage, big data analysis uses various sophisticated algorithms to extract useful information from the massive amount of data generated by the sensors. We discussed the existing implementation and architecture of the IoT and the cognitive-computing-based implementation, using the smart city as an example.

The cognitive IoT enhances the accuracy and capability of complex, sensor-driven systems through learning and embeds more human awareness into the devices and the environment the devices interact with. This enables these intelligent CIoT devices to be integrated into our environment, giving us a human-like interface and comfort from the end-user perspective.

The CIoT can be considered to be a combination of classical IoT systems with cognitive and cooperative abilities; this gives us an increased level of performance and efficiency due to the system's awareness of its environment and its better understanding of the user requirements and how to meet them with higher efficiency, thereby providing better results.

Acknowledgments

I would like to acknowledge the support that Chandramouleeswaran Sankaran received from his family members and his friends, which have been invaluable in this work.

References

[1] Wu Q, Ding G, Xu Y, Feng S, Du Z, Wang J and Long K 2014 Cognitive Internet of Things: a new paradigm beyond connection *IEEE Internet of Things J.* **1** 129–43

[2] Matin Mohammad Abdul 2020 *Towards Cognitive IoT Networks* (Berlin: Springer) pp 12–37

[3] Al-Turjman F 2017 *Cognitive Sensors and IoT: Architecture, Deployment, and Data Delivery* (Boca Raton, FL: CRC Press) pp 18–29

[4] Kumar Sandeep, Raja Rohit, Tiwari Shrikant and Rani Shilpa (ed) 2022 *Cognitive Behavior and Human Computer Interaction Based on Machine Learning Algorithms* (New York: Wiley) pp 2–25

[5] Zhang M, Zhao H, Zheng R, Wu Q and Wei W 2012 Cognitive Internet of Things: concepts and application example *Int. J. Computer Science Issues* **9** 151–8 https://www.ijcsi.org/articles/Cognitive-internet-of-things-concepts-and-application-example.php

[6] Kumar N and Makkar A 2020 *Machine Learning in Cognitive IoT* (Boca Raton, FL: CRC Press) pp 24–64

[7] Qiu R and Wicks M 2014 *Cognitive Networked Sensing and Big Data* (Berlin: Springer) pp 40–86

[8] Hwang K and Chen M 2017 *Big-Data Analytics for Cloud, IoT and Cognitive Computing* (New York: Wiley) pp 5–24 https://www.wiley.com/en-gb/Big+Data+Analytics+for+Cloud%2C+IoT+and+Cognitive+Computing-p-9781119247296

[9] Clark D D, Partridge C, Christopher Ramming J and Wroclawski J T 2003 *A Knowledge Plane for the Internet SIGCOMM'03* pp 1–20

[10] Park Jh, Salim M M, Jo J H, Sicato J C S, Rathore S and Park J H 2019 CIoT-Net: a scalable cognitive IoT based smart city network architecture *Hum. Cent. Comput. Inf. Sci* **9** 29

[11] Pfeifer R and Scheier C 1999 *Understanding Intelligence* (Cambridge, MA: MIT Press) pp 4–21

[12] Fischler M A and Firschein O 1987 *Intelligence: The Eye, the Brain, and the Computer* (Reading, MA: Addison-Wesley) pp 23–43

IOP Publishing

Cognitive Sensors, Volume 1
Intelligent sensing, sensor data analysis and applications
G R Sinha and Varun Bajaj

Chapter 2

Internet of Things-based cognitive wireless sensor networks: applications, merits, and demerits

Tanu Rizvi, Ravi Mishra, Priyanka Nandkishor Chopkar, Anupama Gomkale, Prajakta Upadhye and Devanand Bhonsle

Wireless sensor networks (WSNs), which include huge numbers of sensor networks, have enabled the Internet of Things (IoT) to be built. They perform various operations, viz. the sensing, processing, and observation of physical parameters. Sensor nodes are used to collect data, viz. acoustics, temperature, pressure, movements, etc. from the environment for daily activities. Such data may also be used for military purposes. Conventional sensor nodes in a WSN generally measure the physical parameters of the environment, whereas cognitive wireless sensor networks (CWSNs) are modified versions of conventional WSNs which require the execution of algorithms in the sensor nodes themselves. This chapter discusses various applications and advantages of WSNs. Today, CWSNs are widely used for various military purposes, public security, healthcare, home appliances, real-time surveillance, transportation networks, etc. CWSN provides many benefits such as efficient spectrum utilization, prevention of attacks, global operation, a large communication range, a low node count, good transmission quality, energy efficiency, etc. However, it has some serious issues such as false alarms, the probability of incorrect detection, high power consumption, etc.

2.1 Introduction

In this section, we discuss the basics of the IoT, WSNs, and CWSNs in detail.

2.1.1 The Internet of Things

The IoT consists of the synchronized structured operation of physical devices, computing elements, digital devices, things, machines, living organisms such as

doi:10.1088/978-0-7503-5326-7ch2

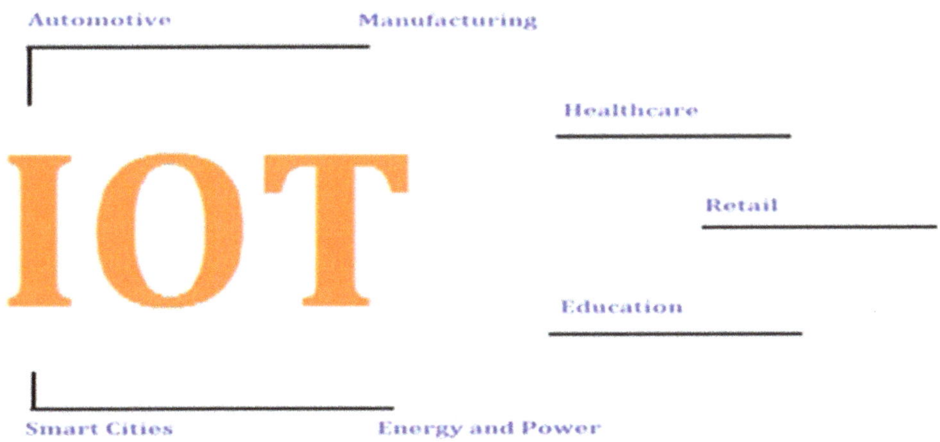

Figure 2.1. Fields in which the IoT is used.

humans and animals, etc. that are interconnected or networked to fulfill end-user requirements using unique identifiers [1, 2]. In the IoT, data can be transferred from one node to another without a physical connection. The subsystems of the IoT comprise networks of sensors, processors, communication devices, and web-enabled smart devices that are employed for data collection, data transfer, and the execution of user commands [3–5]. The data transmitted between sensors and IoT devices are shared via an IoT-enabled gateway called the IoT gateway. The IoT gateway sends the data to cloud or processes them for local analysis. The communication between these devices works on the basis of information received from the various operational devices [6, 7]. The IoT performs a variety of work without any human interaction; however, human contact is sometimes necessary so that humans can provide instructions or access data to interact with them to set them up. Figure 2.1 shows some of the application areas of the IoT [8–11]. Today, the IoT is provided with WSNs and CWSNs and the basic concepts of WSNs and CWSNs, their applications, merits, and demerits have converged in the IoT [12, 13]. The IoT is the main technology that uses wired or wireless networks to connect users to devices. The wireless technologies used include WSNs, near-field communication (NFC), General Packet Radio Service (GPRS), Zigbee, CWSNs, and Bluetooth [14–17]. Many studies and business ideas have been evolving using the IoT.

2.1.2 Wireless sensor networks

A WSN can be defined as a synchronized network consisting of nodes working so as to sense and control their environment. The main sensor system of a WSN is called the perception system. When a WSN is used for information transmission, the sensors then efficiently perceive the external environment and use this information to meet the user requirements [18]. Today, WSNs are believed to be one of the main technologies for intelligent ambience. They are successful in delivering a smart communication environment which sets up intelligent network applications based upon end-user requirements [19–21]. Within the last seven years, wireless technology has sped up by 75% and the total mobile traffic has also grown at a rate of 6.7 exabytes per month [22]. Today, wireless and mobile devices have become one of the most popular consumer technologies. The increasing demands for wireless technology have created significant spectrum challenges for its utilization [23–25].

Since WSNs support highly sophisticated content and complicated structures, it is important to pay attention to their security, because the repercussions of a security breach are unknown. WSNs are essential to the growth of the IoT, and they are also closely tied to how rapidly people can communicate [4–6]. WSN applications can be used to deal with a variety of complex contexts and collect a variety of data and information. In a military setting, they have the ability to detect biological, chemical, and nuclear radiation as well as the placement of troops in hostile territories and other conditions. WSNs can be utilized for data gathering in agricultural environments, following animal footprints, analyzing pollution statuses, and forecasting the appearance of forest fires and mudslides, thereby supporting environmental monitoring and protection. Monitoring crop growth and product flow is a key component of intelligent production in both industry and agriculture [26]. The physiological data of patients can be collected and their problems analyzed in order to provide their medical care, thereby supporting their stability and wellbeing. People's lives are altering because of the evolution of wireless sensor networks. Many different fields, including the military, medical organizations, disaster prevention efforts, and other sectors are concerned about the information security of WSNs [27]. Open wireless communication channels are used by wireless sensor networks to convey data, but without security safeguards, such data is particularly susceptible to both internal and external attacks. However, WSN networks cannot handle the amount of processing required by cryptographic defense techniques [28, 29]. Therefore, selecting an effective encryption technique that can guarantee the security of WSN information is likewise a significant challenge. Figures 2.2(a) and (b) show how sensor node networks and WSNs work.

Figure 2.3 shows the basic components of a WSN. They consist of a power handling unit and a power management unit which initiate the working of the sensor unit, the microcontroller unit, and the transmitter/receiver unit; these manage the sensing, computing, and communication tasks, respectively, depending upon the end-user requirements.

2.1.3 WSNs versus the IoT

The IoT represents a shift in the interconnection of humans and devices that moves toward a completely informed approach. The IoT is based on the application of technical knowledge and intelligence to the surrounding environment. It therefore converts data into valuable smart data [30–32]. A WSN is nothing more than a basic operating component of the IoT. WSNs handle information provided by the IoT. They are employed to keep records of any changes, either physical or technical, and thus they link the systems with gathered information to a central unit [33]. As shown

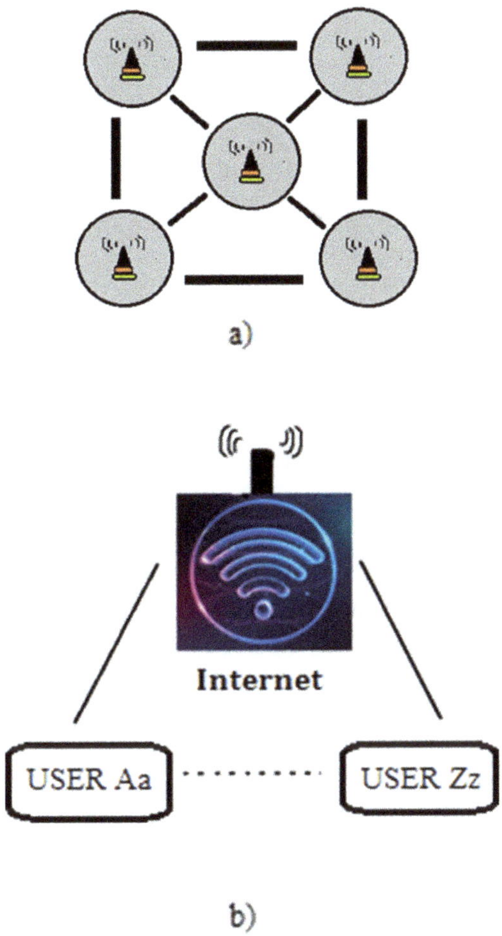

Figure 2.2. (a) A sensor node network. (b) A WSN.

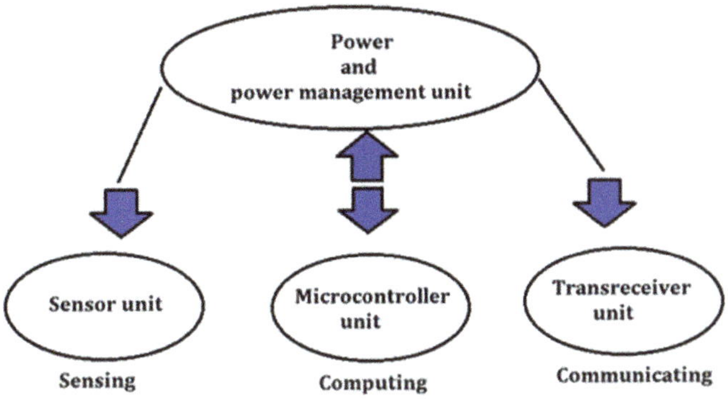

Figure 2.3. Components of a WSN.

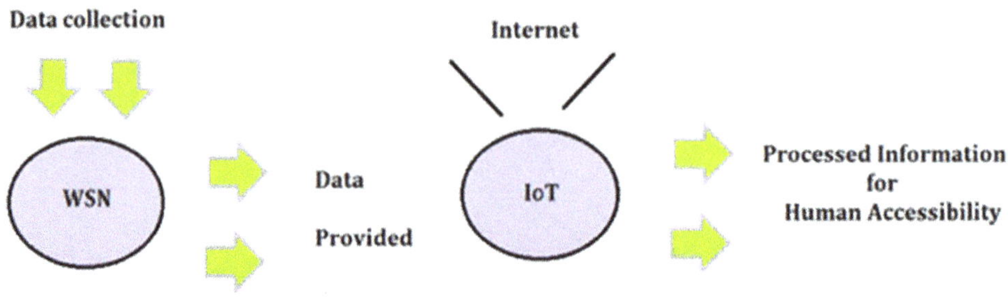

Figure 2.4. A comparison between a WSN and the IoT.

in figure 2.4, it is very clear that WSNs provide data by collecting it, while the IoT is an intelligent operating unit which applies the processed information, thereby intelligently satisfying the end-user requirements by using the Internet [34–37].

2.1.4 CWSNs

The advanced version of a WSN is a CWSN; however, there are many differences between WSNs and CWSNs. CWSN nodes adjust their transmission and reception properties according to the radio environment. In fact, WSNs constitute one of the areas in which cognitive networking is very much in demand [38–40]. WSN resource nodes have restrictions in terms of propagation speed, battery life expectancy, and spectrum availability. For all these reasons, WSNs with a cognitive ability can easily locate a free channel for transmission in the unauthorized spectrum band, or any free spectrum band from the authorized range can also be utilized for effective communication. Instead of using the broadly available 2.4 GHz band, a CWSN might give users access to extra spectrum as well as spectrum with improved propagation properties [41–43]. In a CWSN, selecting a channel with a lower frequency has several advantages, including a greater transmission range, a reduction in the number of sensor nodes required to cover a given region, and lower energy usage [44–46]. Figure 2.5 shows the architecture of a typical distributed CWSN network which is capable of sensing in a cooperative way, illustrating how the sensor nodes and malicious nodes interoperate with the base primary user (PU) station. A WSN is capable of sending, receiving, and idling, whereas a CWSN exhibits all the same properties but is also capable of spectrum sensing, as shown in figure 2.6. The sensing state is an additional stage in a CWSN, which keeps track of all the environmental parameters. As this is directly related to the sending and receiving stages, it therefore consumes more power [47].

For better propagation properties, cognitive technology is used to access new spectrum. A wide range of data rates can be achieved by adaptively adjusting system parameters such as the modulation scheme, transmit power, carrier frequency, and constellation size [48]. This undoubtedly enhances a WSN's reliability, network life, and power usage. There are variations in attack scope between traditional WSNs and CWSNs. In order to to create better security methods, it is very helpful to understand the taxonomy of CWSN assaults [49]. Wireless networks have to face

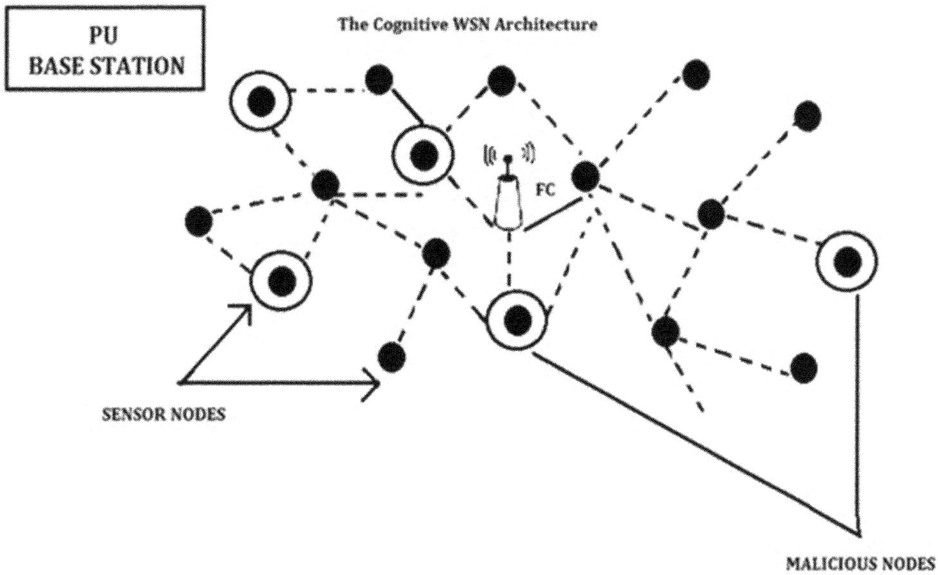

Figure 2.5. Architecture of a CWSN.

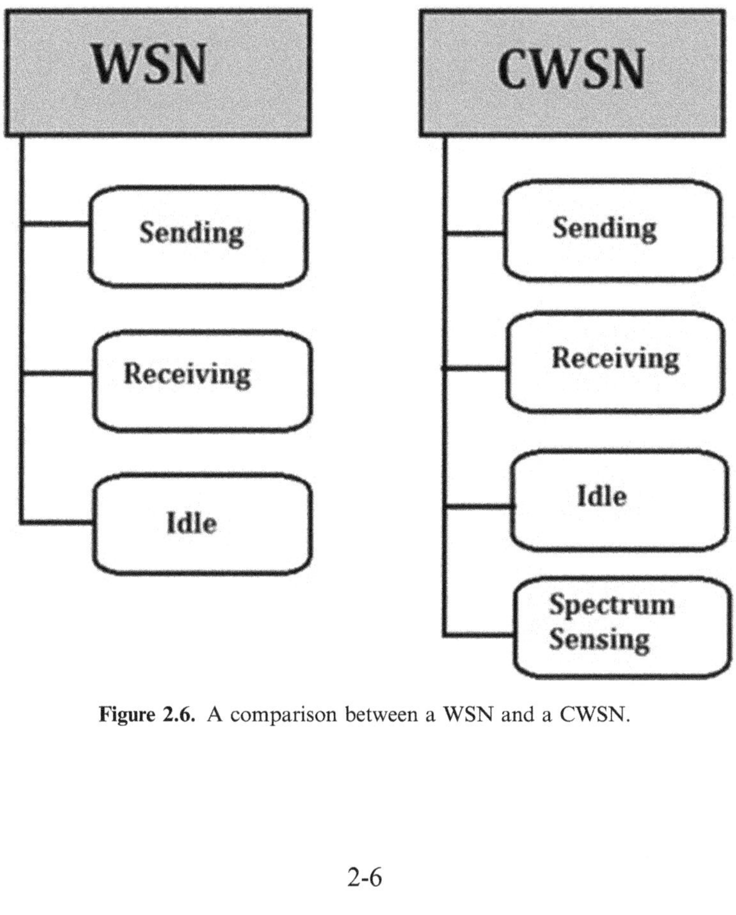

Figure 2.6. A comparison between a WSN and a CWSN.

multiple attack taxonomies and concentrated attacks on WSNs. There is a need to investigate unique network characteristics such as a large transmission range, low energy consumption, low latency, and data reliability, which can make CWSNs more resistant to attacks [50]. The use of radio waves as the transmission method obviously compromises CWSN security. WSNs that have cognition gain many advantages.

Four technical elements constitute cognitive abilities:

- spectrum sensing/monitoring
- analysis, characterization of the environment
- enhancement for an effective communication strategy while considering various restrictions
- adjustment and collaboration strategy [51].

Many benefits are obtained when such cognition capabilities are added to existing WSNs. Following this approach, the CWSN is a novel idea that has been suggested in the research literature; it has the following advantages:

- a large transmission range
- only a few sensor nodes are required to cover a specific scenario
- better spectrum usage
- reduced energy usage
- enhanced communication efficacy
- reduced delays
- enhanced information dependability
- global operability
- utilization of multiple channels.

Three categories can be used to group a CWSN's primary characteristics. These are as follows:

- being aware
- making decisions
- taking the appropriate action.

The primary characteristics of cognitive awareness can be achieved in three stages. The first phase involves uploading all relevant and interesting data to cognitive networks (CNs) [52–54]. These details may include the overall number of nodes and their IDs, the different types of sensor in the sensor networks (SNs), coverage zones, etc. In the second phase, which follows network deployment, CNs should be dynamically updated with the most recent network information, such as the battery life of individual nodes, the data values supplied by the SNs, user preferences, etc. However, the second-phase main characteristic decisions are made using reference tables that the CN also owns [55]. These reference tables should be developed based on the application goals and uploaded prior to deployment. As the result of the cognitive decisions, the CN performs a specific set of activities, which

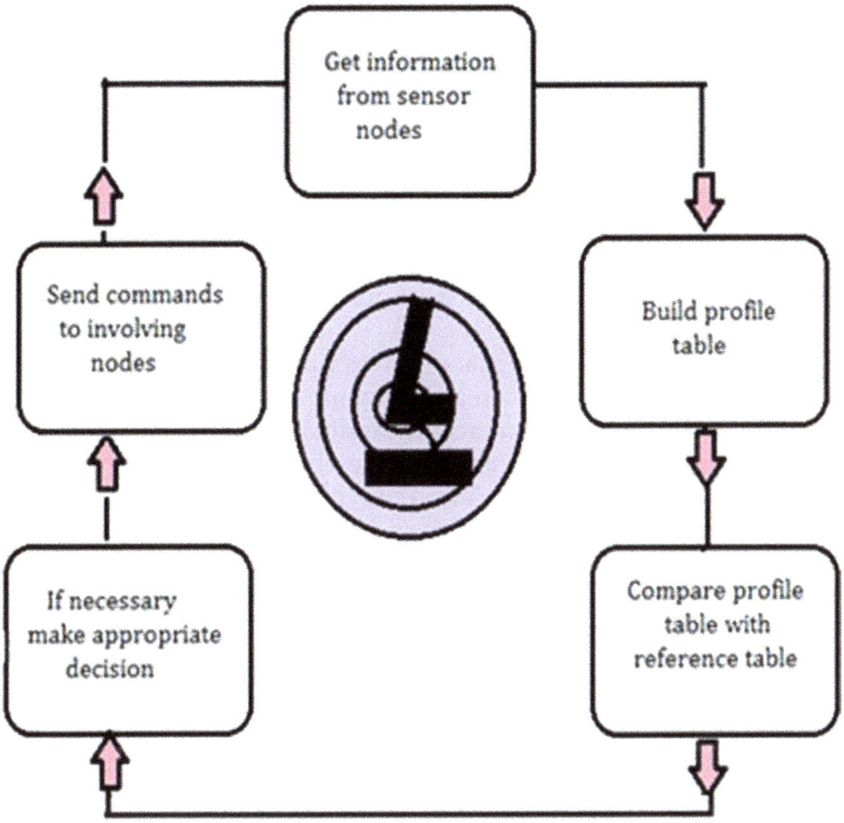

Figure 2.7. Cognitive cycle.

constitute the third primary characteristic [56]. In addition, a CN ought to be capable of communicating with its associated nodes. Figure 2.7 shows the entire cognitive cycle.

Sensor nodes provide information to the cognitive cycle, which is then used to build the profile table. The profile table is then compared with the reference table, and appropriate decisions are made based on the input data; various commands are subsequently signaled to the involved nodes for further action. This is how the entire cognitive cycle works [57]. However, CWSNs also have some flaws. These are as follows:

- short battery life
- inadequate computing power.

These two problems have posed a threat to CWSN network systems. In light of these characteristics, researchers are working to develop a better understanding of the attack mechanisms used against CWSNs and to design more promising defenses for sensor networks [58].

2.2 Literature survey

This section analyzes the literature on the topics of WSNs and CWSNs and the approaches used by various researchers. Researchers have shown that there are many software simulators which can be used to compare the network systems.

V Handziski *et al* (2003) explained that the most popular search term today is 'WSN.' There is no common approach in the field of networking research that is equivalent to the use of Linux on Intel processors, which has achieved popularity regardless of its benefits and constraints. The authors of this paper particularly researched simulation environments, development, hardware and operating systems [59].

K Romer *et al* (2004) explained that a system with different characteristics which requires a WSN system. It has become challenging to find a common discussion of both hardware and software requirements. Systems become more complex in the case of WSNs whose primary requirement is a close collaboration between users, developers, and application experts and their systems. In this article, the author discussed the factors that affect the design space of WSNs, taking the account of the various dimensions that lead to day-to-day applications. Furthermore, they back up our claim by showing how various existing implementations take different approaches in the design space [60].

L Reznik *et al* (2008) presented the idea of cognitive sensor networks and explored the feasibility of implementing them using artificial neural networks. Their paper described a new hierarchical study of tasks that was modeled on the structural topology of brain-like frameworks. These frameworks were composed of artificial neural networks that were dispersed across limited-resource network platforms. The study investigated a cognition concept using signal change detection in its implementation. Novel multilevel neural network architectures were developed and tested using sensor networks built from Crossbow Inc. sensor kits. The outcomes were presented to traditional multilayer perceptron structures due to their functional efficiency and resource consumption [61].

D Cavalcanti *et al* (2008) explained that WSNs operating in unlicensed frequency bands had recently gained popularity. However, there was evidence that the unlicensed spectrum had become overpopulated at that time. Rapid innovations in cognitive radio (CR) technology, on the other hand, enabled a dynamic spectrum access (DSA) model in WSNs to obtain access to less crowded spectrum with possibly better transmission properties. In this paper, researchers introduced a conceptual design for a CR-based WSN, highlighted the major advantages and risks of using CR technology, and proposed potential solutions to tackle the obstacles. As an example, they evaluated the feasibility of using a CR-based WSN for automated production application areas in residential and industrial settings [62].

E Bidra *et al* (2009) explained that the WSN structures used to capture and process vast volumes of data in parallel using tiny, power-limited devices enable WSNs to be used for vigilance, target acquisition, and a variety of other monitoring applications. New ideas for developing CWSNs have recently been proposed to enhance network and environment awareness and helps adaptive decisions to be made based upon specific goals. Compared to the conventional wireless networks, CWSNs are distinctive networks with many limitations. The primary concern, however, is security. The main contribution of these authors was to present the basic concept of CWSNs and also a taxonomy of attacks and countermeasures [63].

A S Zahmati *et al* (2009) explained that cognitive radio networks (CRNs) have many advantages; however, WSNs use a variety of sensors, which is very similar to process formation in WSNs. This paper provided a overview of CWSNs and also discussed emerging themes and recent challenges in the field. The authors discussed the major benefits of CWSNs as well as potential solutions to the challenges. Current WSNs can use the CWSN approach to overcome the restrictions of spectrum insufficiency, which occurs because the spectrum is shared by several relevant systems such as Wi-Fi and Bluetooth. It has been shown that the coexistence of such networks can substantially deteriorate WSN performance. Furthermore, this technology not only provides operation in new spectrum bands but also spectrum regions with enhanced transmission properties [64].

Md Alimul Haque *et al* (2012) explained that the ability of WSNs to capture and process enormous amounts of data concurrently with the aid of tiny, power-restricted devices makes them more effective for surveillance, target detection, and a variety of other monitoring applications. A variety of new theories have been developed for CWSNs, depending upon their development, based on application-based goals, an increased awareness of the system paradigm, and adaptive decisions. A CWSN is a unique type of wireless network that has many limitations compared to other wireless networks. However, security is the main issue. The primary contributions of this research are the taxonomy of attacks, remedies, and the fundamental concepts of cognitive wireless sensor networks [65].

Chijioke Worlu *et al* (2019) stated that one of today's most prominent technological concepts is the IoT, which is based on a vision that users and items can be connected anywhere and at any time using wired and wireless technologies such as Bluetooth, GPRS, LTE, NFC, ZigBee, and WSNs. IoT has attracted a lot of attention in the last ten years from both the business and scientific worlds. Using an IoT approach, application domains may experience a wide range of significant benefits. The goal of this topic of study was to demonstrate a thorough understanding of IoT-based smart environmental monitoring systems. IoT was said to have been dealing with a number of issues including socio-technical trust systems (STTS), reliability, privacy, security, and authentication. Present-day smart environments continue to experience severe IoT setbacks and difficulties in terms of security, privacy, and STTS. One of the key major advances required for the construction of robust structures that will serve to remove uncertainty and technical barriers is the development of an STTS comparison in the IoT. In order to compare results and support its findings, this study offers an overview of security, privacy, and STTS in the IoT. It aims to highlight and explain trust management's usefulness and how the IoT must use it [66].

Zhang Huanan *et al* (2020) describes the IoT, which is currently assisting WSNs to develop quickly. Regardless of time or location limitations, WSNs can provide users with the information they require at any time. The widespread use of WSNs creates a powerful foundation for the advent of the IoT. It is vital to study WSN security in order to reduce security threats and network attacks, since the node installation environment of WSNs is quite complex. The security and use of WSNs are considered in this study [67].

Heejung Yu *et al* (2020) describes IoT networks' recent innovations, extensions, and implementation and explains how they are transforming people's daily lives. The global 5G network uptake is significant for the ongoing development of the IoT. The road to the next-generation IoT networks' consists of WSNs and next-generation cellular networks. Next-generation IoT networks continue to face challenges in lowering end-to-end system latency and boosting throughput without sacrificing reliability. The coexistence of networks using different frequencies is a realistic solution. The major objections, however, are spectrum availability and data bandwidth support. The best technology, which now aims to resolve all these issues and permit the coexistence of the IoT, WSNs, 5G, and post-5G networks, is the CRN [68].

S Amulya *et al* (2021) explained that according to the principles of CR technology, cognitive sensor networks (CSNs) differ significantly from traditional WSNs. CSNs require changes to the transmitter parameters due to interactions with the environment. However, routing is among the crucial elements, and other networks' routing algorithms differ from those of CSNs in that they can take advantage of the spectrum. To construct a trustworthy forwarding path, the routing strategy should comprehend the variable spectrum resources. Buffer overflows and packet drops caused by a lack of spectrum have a significant negative impact on the connectivity between nodes. The prolongation of packet loss affects the network lifetime and the data delivery rate. This flaw is handled during the routing step, which thereby extends the network's life. We have researched alternative routing strategies for cognitive sensor networks based on power dissipation and the packet drop ratios of spectrum links. These approaches help to reduce the loss ratio and ensure that the data can traverse nodes with low drop ratios. As a result, the network's lifespan and energy efficacy will improve [69].

2.3 Methodology

This section discusses the various applications of WSN technologies.

2.3.1 Applications of WSN technologies

It is quite amazing that wireless communication is beginning to be integrated into personal and professional spaces due to the enormous success of wireless voice and informational communication services [53]. Building on this success, researchers have developed an application that focuses on tracking the health of cows and gathers and analyzes information from sensors attached to dairy cattle. It uses a microcontroller to wirelessly control the sensors and GPS to manage cow development. Additionally, it develops a virtual wall application that enables the mobility and space control of the creatures (animals) without the use of permanent buildings

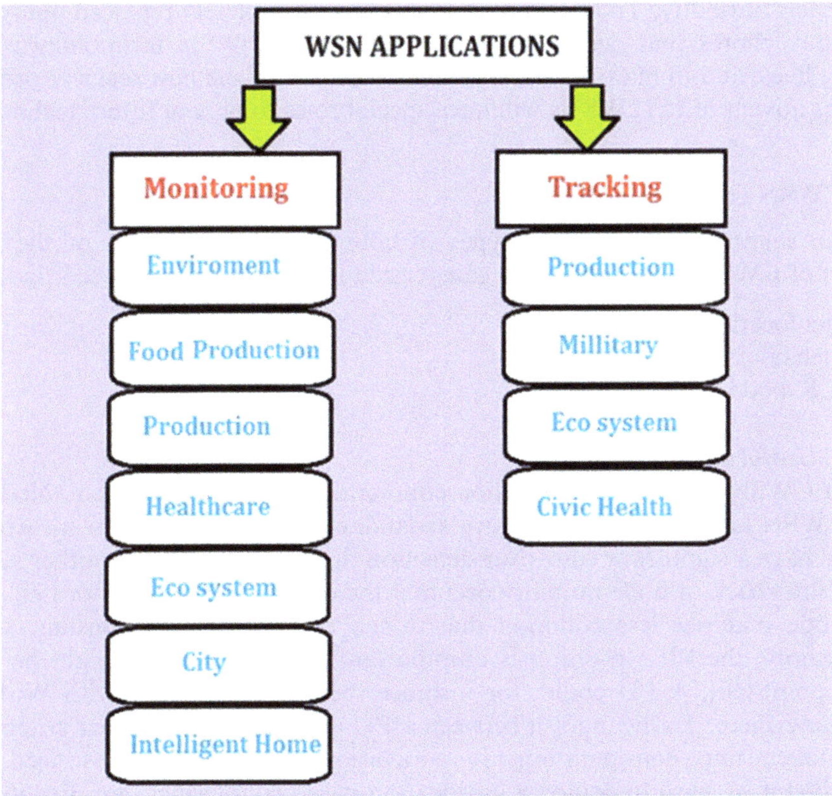

Figure 2.8. Application areas of WSN technology.

constructed by humans. The dairy animals are given a smart necklace that includes a GPS receiver and a sound amplifier.

According to a study, WSN applications make use of many real, useful advancement technologies that have been available for years. As shown in figure 2.8, because of this, the applications that use WSNs are primarily divided into two groups known as:

- monitoring
- tracking.

The monitoring component of WSN usage is defined as the portion that supervises, analyzes, and diligently monitors a system's movement in real time. The tracking part is defined as the component that monitors an overall utilization in order to continue altering an event, a person, an animal, etc [54, 55]. According to a different study, monitoring applications are available for biomedical or health monitoring, industrial monitoring, precision agriculture, internal and external environmental monitoring, electrical network monitoring, military location tracking, etc [56]. WSNs have become an important part of day-to-day human activities due to their flexibility. Today, smart gadgets have completely replaced many of the day-to-day chores that need a physical presence [57]. WSN technology is transforming lives, and in many cases, it is seen to be one of the best security providers. With the advent of IoT, WSNs will have special roles to play in future technologies.

2.3.2 CWSN countermeasures

Here we suggest three different types of countermeasure in light of the unique features of CWSNs [58]. These are categorized into three groups based upon:

- geolocation
- behavior
- CR node trust factors.

2.3.2.1 Geolocation

In most CWSN instances, geolocation countermeasures are ineffective. Nodes in a typical WSN have the ability to move around; even attackers can do so. Attackers actually have a significant edge over detection during movement. Another security-related drawback of node mobility occurs if the objective is to monitor [59]. In this case, node batteries less common due to the requirement for constant sensing. Furthermore, the PU's position is unimportant for security if it might be at any spatial point [60]. A PU could, for instance, be a mobile phone with Wi-Fi that travels anywhere. To distinguish between a PU and an attacker, other criteria need to be taken into consideration. In conclusion, some restrictions need to be established if we wish to deploy a geolocation-based countermeasure. For instance, a scenario can include fixed numbers of PUs or restricted regions for attackers [61].

2.3.2.2 Behavior

Defenses based on behavior aim to simulate the primary user (PU) and work in a similar way to geolocation countermeasures. This concept is used to search for distinctions between an attacker and a PU. For instance, it employs a few radio factors to determine whether a transmitter is an attacker or an incumbent transmitter [62]. Those criteria are signal characteristics, transmission power, and location. In CR, a typical television setting, the PU model might be extremely accurate. The earlier research does not apply to CWSNs, however, as it does not work for geolocation under countermeasures [63, 64]. Unfortunately, there is still no PU model in CWSNs. The PU is typically more erratic than in the earlier cases. However, the PU can be particularly identified if we concentrate our CWSN in constrained circumstances, such as environmental intelligence about a house or building. It is possible to identify parameters such as power transfer, spectrum time occupancy, and frequency usage. Algorithms based on genetic or self-organizing maps may be used to identify the behavior of PUs and distinguish them from attackers. These algorithms are a suitable solution to this issue, since they can spot patterns and changes in behavior [65–67]. However, it is important to consider the cost in terms of the computing resources used and battery life.

2.3.2.3 CR node trust factors

Two distinct categories of defense are connected to the scenario and its behaviors. The third group is an addition to the earlier techniques that might enhance assault detection. In WSNs, reputation systems are fairly widespread [68]. Reputation systems benefits from the redundancy and adaptability that are inherent to WSNs. A network of sensors is typically formed, and information is returned. Repetition can be utilized to identify and separate compromised or defective nodes [69].

2.4 Results and discussion

Several attacks are encountered by CWSNs. They can be categorized into the main functional blocks shown in figure 2.9. The well-known CWSN attack methods are:

- Communication attacks
- Node-targeted attacks
- Power consumption attacks
- Cryptographic attacks
- Policy attacks
- Privacy attacks.

Communication attacks very often have the direct aim of affecting the transmission between associated nodes. The aim could be node isolation, which hampers the performance of the whole network and is subcategorized into various types. Node-targeted attacks are very important, as information transmission is very important for the operation of CWSNs [70]. Nodes that are under an attacker's control can be utilized to work in opposition to their normal actions. This attack can hamper the functionality of the whole network. These networks are vulnerable to power attacks because the node sizes and batteries are very small size. These attacks

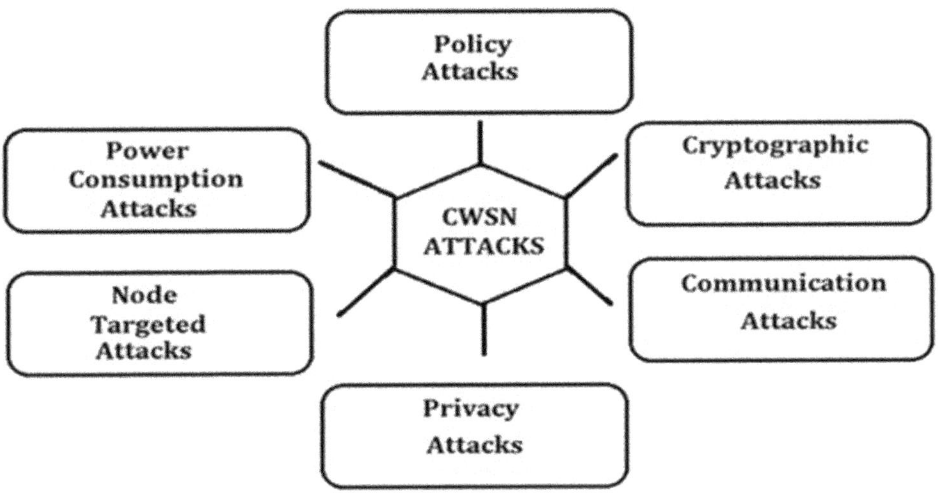

Figure 2.9. CWSN attacks.

may result in loss of battery power, since the nodes are engaged in unnecessary communications defined by the attackers [71]. If attackers can discover a weakness in the code or the key management system, this may enable cryptographic attacks to take place. Policy and security attacks are synonymous, since both are affected by the defined rules [72]. CWSNs share their resources to set up communication channels, which can enable privacy attacks to take place. Node information can be extracted through these communication channels, resulting in privacy or security attacks.

Researchers have devised approaches that can be used to simulate and model attacks using performance analysis. They have found that the best-known network simulators include NS-2, NS-3, Cooja, Castalia, OMNET++, GloMoSim, TOSSIM, and Avrora. The applications used to simulate discrete event networks include OMNET++, NS-2, and NS-3. In this study, we have analyzed the well-known NS-2 open source technology by simulating the SUNSET undersea sensor network. Additionally, it has been found that in the case of radio access in particular, OMNET++ can test the protocols and distributed algorithms used in actual wireless channels and radio models as well as node behavior. Reputation systems can indicate whether the primary and secondary users (SUs) are behaving as expected in CWSNs, where information is crucial for cognitive behavior and sharing information is practically required [72]. The reputation system can alter every node's reputation and trust thanks to the large amount of information it receives. The greatest benefit of reputation-based systems is their adaptability. Their counter-measures might be employed in conjunction with other attacks and could be deployed in any device, including small sensors with few resources. Simulation findings and hardware testbed results have been mentioned by various researchers. This enables simulation-based measures to be combined with real-world testbed results [73]. They also emphasized how crucial it is to use several metrics in order to better understand the process being studied. Although simulations by themselves are unable to portray the dynamism and intricacies of the real world, they demonstrate the system's consistency and dependability. The displayed system shown in figure 2.9 offers a few highlights that will be helpful in real WSN deployments in the health sector [74]. Considering low energy consumption, good precision, security, truthfulness, and accessibility as examples, table 2.1 elaborates the simulator results obtained using different simulators for various parameters such as the hardware (HW) platform support, operating system (OS) support, power usage, security, and the limitations of the various simulators. Figure 2.10 shows the favorable simulators for the various attack parameters [75].

These simulators are significant because they promote hazard evaluation, system management, and data security while also encouraging risk assessment. According to the results obtained by researchers, the simulations closely mirrored the results of deployments in the real world. However, they also involve determining the appropriate malicious node ID(s), detecting false positives or negatives, and creating a new routing path [76, 77]. They support heterogeneous networks in terms of both the hardware and software used for simulation. According to extensive analysis it was said that the study's results were approximated by simulations, and that the procedures described using the cluster topology were energy efficient. Furthermore,

Table 2.1. Simulator result.

Simulation	Parameters HW platform support	OS support	Power usage	Security	Limitations
UW SIM	Good	None	None	None	For water networks
Avrora	Limited	None	Good	None	For Mica2 sensor networks
Castalia	None	None	Good	None	Not a sensor-specific platform
TOSSIM	None	TinyOS	Good	None	TinyOS code only
NS-2	None	None	Good	None	No real traffic
NS-3	None	None	Good	None	No real traffic

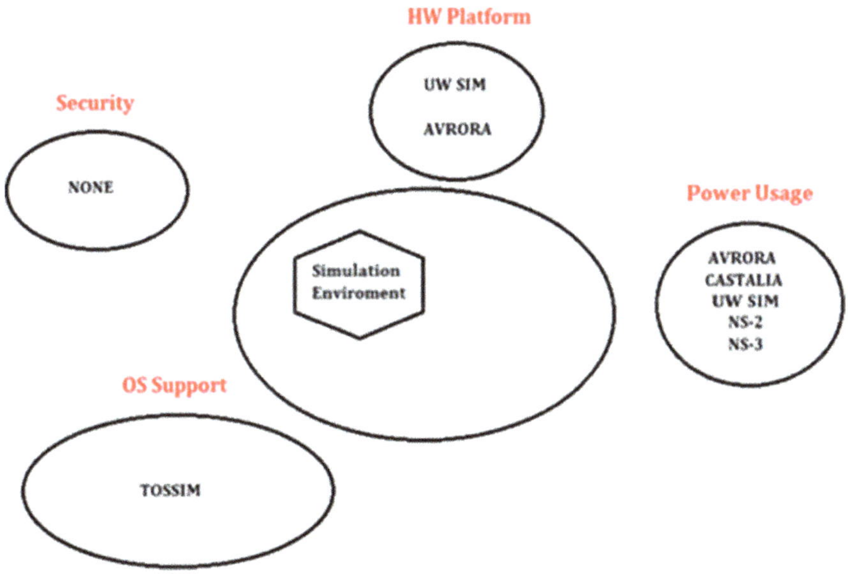

Figure 2.10. The preferred simulators for various attack parameters.

papers show that according to extensive simulations of sensor nodes placed at random, the proposed technique is found to be more favorable and supportive of various IoT uses than the old-style WSN concept [78–80].

According to studies, the simulation results show that CWSNs are more flexible and energy efficient than traditional WSN concepts and may thus be implemented for successful communication in the IoT [81–83]. According to the findings, researchers have proved that it is very important to select a travel option based on the suggested algorithm. This is accomplished by using a linear optimization that calculates an acceptable area and enforces a certain energy use within that area, consistently pushing the network node to function for the longest possible duration [66, 84].

2.5 Conclusions and future scope

Rapid developments in sensor and communication technologies led to the use of WSNs and thus CWSNs. CWSN technology will soon be much deeper and wider. It can be employed as basic security for end-user services. The use of CWSNs for the IoT is also steadily growing within industry. Through the massive growth of surrounding devices, IoT has the great potential to continually fuse the real and virtual worlds together, creating new, intriguing, and difficult opportunities for both academia and industry. This chapter primarily analyzed many IoT, WSN, and CWSN characteristics and described how the IoT will be the main focus of upcoming technological advancements. It also highlighted many obstacles that must be overcome in the use of the IoT. Taking CWSN smart environment monitoring systems as an example, a comparison of the findings for simulated security and privacy concerns was researched and presented. The most significant application fields were highlighted, and different use cases were identified. This chapter also analyzed the various features that CWSN architectures can offer, particularly in relation to intelligent environmental monitoring systems. This research supports observers and experts in this field of study by helping them to recognize the vast functionality of CWSNs, the critical issues to be addressed, and methods for implementing creative and immediate solutions that are crucial and can be used to convert CWSNs from a highly hypothetical abstract concept into an actuality within the real world.

References

[1] Lee I and Lee K 2015 The internet of things (IoT): applications, investments, and challenges for enterprises *Bus. Horiz.* **58** 431–40

[2] Butun I and Kocakulak M 2017 An overview of Wireless Sensor Networks towards internet of things *IEEE 7th Annual Computing and Communication Workshop and Conference (CCWC)* (Piscataway, NJ: IEEE)

[3] Hoyer V and Fischer M 2008 Market overview of enterprise mashup tools *Service-Oriented Computing – ICSOC 2008* (*Lecture Notes in Computer Science* vol 5364) pp 708–21

[4] Qin Y, Sheng Q Z, Falkner N J G, Dustdar S, Wang H and Vasilakos A V 2016 When things matter: a survey on data-centric internet of things *J. Netw. Comput. Appl.* **64** 137–53

[5] Cero E, Husic J B and Barakovic S 2017 IoT's tiny steps towards 5G: Telco's perspective *Symmetry* **9** 213

[6] Guillen E, Sanchez J and Ramos C O 2017 A model to evaluate the performance of IoT applications *Int. MultiConference of Engineers and Computer Scientists* (Hong Kong: International Association of Engineers) pp 15–8 http://www.iaeng.org/publication/IMECS2017/IMECS2017_pp147-150.pdf

[7] Sundmaeker H, Guillemin P, Friess P and Woelffle S 2010 *Vision and Challenges for Realising the Internet of Things* (Brussels: European Commission)

[8] Ning H and Wang Z 2011 Future internet of things architecture: like mankind neural system or social organization framework? *IEEE Commun. Lett.* **15** 461–63

[9] Perera C, Zaslavsky A, Christen P and Georgakopoulos D 2012 CA4IOT: Context Awareness for Internet of Things *2012 IEEE Int. Conf. on Green Computing and Communications* (Piscataway, NJ: IEEE) pp 775–82

[10] Vermesan O *et al* 2009 Internet of things strategic research roadmap *Internet Things- Global Technol. Soc. Trends* **1** 9–52 http://www.internet-of-things-research.eu/pdf/IoT_Cluster_Strategic_Research_Agenda_2011.pdf
[11] Yu L, Lu Y and Zhu X 2012 Smart hospital based on internet of things *J. Networks* **7** 1654–61
[12] Kaur S and Singh I 2016 A survey report on internet of things applications *Int. J. Computer Science Trends and Technology* **4** 330–35 http://ijcstjournal.org/Vol4Issue2No1.html
[13] Aly H, Elmogy M and Barakt S 2015 Big data on internet of things: applications, architecture, technologies, techniques, and future directions *Int. J. Computer Science Engineering* **4** 300–13 https://www.ijcse.net/docs/IJCSE15-04-06-040.pdf
[14] Vermesan O and Friess P 2014 *Internet of Things From Research and Innovation to Market Deployment* (Aalborg: River Publishers) https://www.riverpublishers.com/pdf/ebook/RP_E9788793102958.pdf
[15] Lee I and Lee K 2015 The internet of things (IoT): application, investment, and challenges for enterprises *Bus. Horiz.* **58** 431–40
[16] Gubbi J, Buyya R, Marusic S and Palaniswami M 2013 Internet of things (IoT): a vision, architectural elements, and future directions *Futur Gener Comput Syst* **29** 1645–60
[17] Bi Z, Da Xu L and Wang C 2014 Internet of things for enterprise systems of modern manufacturing *IEEE Trans. Ind. Informatics* **10** 1537–46
[18] Prathibha S R, Hongal A and Jyothi M P 2017 IoT based monitoring system in smart agriculture *2017 Int. Conf. on Recent Advances in Electronics and Communication Technology (ICRAECT)* (Piscataway, NJ: IEEE) pp 81–4
[19] Ejaz W and Ibnkahla M 2018 Multiband spectrum sensing and resource allocation for IoT in cognitive 5Gnetworks *IEEE Internet Things J.* **5** 150–63
[20] Henkel J, Pagani S, Amrouch H, Bauer L and Samie F 2017 Ultra-low power and dependability for IoT devices (invited paper for IoT technologies) *Design, Automation & Test in Europe Conference & Exhibition (DATE), 2017* (Piscataway, NJ: IEEE) pp 954–59
[21] Grønbæk I 2008 Architecture for the internet of things (IoT): API and interconnect *2008 Second Int. Conf. on Sensor Technologies and Applications (sensorcomm 2008)* (Piscataway, NJ: IEEE) pp 802–7
[22] Sadeghi A R, Wachsmann C and Waidner M 2015 Security and privacy challenges in industrial internet of things *DAC '15: The 52nd Annual Design Automation Conf. 2015* pp 1–6
[23] Mikko H and Nyman L 2017 The internet of (vulnerable) things: on Hypponen's law, security engineering, and IoT legislation *Technology Innovation Management Review* **7** 5–11
[24] Shin D 2014 Telematics and informatics a socio-technical framework for internet-of-things design: a human centered design for the internet of things *Telemat, Informatics* **31** 519–31
[25] Shin D H and Park Y J 2016 Understanding the internet of things ecosystem: multi-level analysis of users, society, and ecology *Digit. Policy, Regul. Gov.* **19** 77-100
[26] Vijay G, Elyes B A B and Ibnkahla M 2011 Congnition in Wireless Sensor Networks: A Perspective *IEEE Sensors J.* **11** 582–592
[27] Mitola J 2000 Cognitive radio: an integrated agent architecture for software defined RADIO *Doctoral Theses* Royal Institute of Technology, Stockholm, Sweden no. 2000-9566

[28] Aalamifar F, Vijay G, Khozani P A and Ibnkahla M 2011 *Cognitive wireless sensor networks for highway safety MSWiM '11: The 14th ACM Int. Conf. on Modeling, Analysis and Simulation of Wireless and Mobile Systems* (New York: ACM)

[29] Boukerche A, Oliveira H A B F, Nakamura E F and Loureiro A A F 2008 Secure localization algorithms for wireless sensor networks *IEEE Commun. Mag.* **46** 96–101

[30] Isaac S J, Hancke G P, Madhoo H and Khatri A 2011 A survey of wireless sensor network applications from a power utility's distribution perspective *IEEE Africon'11* (Piscataway, NJ: IEEE) pp 1–5

[31] Stabellini L and Zander J 2010 Energy-aware spectrum sensing in cognitive wireless sensor networks: a cross layer approach *IEEE Proc. Wireless Communications & Networking Conf. (WCNC 2010)* (Piscataway, NJ: IEEE)

[32] Ishizu K, Murakami H, Miyamoto G, Tran H N, Filin S, Hasegawa M, Murata Y, Harada H and Kato S 2008 *Design and Implementation of Cognitive Wireless Network based on IEEE P1900.4 2008 5th IEEE Annual Communications Society Conference on Sensor, Mesh and Ad Hoc Communications and Networks Workshops* (Piscataway, NJ: IEEE) pp 1–8

[33] Alhumud H and Zohdy M 2018 Managing energy consumption of wireless sensors networks in multiple greenhouses *Wireless Engineering and Technology* **99** 11–9

[34] Wang N, Huang Y and Liu W 2008 A fuzzy-based transport protocol for mobile ad hoc networks *2008 IEEE International Conference on Sensor Networks, Ubiquitous, and Trustworthy Computing (sutc 2008)* (Piscataway, NJ: IEEE) pp 320–25

[35] Zeng B and Yan D 2014 An energy efficient harmony search based routing algorithm for small-scale wireless sensor networks *IEEE 17th Int. Conf. on Computational Science and Engineering, Chengdu* pp 362–67

[36] Lee H M 2016 Optimal cost design of water distribution networks using a decomposition approach *Eng. Optim.* **48** 2141–56

[37] Hoang D C 2017 Real-time implementation of a harmony search algorithm based clustering protocol for energy-efficient wireless sensor networks *IEEE Trans. Ind. Inf.* **10** 774–83

[38] Parenreng J M and Kitagawa A 2017 A model of security adaptation for limited resources in wireless sensor network *J. Computer and Communications* **05** 10–23

[39] Vijayarajeswari R, Rajivkannan A and Santhosh J 2016 A simple steganography algorithm based on lossless compression technique in WSN *Circuits and Systems* **7** 1341–51

[40] Saravanaselvan A and Paramasivan B 2016 Implementation of an efficient light weight security algorithm for energy-constrained wireless sensor nodes *Circuits and Systems* **7** 2234–41

[41] Savoine M M, Menezes M O D and Andrade D A D 2018 Proposal of a methodology for the assessment of security levels of IoT wireless sensor networks in nuclear environments *World J. Nuclear Sci. Technol.* **8** 78–85

[42] Liu Y and Morgan Y 2018 Security analysis of subspace network coding *J. Information Security* **9** 85–94

[43] Rawat P, Singh K D, Chaouchi H and Bonnin J M 2014 Wireless sensor networks a survey on recent developments and potential synergies *J. Supercomput.* **68** 1–48

[44] Chen D B, Zhang N L, Zhang M G, Wang Z H and Zhang Y 2017 Study on remote monitoring system of crossing and spanning tangent tower *IOP Conf. Ser.: Mater. Sci. Eng.* **199** 012038

[45] Sicari S, Cappiello C, Pellegrini F D, Miorandi D and Porisini A C 2016 A security-and quality-aware system architecture for Internet of Things *Information Systems Frontiers* **18** 665–77

[46] Miorandi D, Sicari S, De Pellegrini F and Chlamtac I 2012 Ad hoc networks internet of things: vision, applications and research challenges *Ad Hoc Networks* **10** 1497–516
[47] Pirbhulal S, Zhang H, Alahi M E, Ghayvat H, Mukhopadhyay S, Zhang Y T and Wu W 2017 A novel secure IoT-based smart home automation system using a wireless sensor network *Sensors* **17** 69
[48] Hodge V J, Keefe S O, Weeks M and Moulds A 2015 Wireless sensor networks for condition monitoring in the railway industry: a survey *IEEE Trans. Intell. Transp. Syst.* **16** 1088–106
[49] Sidek H H A B, Jamal A A, Makhtar M and Fadzli S A 2018 Smart bicycle sharing system for non commercial purposes using time-based one-time password algorithm *Int. Journal of Engineering and Technology (UAE)* **7** 275–77
[50] Viani F, Robol F, Bertolli M, Polo A, Massa A, Ahmadi H and Boualleague R 2016 A wireless monitoring system for phytosanitary treatment in smart farming applications *IEEE Int. Symposium on Antennas and Propagation* pp 2001–2
[51] Wadud Z, Javaid N, Khan M A, Alrajeh N, Alabed M S and Guizani N 2017 Lifetime maximization via hole alleviation in IoT enabling heterogeneous wireless sensor networks *Sensors* **17** 1677
[52] Mambou E N, Nlom S M, Swart T G, Ouahada K, Ndjiongue A R and Ferreira H C 2016 Monitoring of the medication distribution and the refrigeration temperature in a pharmacy based on internet of things (IoT) technology *IEEE 18th Mediterranean Electrotechnical Conf.* pp 1–5
[53] Butler Z, Corke P, Peterson R and Rus D 2004 Virtual fences for controlling cows *IEEE Int. Conf. on Robotics and Automation* 4429–36
[54] Fei Z, Li B, Yang S, Xing C, Chen H and Hanzo L 2016 A survey of multi-objective optimization in wirelesss sensor networks: metrics, algorithms, and open problems *IEEE Commun. Surv. Tutorials* **19** 1677
[55] Arampatzis T, Lygeros J, Member S and Manesis S 2005 A survey of applications of wireless sensors and wireless sensor networks *IEEE Int. Symp. on Mediterranean Conf. on Control and Automation Intelligent Control* pp 719–24
[56] Ovsthus K and Kristensen L M 2014 An industrial perspective on wireless sensor networks-A survey of requirements, protocols, and challenges *IEEE Communications Surveys and Tutorials* **16** 1391–412
[57] Zhu Y, Song J and Dong F 2011 Applications of wireless sensor network in the agriculture environment monitoring *Procedia Engineering* **16** 608–14
[58] Milenkovic A, Otto C and Jovanov E 2006 Wirelesssensor networks for personal health monitoring: Issues and an implementation *Comput. Commun.* **29** 2521–33
[59] Handziski V, Kopke A, Karl H and Wolisz A 2003 *A common wireless sensor network architecture* TKN-03-012 Technische Universitat Berlin pp 10–7 http://www.eecs.tu-berlin.de/fileadmin/fg112/Papers/sensornetz_architecture.pdf
[60] Romer K and Mattern F 2004 The design space of wireless sensor networks *IEEE Wirel. Commun.* **11** 54–61
[61] Reznik L and Von Pless G 2008 Neural networks for cognitive sensor networks *IEEE Int. Joint Conf. on Neural Network, IJCNN* pp 1235–41
[62] Cavalcanti D, Das S, Wang J and Challapali K 2008 Cognitive radio based wireless sensor networks *Proc. 17th Int. Conf. on Computer Communications and Networks* vol 1 (Piscataway, NJ: IEEE) pp 1–6

[63] Bidra E and Ibnkahla M 2009 *Performance Modeling of Cognitive Wireless Sensor Networks Applied to Environmental Protection GLOBECOM 2009 - 2009 IEEE Global Telecommunications Conf.* (Piscataway, NJ: IEEE) pp 1–6

[64] Zahmati A S, Hussain S, Fenando C and Grami A 2009 Cognitive Wireless Sensor Networks: Emerging topics and recent challenges *2009 IEEE Toronto Int. Conf. Science and Technology for Humanity (TIC-STH)* pp 593–96

[65] Alimul Haque M, Faizanuddin M and Singh N K 2012 A study of cognitive wireless sensor networks: taxonomy of attacks and countermeasures *World Applied Programming J.* **2** 477–84

[66] Worlu C, Jamal A A and Mahiddin N A 2019 Wireless sensor networks, internet of things and their challenges *Int. J. Innovative Technology and Exploring Engineering (IJITEE)* **8** 556–66

[67] Huanan Z, Suping X and Jiannan W 2021 Security and application of wireless sensor network *Elsevier 10th Int. Conf. of Information and Communication Technology (ICICT-2020)* **vol 183** 486–92

[68] Yu H and Zikria Y B 2020 Cognitive radio networks for internet of things and wireless sensor networks *Sensors* **2020** 5288

[69] Amulya S and Madhuri M 2021 Survey on improving QoS of cognitive sensor networks using spectrum availability based routing techniques *Turkish J. Comput. Math. Edu.* **12** 2206–25 https://turcomat.org/index.php/turkbilmat/article/view/6211/5150

[70] Xia N and Yang C S 2016 Recent advances in machineto machine communications *J. Comput. Commun.* **4** 107–11

[71] Song Q, Nuaymi L and Lagrange X 2016 Survey of radio resource management issues and proposals for energy efficient cellular networks that will cover billions of machines *Eurasip J. Wirel. Commun. Netw.* **2016** 140

[72] Zaslavsky A, Perera C and Georgakopoulos D 2013 Sensing as a service and big data *Int. Conf. on Advances in Cloud Computing* pp 1–8

[73] Potdar V, Sharif A and Chang E 2009 Wireless sensor networks: a survey *2009 International Conference on Advanced Information Networking and Applications Workshops* (Piscataway, NJ: IEEE) pp 636–41

[74] Lazarescu M T 2013 Design of a WSN platform for long term environmental monitoring for IoT applications *IEEE J. Emerg. Sel. Top. Curcuits Syst.* **3** 45–54

[75] Cabra J, Castro D, Mendez J C D and Trujillo L 2017 An IoT approach for wireless sensor networks applied to e-health environmental monitoring *IEEE Int. Conf. on Internet of Things and IEEE Green Computing and Communications and IEEE Cyber, Physical and Social Computing and IEEE Smart Data* (Piscataway, NJ: IEEE) pp 578–83

[76] Chi Q, Yan H, Zhang C, Pang Z and Xu L D 2014 A reconfigurable smart sensor interface for industrial WSN in IoT environment *IEEE Trans. Ind. Informatics* **10** 1417–25

[77] Fang S, Xu L, Pei H, Liu Y, Liu Z, Zhu Y, Yan J and Zhang H 2014 An integrated approach to snowmelt flood forecasting in water resource management *IEEE Trans. Ind. Informatics* **10** 548–58

[78] Nishiyama Y, Ishino M, Koizumi Y, Hasegawa T, Sugiyama K and Tagami A 2016 Proposal on routing-based mobility architecture for ICN-based cellular networks *IEEE Conf. on Computer Communications Workshops* pp 467–72

[79] Mondal A and Bhattacharjee S 2017 A reliable, multi-path, connection oriented and independent transport protocol for IoT networks *IEEE 9th Int. Conf. on Communication Systems and Networks* (Piscataway, NJ: IEEE) pp 590–91
[80] Sicari S, Rizzardi A, Grieco L A and Porisini A C 2017 A secure ICN-IoT architecture *IEEE Int. Conf. Commun. Work.* (Piscataway, NJ: IEEE) pp 259–64
[81] Diaz A and Sanchez P 2016 Simulation of attacks for security in wireless sensor network *Sensors* **16** 1932
[82] Mathur A, Newe T and Rao M 2016 Defence against black hole and selective forwarding attacks for medical WSN sin the IoT *Sensors* **16** 118
[83] Rani S, Talwar R, Malhotra J, Ahmed S H, Sarkar M and Song H 2015 A novel scheme for an energy efficient Internet of Things based on wireless sensor networks *Sensors* **15** 28603–26
[84] Zhang T, Chen Z, Ouyang Y, Hao J and Xion Z 2009 An improved RFID-based locating algorithm by eliminating diversity of active tags for indoor environment *Comput. J.* **52** 902–9

IOP Publishing

Cognitive Sensors, Volume 1
Intelligent sensing, sensor data analysis and applications
G R Sinha and Varun Bajaj

Chapter 3

Psychiatric disorders and cognitive impairment following COVID-19: a comprehensive review and its implications for smart healthcare design

Tayebe Azimi, Amirhossein Koochekian, Hamid Reza Marateb, Mehdi Shirzadi, Mónica Rojas-Martínez, Joan F Alonso, Alejandro Bachiller Matarranz, Marjan Mansourian, Manuel Rubio Rivas, Sergio Romero Lafuente and Miquel Angel Mañanas

More than 80% of hospitalized COVID-19 patients experience post-COVID symptoms. Cognitive impairment, fatigue, depression, and dyspnea are some of the post-acute symptoms. The monitoring of such high-risk individuals is critical to avoid the development of additional severe long-term disorders. Artificial intelligence (AI)-based integration and analysis of the input symptoms at an early stage of the disease help in the provision of appropriate treatment. Various questionnaires, including the General Health Questionnaire (GHQ-12), the Connor-Davidson Resilience Scale, the Life Events Checklist for DSM-5 (LEC-5), the PTSD Checklist for DSM-5 (PCL-5), the Kessler Psychological Distress Scale (K10), the Brief Illness Perception Questionnaire, and the COVID-19 Peritraumatic Distress Index (CPDI) can be used to extract features for such predictions. Knowing that mild or moderate COVID-19 infection can be linked to cognitive impairments, such prediction systems are promising for smart healthcare design. This chapter discusses post-COVID symptoms and the development of related computer-aided diagnosis and prognosis systems.

3.1 Introduction

The coronavirus disease 2019 (COVID-19) emerged in Wuhan, China, in December 2019 and caused a global pandemic [1, 2]. It has affected 279 114 972 individuals worldwide and led to 5 397 580 deaths up until December 2021, according to the World Health Organization (WHO) [2, 3]. At first, COVID-19 was mainly known as a respiratory illness [1], However, observations have shown that its severity level differs from asymptomatic to severe [4, 5], and it may involve various body parts, such as the

doi:10.1088/978-0-7503-5326-7ch3

heart, lungs, muscles, and nervous system [6]. Lockdown and social distancing guidelines implemented by countries to reduce person-to-person contact to control the spread of COVID-19 have had a massive impact on people's mental health [7, 8]. These policies led to fear, depression, fatigue, weakness, and cognitive symptoms in the post-acute sequelae of SARS-COV-2 for at least four weeks after infection [6,9–18]. People with preexisting mental disorders and vulnerable populations, such as older adults with type two diabetes or cardiovascular disease, are at high risk of deteriorating mental health [19–21]. This may primarily happen due to their understanding of poor COVID-19 outcomes or concerns about their access to doctors [21].

Post-COVID sequelae were experienced by more than 80% of hospitalized COVID-19 patients [22]. Based on 103 papers published during the COVID-19 outbreak, anxiety prevalence reached 39.6% and 27.3% for COVID-19 patients and the general population, respectively [23]. It has also been shown that the prevalences of schizophrenia among patients infected with COVID-19 and the general population were 3.6% and 0.66%, respectively [24–26]. Studies have shown that symptoms related to cognitive impairment and psychiatric disorders are elements of 'long COVID-19' associated with the central nervous system [27]. Cognitive performance can be influenced by a number of variables such as age [28, 29], inflammatory processes due to cytokines released by immune cells [30, 31], lifestyle factors [32–37], genetic variants [38, 39], and infection variants [40, 41].

The development of infrastructure is needed, together with a comprehensive understanding, in order to prepare a post-acute COVID-19 timetable for clinics. This timetable includes symptoms from the first week to 12 weeks after infection (figure 3.1) [6]. The early diagnosis of these symptoms will decrease the symptoms and prevent the onset of new symptoms [42].

In this chapter, in addition to gathering information about post-acute COVID-19 syndrome (PCS) from multiple studies, we seek AI algorithms that can be used to characterize patients who have recovered from COVID-19.

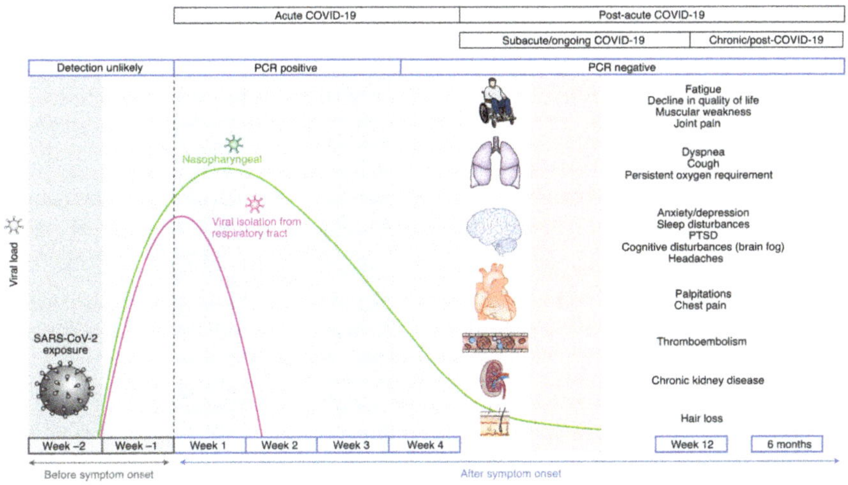

Figure 3.1. Timetable for post-acute COVID-19 (reprinted from [6] by permission from Springer Nature, copyright (2021)).

3.2 Psychiatric disorders and post-COVID symptoms

Individuals who recovered from COVID-19 due to particular guidelines for hospitalization and who feared dying might feel hopeless, depressed, or despaired. Patients had to stay alone during the infection for some time, which increased the chance of depression [43].

3.2.1 Potential tools for analyzing the psychiatric outcomes of post-COVID patients

Various questionnaires have been used in the literature to evaluate the psychopathological status of patients infected with COVID-19, some of which are listed below [1, 42, 44].

- Impact of Event Scale-Revised (IES-R) [45–47]
- PTSD checklist for DSM-5 (PCL-5) [48]
- Zung Self-Rating Depression Scale (ZSDS) [49]
- Thirteen-item Beck Depression Inventory (BDI-13) [50]
- State-Trait Anxiety Inventory From Y (STAI-Y) with 20 items for measuring traits and anxiety [51]
- Women's Health Initiative Insomnia Rating Scale (WHIIRS) [52]
- Obsessive-Compulsive Inventory (OCI) [53]
- The Mini-Mental State Examination (MMSE) [54]
- Montreal Cognitive Assessment (MoCA) [55]
- The Yale–Brown Obsessive-Compulsive Scale (Y-BOCS) [56]
- The Brown Assessment of Beliefs Scale (BABS) [57]
- The Beck Depression Inventory-II (BDI-II) [58]
- Zung Self-Rating Anxiety Scale (SAS) [59]
- King's Brief Interstitial Lung Disease (K-BILD) [60]
- Insomnia Severity Index (ISI) [61]

Various studies have used different cutoffs to consider the presence of psychiatric disorders. In a study of persistent psychopathology and neurocognitive impairment in COVID-19 patients, individuals who had a ZSDS index $\geqslant 50$, a PCL-5 $\geqslant 33$, a STAI-state $\geqslant 40$, an IES-R $\geqslant 33$, a WHIIRS $\geqslant 9$, a BDI-13 $\geqslant 9$, or an OCI $\geqslant 21$ in at least one questionnaire in three months were considered to be patients with psychiatric disorders [44]. Another study used supervised machine learning techniques based on the outputs of questionnaires (Y-BOCS, BABS, BDI-II, STAI-Y) and some of the clinical risk factors in its prediction technique [42]. Random forest and support vector machine (SVM) were utilized to predict psychiatric symptoms such as anxiety, depression, obsessive-compulsive symptoms, and belief symptoms. The results of these two methods are shown in table 3.1 [42].

Figure 3.2 shows the critical markers used to predict anxiety based on these two machine learning methods [42].

In another study of a multidimensional protocol used to assess the psychiatric burden after three months, patients' socio-demographic and psychological data were collected from online structured electronic data. They were then asked to evaluate

Table 3.1. Results of the prediction model (reproduced from [42] under the CC BY 4.0 license).

	Random Forest				SVM			
Symptom	Precision	Recall	F1_score	Accuracy	Precision	Recall	F1_score	Accuracy
Depression symptoms	0.60	0.68	0.63	0.68	0.56	0.66	0.59	0.66
Anxiety	0.68	0.75	0.70	0.75	0.61	0.67	0.63	0.67
Obsessive-compulsive symptoms	0.65	0.70	0.67	0.70	0.65	0.71	0.67	0.71
Belief symptoms	0.90	0.90	0.89	0.90	0.90	0.92	0.90	0.92

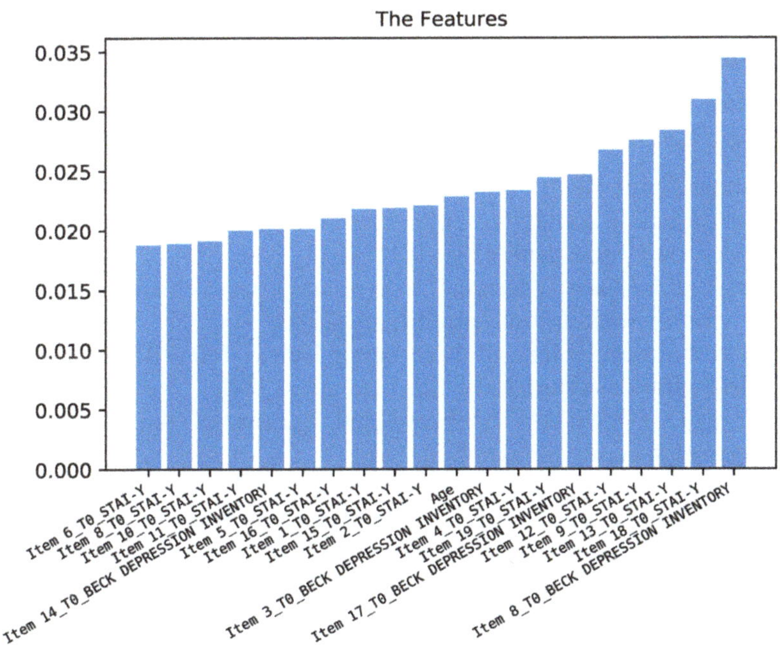

Figure 3.2. Important markers used to predict anxiety (reproduced from [42] under the CC BY 4.0 license).

their satisfaction with family support, changes in their quality of life, and worries about getting infected with COVID-19 again, using scores from zero to ten. This score indicated individuals' depression symptoms, anxiety, insomnia, and health impairment [62]. Zung's Self-Rating Anxiety Scale (SAS) uses twenty items to evaluate anxiety symptoms and produces scores ranging from 20 to 80, with a cutoff of 40 [63]. K-BILD contains 15 items, the IES-R questionnaire uses 22 items, the ISI questionnaire has seven items, and the BDI-II questionnaire uses 21 items to evaluate health status, current subjective distress, insomnia symptoms, and depression symptoms. Based on these cutoffs, patients were analyzed for their mental health symptoms [62].

3.3 Cognitive impairment and post-COVID symptoms

Cognitive impairment, which can take the form of memory loss, attention problems, difficulties in language and reasoning, emotional expression difficulties, intellectual disabilities, or brain injuries, is related to the central nervous system (CNS) and is one of the post-acute COVID-19 symptoms (also known as COVID-19 fog (CF)) [27, 64]. In an analysis of COVID-19 convalescents over five months, processing speed deficits, delayed verbal recall deficit, and problems in processing speed and verbal memory were observed at percentages of 42.1%, 26.3%, and 21%, respectively [65]. The occurrence of depression and anxiety was high in patients with severe acute respiratory syndrome coronavirus 2 [44]. Since cognitive disorders result in poor life and decreased quality of life [27], determining COVID-19 patients' cognitive problems and disease characteristics is essential [66].

3.3.1 Potential tools for analyzing the cognitive outcomes of post-COVID patients

Several studies have used various questionnaires to evaluate patients' cognitive functions [44, 67–71].

- The Short Form Health Survey 36 (SF-36) contains 36 items that are used to analyze physical activities, social interactions, emotional problems, mental disorders, pain, and general health [72].
- The Barthel Index was designed to evaluate improvement after rehabilitation in individuals with chronic disabilities [73].
- The Psychological General Well-being Index (PGWBI) contains 22 items used to evaluate subjective health or distress [74].
- The EuroQol measures health-related quality of life (HRQoL) without considering specific diseases [75].
- The Pittsburgh Sleep Quality Index (PSQI) evaluates sleep quality [76].
- The Mini-Mental Test assesses various areas of brain function such as calculation, memory, and attention using simple questions and graphical tasks [77].
- The Brief Pain Inventory (BPI) measures the severity level of pain [78].
- The Post-Traumatic Stress Syndrome 14 (PTSS-14) Questions Inventory identifies the level of patient post-traumatic stress disorder [79].
- The Hospital Anxiety and Depression Scale (HADS) analyzes the severity rate of anxiety and depression [80].
- The Mini Nutritional Assessment (MNA) is used to evaluate nutritional status [81, 82]
- The Activities of Daily living (ADLs) [83]
- The Auditory Verbal Learning Test (AVLT) [84]
- The Trail Making Test (TMT) [85]
- The Depression Symptom Inventory [86]

In addition, some papers have been published on the effectiveness of inflammatory biomarkers in predicting cognitive disorders with clinical features. The neutrophil/lymphocyte ratio (NLR), C-reactive protein (CRP), and monocyte/

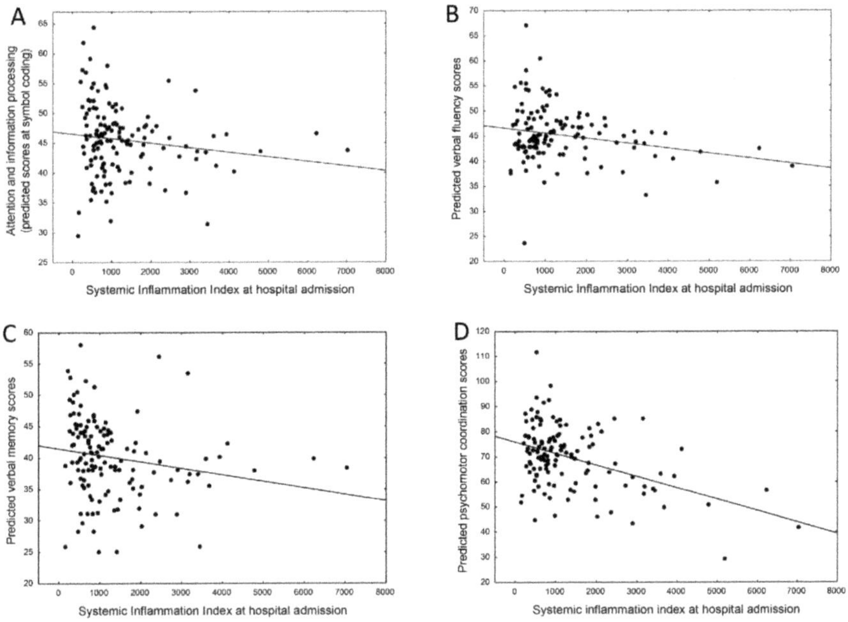

Figure 3.3. Assessments of the effects of systemic inflammation (as measured by the SSI) on attention and processing speed, verbal fluency, psychomotor coordination, and verbal memory (reprinted from [44], copyright (2021), with permission from Elsevier).

lymphocyte ratio (MLR) have been used as inflammatory markers to calculate the systemic immune-inflammation index (SII) [44]. Figure 3.3 shows the effect of the SSI on different psychiatric symptoms.

$\text{SSI} = \text{Platelets} \times \frac{\text{neutrophils}}{\text{lymphocytes}}$ [44].

This study evaluated verbal memory, fluency, working memory, attention, and processing speed using the Brief Assessment of Cognition in Schizophrenia (BACS) scale [87]. In this assessment, scores of two, three, and four represent good performance. However, zero and one denote poor performance. The general linear model (GLM) was used to investigate the effect of predictors on the outcome and evaluate changes over time. Repeated measures ANOVA was utilized [44]. The generalized linear model (GLZM) has been used to analyze the impact of systemic inflammation on the level of severity of psychopathology and neurocognitive impairment [44].

Since traditional cognitive assessments have many limitations, a new rehabilitation method based on brain–computer interface-virtual reality (BCI-VR) was designed that merges the detection characteristics of the brain–computer interface (BCI) with virtual reality (VR). Using this approach, the authors prepared a one-step cognitive rehabilitation and cognitive impairment assessment system for COVID-19 patients (figure 3.3) [88]. The brain specificity of individuals with cognitive disorders was detected using an EEG signal, which is easy to use even at home [89–91]. Moreover, VR technology improves patient participation and training effectiveness regarding cognitive damage [92–96]. This study provided a

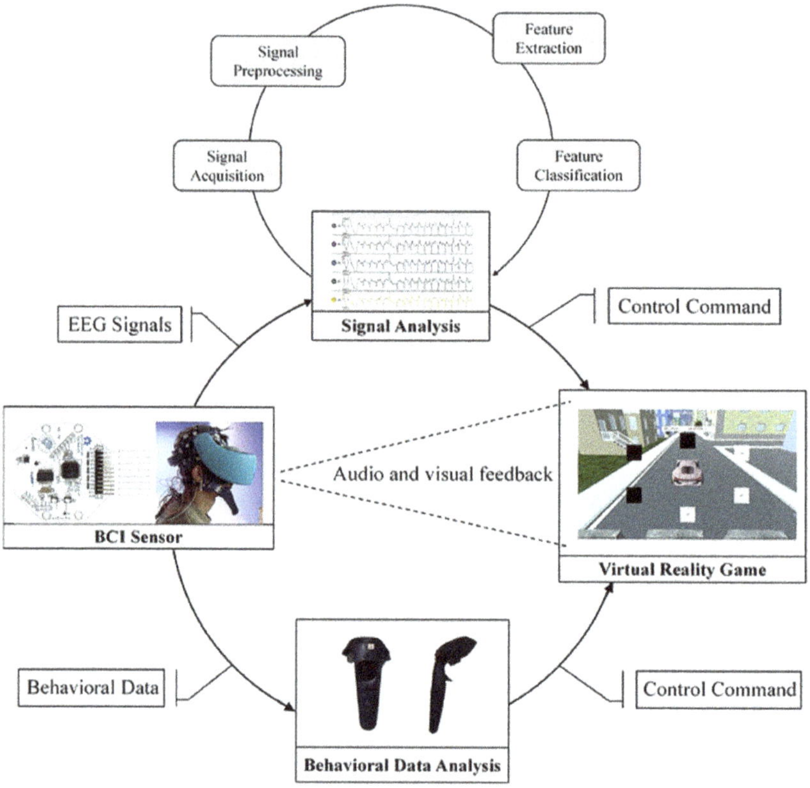

Figure 3.4. Diagram of BCI-VR (reproduced from [88] under the CC BY 4.0. license).

proof of concept for assessing cognitive impairment, rehabilitating patients at home, and meeting their needs [88]. VR wearable devices and electroencephalography (EEG) acquisition instruments are needed for implementation (figure 3.4). They are comparatively light compared to hospital medical devices [88].

It can be beneficial to assess the risk of cognitive impairment using questionnaires. The authors of a study used a questionnaire known as the Depression Symptom Inventory, which has 16 items, to their mood stability and residents' demographics [86]. Cronbach's alpha was used to examine the internal consistency of the questionnaire by evaluating the correlation coefficient of the answers; the result, 0.89, confirms that it has good internal consistency [86]. In this questionnaire, one to four points represent low risk, five to nine points indicate moderate risk, and more than ten points denote a high risk of depression [86].

3.4 Comprehensive literature review

Table 3.2 is a literature review of published papers on COVID-19 cognitive impairment and psychiatric disorders from 2020 to 2022 (table 3.2). The

Table 3.2. Literature review of articles on the post-acute symptoms of COVID-19 published from 2020 to 2022.

Reference	Sample Size	Study country and period	Outcome	Population characteristics (numbers, gender	Method of evaluation	Control group	Performance index
Singh *et al* 2020 [97]	1031 participants	UK British thoracic society 9 April 2020 to 15 April 2020	Functional impairment Psychological disturbances	71% of respondents were physiotherapists 84% female 35–44 years (34%) 45–54 years (27%) 25–34 years (22%) Respondents were from Europe	Online Survey	Ten healthy individuals with similar age	Acc = 90%
Woo *et al* 2020 [98]	28 participants (18 patients and 10 control group) Individuals were recruited until 14 July 2020	Germany University Medical Center Hamburg-Eppendorf (UKE)	Neurocognitive deficits (including concentration, short-term memory, and attention)	18 patients (10 females), 17–71 years	Questionnaire (phone or direct interview)		
Janiri *et al* 2020 [99]	61 patients	Italy Universitario Agostino Gemelli IRCCS	Psychological distress	61 patients Patients aged >60 years	Self-rating questionnaire Kessler questionnaire K10		
Vance *et al* 2021 [100]		USA	Neurological symptoms Cognitive impairment Dysautonomia Anosmia Chronic fatigue		Extensive literature review		
De Lorenzo *et al* 2020 [101]	195 patients (185 were included)	Italy San Raffaele University Hospital in Milan, Italy 7 April 2020 up until 7 May 2020	Residual dysfunction Cognitive impairment Post-traumatic stress disorder (PTSD) Respiratory dysfunction Uncontrolled blood pressure	185 patients aged ⩾ 18 years	Prospective and retrospective cohort study		

Korompoki *et al* 2021 [102]	Populations were from Asia, Europe, and the USA. Eight prospective studies, four retrospective studies. Nine studies out of 12 were included in the study	April 2021	Post-COVID-19 Symptoms: Chronic fatigue, Cognitive dysfunction, Pain, etc.	A narrative review on COVID-19 post-acute phase. Eight retrospective and four prospective studies.	Comprehensive literature review	
Cabrera Martimbianco *et al* 2021 [103]	25 observational studies 5440 participants	1 February 2021	Post-COVID-19 symptoms	22 360 electronic searches, 25 included studies, and 5440 subjects	Systematic review databases are CENTERAL, Embase, Epistemonikos, Health System Evidence, LILACS, and MEDLINE.	
Yun-Kuan Thye *et al* 2022 [104]	3470 articles		PTSD Anxiety Depression	Records collected from three database searches (n=3434) Additional records identified through Google scholar (n=36) Studies included for qualitative analysis (n=73)	Narrative review study	
Coleman *et al* 2021 [105]	7 139 696 patients	USA 20 October 2021	Unfavorable mental health significances	Electronic health records from 65 clinical organizations in the US. Population was 7 139 696 patients out of which 1 834 913 were COVID-19 positive	Retrospective cohort study	Patients with a similar health event

(*Continued*)

Table 3.2. (*Continued*)

Reference	Sample Size	Study country and period	Outcome	Population characteristics (numbers, gender	Method of evaluation	Control group	Performance index
Akbarialiabad *et al* 2021 [106]	120 included papers 290 publications		Arthralgia, Fatigue, Breathlessness, Sleep difficulties, Mental health, and Chest pain	Electronic databases resulted in 66 eligible studies 120 papers 290 publications	Comprehensive systematic scoping review. Databases: Google Scholar, PubMed, Cochrane Library, PsyclNFO, and Embase.		
Lopez-Leon *et al* 2021 [107]	18 251 publications 47 910 patients (aged 17–87)		Long-term COVID-19 symptoms, including attention disorder and dyspnea	18 251 publications 47 910 patients (aged 17–87) Records identified from: LitCOVID (n=24 301) and Embase (n=14 255)	Meta-analysis and systematic review		
Crispo *et al* 2021 [69]	4212 studies		Headaches Difficulties with attention, concentration, and memory Fatigue Insomnia Depression PTSD Cognitive impairment	Records identified through database searches (n=4212) Full-text articles included (n=18)	Narrative review study databases: PubMed, Embase, and Scopus		
Leth *et al* 2021 [108]	71 patients	Denmark Department of Infection Diseases, Aarhus University Hospital, Denmark	Fatigue Dyspnea Subjective difficulties concentrating Impaired smell and taste	71 patients (between March 11 and May 15) 49 patients included in study. The median age of the 49 included patients was 58 (43–78); 28 were female	Follow-up consultations, using telephone or personal visit		
Shanbehzadeh *et al* 2021 [109]	7398 studies, among which 34 were included (8932 participants)		Mental and physical health complications post-COVID-19, including depression, anxiety, and PTSD	7398 studies	Scoping review from 1 January to 7 November 2020		

Salamanna *et al* 2021 [110]	11 361 publications		Long-term COVID-19 symptoms (neurocognitive impairment, …)	11 361 articles identified 145 articles included	Systematic review	
Venturelli *et al* 2021 [111]	2965 patients 767 included patients		Anxiety, PTSD, depression, and cognitive impairment 2965 patients (>18 years)	767 adult patients were included	2965 patients met the criteria, 767 patients were included in study Exclusion of pediatric patients (<18 years)	
Moreno-Pérez *et al* 2021 [112]	Total population was 422 patients 277 patients were included		Dyspnea Fatigue Memory disorder Headaches Cognitive impairment	422 patients were considered (277 participants) in the emergency department in 2020	Prospective cohort study	
Schou *et al* 2021 [113]	1725 publications		Anxiety Depression PTSD Cognitive deficits Sleep disturbances Obsessive-compulsive disorder Psychotic episodes	1725 unique studies were identified	Systematic review databases: PubMed and Embase	
Poyraz *et al* 2021 [114]	239 patients	Turkey 1 June and 1 July, 2020	PTSD Depression Anxiety Poor sleep	239 patients239 responses received. 79 additional paper surveys completed by the volunteering patients. 284 surveys were included Mean age: 39.7±12.7 Female-to-male ratio = 0.99, p˃0.05	Cross-sectional survey study Online survey (survey contained three parts and 128 questions)	Comparison groups: Normal PTSD symptoms Mild PTSD symptoms
Miskowiak *et al* 2022 [115]	150 total patients 25 were included in study	Denmark Bispebjerg Hospital in Denmark	Cognitive impairment Functional implication	150 patients admitted to the Respiratory Department at Bispebjerg Hospital 25 patients included in cognitive assessment at one year	Longitudinal study	Compared to matched healthy control group

(*Continued*)

Table 3.2. (*Continued*)

Reference	Sample Size	Study country and period	Outcome	Population characteristics (numbers, gender	Method of evaluation	Control group	Performance index
Reeves and Willoughby 2022 [116]	51-year-old woman	USA	Cognitive impairment	51-year-old woman	Case report		
Johanna Josepha op't Hoog *et al* 2022 [117]	5052 articles		Physical impairments Cognitive impairment Psychological health	5052 articles, five included studies	Systematic review and meta-analysis		
Puthucheary *et al* 2021 [118]	314 patients	UK	Cognitive Problems Depression Anxiety PTSD Psychological problems	314 patients from 26 hospitals around England 139 patients had at least one complete Post-ICU presentation screen (PICUPS)	Clinical and workforce data		
Albu *et al* 2021 [119]	41 patients	Spain	Fatigue Dyspnea Subjective cognitive impairment Neurological sequelae	41 patients 30 patients were included in study Median age was 54; 11 female 19 male	Cross-sectional observational study		
Schoeps *et al* 2022 [120]	6283 adolescents and young adults		Psychiatric disorder (depression, anxiety, …)	6283 Patients Age: 12–30 years old (M= 18.79; SD=3.48) 83.7% female	Cross-cultural mediation study		
Hicks *et al* 2022 [121]	261 patients with dementia	UK	Dementia Cognitive impairment	261 people with dementia	Cohort Study DETERMIND-C19 study questionnaire		
Nanwani-Nanwani *et al* 2022 [122]	1093 patients, among which 186 were included	Spain	Psychological (anxiety, depression, PTSD) Physical Functional Cognitive	1093 patients admitted into ICU 186 patients assessed in follow-up clinics Mean age: 59±12 years old 32% of patients were female	Ambispective, observational study		
Allen Christensen *et al* 2022 [5]	11 patients	Denmark	Cognitive status Ability to perform ADLs	11 adult patients	Prospective multiple case study		
Latronico *et al* 2022 [123]	114 C-ARDS survivors	Italy	Mental impairment Physical impairment	224 patients admitted to ICU114 C-ARDS survivors were included in study	Prospective longitude study		

Beghi *et al* 2022 [124]	2578 psychiatric assessments	Italy	Psychological distress	2578 psychiatric assessments (1220 in 2019 and 1358 in 2020)	Retrospective longitude observational study
Giordano *et al* 2022 [125]	49 patients	Italy	Cognitive and psychopathological disturbances	49 consecutive adult patients	Observational study
Klimkiewicz *et al* 2022 [27]	17 patients	Poland Hospital of the Military Institute of Medicine, Warsaw, Poland	Cognitive impairment	17 individuals 10 female 7 male Aged 65±14 years	Preliminary report
Vannorsdall *et al* 2022 [126]	82 patients	USA Johns Hopkins Post-Acute COVID-19 Team (JH PACT)	Fatigue Dysautonomia Dyspnea Pain Cognitive complaints	82 PACT patients 59% female Mean (SD) age: 54.5 (14.6) years old	Clinical telephone-based assessment
Mattioli *et al* 2022 [127]	215 patients	Italy University Hospital of Brescia	Neurological and cognitive impairment	215 patients	Prospective observational study
Oluyinka Elugbadebo and Baiyewu 2022 [128]	178 dyads of patients and their caregivers	Nigeria University of Ibadan/ University College Joint Ethical Committee	Mild anxiety and depression disorders	178 dyads of patients and their caregivers	Cross-sectional study administered a semi-structured questionnaire
Bonizzato *et al* 2022 [1]	12 patients	Italy	Cognitive Impairment Anxiety Depression	12 patients aged between 47 and 85 Mean age: 71.33 ± 10.08	Observational study
Arica-Polat *et al* 2022 [66]	176 patients	Turkey	Cognitive deficit	176 patients Mean age 66.09 ± 13.96 years 100 (56.8%) Female 76 (43.2%) Male	Cross-sectional observational study

(*Continued*)

Table 3.2. (*Continued*)

Reference	Sample Size	Study country and period	Outcome	Population characteristics (numbers, gender	Method of evaluation	Control group	Performance index
Ceban *et al* 2022 [129]	Records identified through databases and other sources (n= 10 979)		Fatigue and cognitive impairment	Records identified through databases (n=10 923) Records identified through other sources (n=56) Articles included in study (n=74)	Systematic review and meta-analysis Meta-analysis of observational studies in epidemiology (MOOSE)		
Nishimi *et al* 2022 [130]	845 adults	USA	PTSD symptoms	845 adult participants 77.5% female with mean age: 37.0 (SD=11.0) years old	Cross-sectional online US study		
Nishimi *et al* 2022 [131]	263 697 patients	USA administrative and electronic health records of US Department	Psychiatric disorders	263 697 patients Mean [SD] age, 66.2 [13.8] years; 239 539 men [90.8%]	Retrospective cohort study		
Aaltonen *et al* 2022 [132]	59 identified persons	Finland	Psychiatric disorders	59 identified persons 57 home-quarantined participants were included in study Participation rate 97%; 43 with exposure and 14 with Sars-Cov-2 infection	Cohort study questionnaire-9 clinical outcomes in routine evaluation – Outcome	The study included a control group	
Saali *et al* 2022 [133]	147 third-year medical students	USA	PTSD Major Depressive Disorder (MDD) Generalized Anxiety Disorder (GAD)	147 third-year medical students; 110 (75%) participated in study, 108 were included in the final analysis	Survey-based study		
Cristillo *et al* 2022 [134]	168 patients	Italy	Cognitive impairment Dementia	168 total patients 106 patients were included in study Mean age: 64.9 years old 73.3% Female	Observational cohort study		
Stickley and Ueda 2022 [135]	9000 people	Japan	Mental health (depressive and anxiety symptoms)	9000 people	Logistic regression analysis Online survey		

Valdes *et al* 2022 [136]	382 patients	USA	Neurologic and psychiatric symptoms	382 Patients 215 patients (56%) completed the t-Montreal Cognitive Assessment (MoCA) and were included in study	Retrospective analysis Multivariable logistic regression	Included comparison group
Cortés Zamora *et al* [137]	953 residents	Spain	Depression Anxiety PTSD Sleep disturbances	953 residents 215 residents ⩾ 65 years were included in study	Cohort longitudinal study	
Costas-Carrera *et al* 2022 [138]	102 patients	Spain	Neuropsychological performance Cognitive impairment Anxiety		Prospective cohort study	
Tang *et al* 2022 [139]			Cognitive impairment Depression Anxiety Concentration impairment	102 severe SARS-COV-2 survivors 58 patients were included in study 29% female Aged between 37 and 81 years old Mean Age: 65±9.32	Literature review study on PubMed	
Tabacof *et al* 2022 [140]	386 patients	USA Mount Sinai's PACS clinic	Cognitive function Physical function Fatigue Brain fog Headache, …	386 total patients 156 (48%) patients responded to survey	Cross-sectional observational study	
Liu *et al* 2022 [141]	3233 COVID-19 survivors	China	Cognitive impairment	3233 COVID-19 survivors aged ⩾60 years old 466 uninfected spouses as a control group 1438 participants and 438 uninfected control individuals	Longitudinal cohort study	466 uninfected spouses, 438 uninfected control individuals
Vagheggini *et al* 2022 [62]	21 patients	Italy May and July 2020	Cognitive Impairment PTSD Moderate depressive symptoms Clinical insomnia	21 patients Mean age: 57.05 years old SD = 11.02	Single arm study	
Hamlin *et al* 2022 [142]		Sweden	Psychiatric disorder		Register-based study	Data collected during the first two COVID-19 waves were compared to control periods between 2018 and 2019

(*Continued*)

Table 3.2. (*Continued*)

Reference	Sample Size	Study country and period	Outcome	Population characteristics (numbers, gender	Method of evaluation	Control group	Performance index
Crivelli *et al* 2022 [143]	6202 articles		Cognitive impairment	6202 articles, 27 studies (2049 subjects) were included	Systematic review and meta-analysis	There were 506 healthy controls and 2103 patients	
Henneghan *et al* 2022 [144]	105 adults	USA	Cognitive impairment	105 adults 79 adults were included in study Aged 22 to 65 years old	Cross-sectional study		
Górski *et al* 2022 [86]	273 residents	Poland	Cognitive impairment Depression	273 residents 51.6% (*N*=141) female 48.4% (*N*=132) male Mean age 80.81±8.17 years	Custom questionnaire		
Badenoch *et al* 2022 [145]	4385 database records identified	UK	Objective cognitive impairment Anxiety PTSD Fatigue Sleep disturbance	4385 records identified through database searches 51 studies were included in meta-analysis 18 917 patients	Systematic review and meta-analysis Study databases were Embase, MEDLINE		
Ferrucci *et al* 2022 [146]	76 patients	Italy	Cognitive impairment	76 patients Mean age: 56.2±12.08	Observational study		
Crivelli *et al* 2022 [71]	45 patients and 45 healthy controls		Cognitive impairment	45 patients (age M=50 (43,63) years), 22(49%) female 45 healthy controls (age M=57 (46,64) years, 20 (44%) female)	Cohort study	45 healthy controls	
D'Urso *et al* 2022 [42]	Healthy Subjects OCD patients Alzheimer's disease (AD) patients	Italy	Anxiety symptoms Depression symptoms	Healthy subjects OCD patients AD patients	Supervised machine learning method		92% acc
Na *et al* 2022 [147]	3078 US veterans	USA	Major depression PTSD Generalized anxiety	3078 veterans Average age of the sample 62.2 years old 233 of them reported having COVID-19	Prospective cohort study		

specifications of the studies are provided, including the setting, outcome, control group, and performance indices.

More than 38% of studies focused on both psychiatric disorders and cognitive impairments, which were analyzed exclusively, accounting for 30% and 31.7%, respectively, of the articles listed in table 3.2. On the other hand, smart healthcare design can be seen in less than 17% of the papers (figure 3.5). Among those focused on smart healthcare design, only 70% reported their performance based on any evaluation criteria.

3.5 Smart cities and post-COVID symptoms

The adoption of smart technologies for COVID-19 pandemics is shown diagrammatically in figure 3.6. Telemedicine and big data are two solutions that could be used for post-COVID pandemic management. Such solutions could be implemented

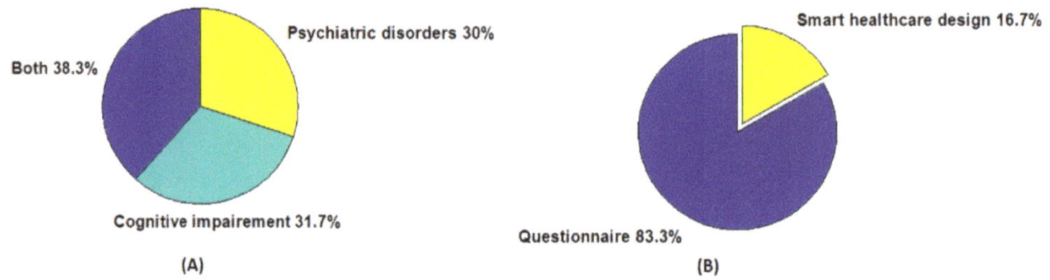

Figure 3.5. The percentage of the papers listed in table 3.2 for (a) psychiatric disorders, cognitive impairment, and both topics, and (b) the percentage of the papers related to smart healthcare design and questionnaire analysis.

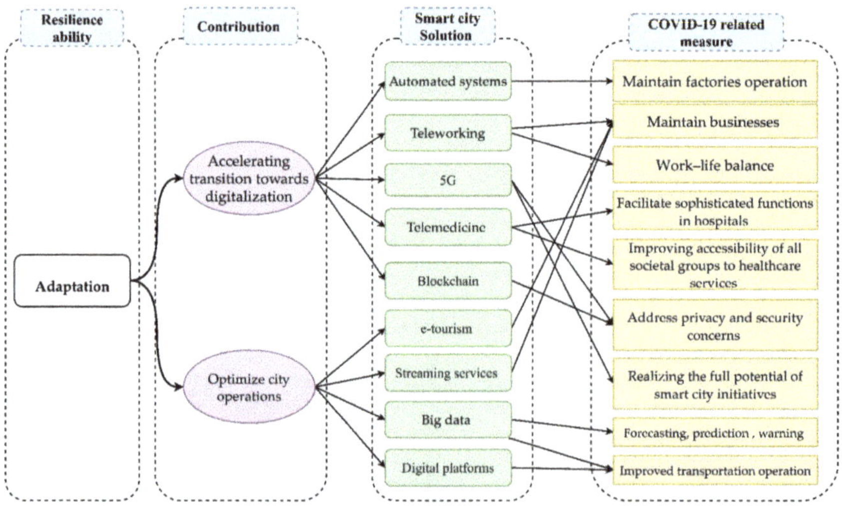

Figure 3.6. The contributions of various smart solutions to COVID-19 pandemic management (reproduced from [154] under the CC BY 4.0 license).

for 'neurocovid,' which affects the human brain [148]. However, as shown in figure 3.5, less than 17% of the studies on COVID-19 neurocovid focused on smart systems. The reason for this is mainly that smart systems are not usually used in psychological and mental health studies [149]. Even if they are used, the validation of such systems is not aligned with clinicians' expectations. We have discussed this gap previously for various clinical problems [150], including COVID-19 [151, 152]. In addition to complying with medical regulations related to digital healthcare solutions, it is necessary to provide updated standardizations [153].

3.6 Conclusions and future scope

Although smart healthcare methodologies have the prognostic potential to reveal high-risk patients and the association between symptoms and the post-COVID period, some of the articles reviewed in this chapter used follow-up data collected by telephone interview, making them unreliable. In addition, almost all of the studies were based on a small sample size; therefore, such results cannot be applied to the whole population. Multicenter international cohorts and randomized controlled trials are thus required to prove the suitability of such digital solutions for addressing cognitive impairment and psychiatric distress in COVID-19 patients.

Acknowledgments
This work was supported by the Beatriu de Pinós post-doctoral program of the Office of the Secretary of Universities and Research of the Ministry of Business and Knowledge of the Government of Catalonia (Grant number: #2020 BP 00261), the TecnioSpring INDUSTRY fellowship (ACCIO, H2020-EU—EXCELLENT SCIENCE—Marie Skłodowska-Curie Actions, #ACE026/21/000035), the Spanish Ministry of Science and Innovation (project PID2020-117751RB-I00), and the Serra Húnter and María Zambrano fellowships.

References

[1] Bonizzato S, Ghiggia A, Ferraro F and Galante E 2022 Cognitive, behavioral, and psychological manifestations of COVID-19 in post-acute rehabilitation setting: preliminary data of an observational study *Neurol. Sci.* **43** 51–8

[2] Haddad C, Dib J E, Akl N, Hallit S and Obeid S 2022 COVID-19 and psychosis, depression, obsession and quality of life in Lebanese patients with schizophrenia: any changes after 5 months of quarantine? *BMC Psychol.* **10** 32

[3] Tuna Ö, Ermis C, Darcin A E, Dagistan E and Salman S 2022 Comparison of inflammation markers and severity of illness among patients with COVID-19, acute psychiatric disorders and comorbidity *Eur J. Psychiatry* https://search.bvsalud.org/global-literature-on-novel-coronavirus-2019-ncov/resource/en/covidwho-1664884

[4] Ji D *et al* 2020 Prediction for progression risk in patients with COVID-19 pneumonia: the CALL score *Clin. Infect. Dis.* **71** 1393–9

[5] Christensen K A, Christensen J and Eskildsen S J 2022 Exploring the ability to perform activities of daily living and cognitive status after hospitalization with COVID-19: a multiple case study *Occup. Ther. Int.* **2022** 4605989

[6] Nalbandian A *et al* 2021 Post-acute COVID-19 syndrome *Nat. Med.* **27** 601–15

[7] Torales J, O'Higgins M, Castaldelli-Maia J M and Ventriglio A 2020 The outbreak of COVID-19 coronavirus and its impact on global mental health *Int. J. Soc. Psychiatry* **66** 317–20

[8] Horton R 2020 Offline: 2019-nCoV-'A desperate plea' *Lancet* **395** 400

[9] Paneroni M, Simonelli C, Saleri M, Bertacchini L, Venturelli M, Troosters T, Ambrosino N and Vitacca M 2021 Muscle strength and physical performance in patients without previous disabilities recovering from COVID-19 pneumonia *Am. J. Phys. Med. Rehabil.* **100** 105–9

[10] Huang C *et al* 2021 6-Month consequences of COVID-19 in patients discharged from hospital: a cohort study *Lancet* **397** 220–32

[11] Moghimi N *et al* 2021 The neurological manifestations of post-acute sequelae of SARS-CoV-2 infection *Curr. Neurol. Neurosci. Rep.* **21** 44

[12] Nersesjan V *et al* 2021 Central and peripheral nervous system complications of COVID-19: a prospective tertiary center cohort with 3-month follow-up *J. Neurol.* **268** 3086–104

[13] Wiertz C M H, Vints W A J, Maas G J C M, Rasquin S M C, van Horn Y Y, Dremmen M P M, Hemmen B and Verbunt J A 2021 COVID-19: patient characteristics in the first phase of postintensive care rehabilitation *Arch. Rehabil. Res. Clin. Transl.* **3** 100108

[14] Anon 2020 *COVID-19 rapid guideline: managing the long-term effects of COVID-19* NG188 National Institute for Health and Care Excellence (NICE) COVID-19 rapid guideline: managing the long-term effects of COVID-19 https://www.nice.org.uk/guidance/ng188

[15] Halpin S J *et al* 2021 Postdischarge symptoms and rehabilitation needs in survivors of COVID-19 infection: a cross-sectional evaluation *J. Med. Virol.* **93** 1013–22

[16] Del Brutto O H, Wu S, Mera R M, Costa A F, Recalde B Y and Issa N P 2021 Cognitive decline among individuals with history of mild symptomatic SARS-CoV-2 infection: a longitudinal prospective study nested to a population cohort *Eur. J. Neurol.* **28** 3245–53

[17] Fernández-de-Las-Peñas C, Palacios-Ceña D, Gómez-Mayordomo V, Florencio L L, Cuadrado M L, Plaza-Manzano G and Navarro-Santana M 2021 Prevalence of post-COVID-19 symptoms in hospitalized and non-hospitalized COVID-19 survivors: a systematic review and meta-analysis *Eur. J. Intern. Med.* **92** 55–70

[18] Ali A M, Alkhamees A A, Abd Elhay E S, Taha S M and Hendawy A O 2021 COVID-19-related psychological trauma and psychological distress among community-dwelling psychiatric patients: people struck by depression and sleep disorders endure the greatest burden *Front. Public Health* **9** 799812

[19] Hassan L, Peek N, Lovell K, Carvalho A F, Solmi M, Stubbs B and Firth J 2022 Disparities in COVID-19 infection, hospitalisation and death in people with schizophrenia, bipolar disorder, and major depressive disorder: a cohort study of the UK Biobank *Mol. Psychiatry* **27** 1248–55

[20] Iasevoli F, Fornaro M, D'Urso G, Galletta D, Casella C, Paternoster M, Buccelli C, de Bartolomeis A and COVID-19 in Psychiatry Study Group 2021 Psychological distress in patients with serious mental illness during the COVID-19 outbreak and one-month mass quarantine in Italy *Psychol. Med.* **51** 1054–6

[21] Kowall B, Kostev K, Landgraf R, Hauner H, Bierwirth R and Rathmann W 2022 Effects of the COVID-19 pandemic on clinically diagnosed psychiatric disorders in persons with type 2 diabetes *Diabet. Med.* **39** e14852

[22] Huang Y *et al* 2021 COVID symptoms, symptom clusters, and predictors for becoming a long-hauler: looking for clarity in the haze of the pandemic *Clinical Nursing Res.* https://doi.org/10.1101/2021.03.03.21252086

[23] Pashazadeh Kan F *et al* 2021 A systematic review of the prevalence of anxiety among the general population during the COVID-19 pandemic *J. Affect. Disord.* **293** 391–8
[24] Ji W *et al* 2020 Effect of underlying comorbidities on the infection and severity of COVID-19 in Korea: a nationwide case-control study *J. Korean Med. Sci.* **35** e237
[25] Cho S J, Kim J, Kang Y J, Lee S Y, Seo H Y, Park J E, Kim H, Kim K-N, Lee J Y and Sohn J H 2020 Annual prevalence and incidence of schizophrenia and similar psychotic disorders in the republic of Korea: a national health insurance data-based study *Psychiatry Investig.* **17** 61–70
[26] Wang Q, Xu R and Volkow N D 2021 Increased risk of COVID-19 infection and mortality in people with mental disorders: analysis from electronic health records in the United States *World Psychiatry* **20** 124–30
[27] Klimkiewicz J, Pankowski D, Wytrychiewicz-Pankowska K, Klimkiewicz A, Siwik P, Klimczuk J and Lubas A 2022 Analysis of the relationship among cognitive impairment, nutritional indexes and the clinical course among COVID-19 patients discharged from hospital-preliminary report *Nutrients* **14** 1580
[28] Verhaeghen P and Cerella J 2002 Aging, executive control, and attention: a review of meta-analyses *Neurosci. Biobehav. Rev.* **26** 849–57
[29] Salthouse T A 2019 Trajectories of normal cognitive aging *Psychol. Aging* **34** 17–24
[30] Marsland A L, Gianaros P J, Kuan D C-H, Sheu L K, Krajina K and Manuck S B 2015 Brain morphology links systemic inflammation to cognitive function in midlife adults *Brain Behav. Immun.* **48** 195–204
[31] Sartori A C, Vance D E, Slater L Z and Crowe M 2012 The impact of inflammation on cognitive function in older adults *J. Neurosci. Nurs.* **44** 206–17
[32] Ballesteros S, Kraft E, Santana S and Tziraki C 2015 Maintaining older brain functionality: a targeted review *Neurosci. Biobehav. Rev.* **55** 453–77
[33] Greenwood P M and Parasuraman R 2010 Neuronal and cognitive plasticity: a neurocognitive framework for ameliorating cognitive aging *Front. Aging Neurosci.* **2** 150
[34] Gajewski P D and Falkenstein M 2011 Neurocognition of aging in working environments *J. Labour Mark. Res.* **44** 307–20
[35] Gajewski P D and Falkenstein M 2016 Lifestyle and interventions for improving cognitive performance in older adults *Performance Psychology* (Amsterdam: Elsevier) pp 189–203
[36] Gajewski P D and Falkenstein M 2018 *Physical Activity and Inactivity Impacts on Cognitive and Emotional Functioning in Later Life* (Oxford: Oxford University Press)
[37] Lupien S J, McEwen B S, Gunnar M R and Heim C 2009 Effects of stress throughout the lifespan on the brain, behaviour and cognition *Nat. Rev. Neurosci.* **10** 434–45
[38] Davies G *et al* 2018 Study of 300,486 individuals identifies 148 independent genetic loci influencing general cognitive function *Nat. Commun.* **9** 2098
[39] Erickson K I, Kim J S, Suever B L, Voss M W, Francis B M and Kramer A F 2008 Genetic contributions to age-related decline in executive function: a 10-year longitudinal study of COMT and BDNF polymorphisms *Front. Hum. Neurosci.* **2** 11
[40] Colzato L, Zhang W, Beste C and Stock A-K 2021 Dissociating direct and indirect effects: a theoretical framework of how latent toxoplasmosis affects cognitive profile across the lifespan *Neurobiol. Aging* **102** 119–28
[41] Pistarini C, Fiabane E, Houdayer E, Vassallo C, Manera M R and Alemanno F 2021 Cognitive and emotional disturbances due to COVID-19: an exploratory study in the rehabilitation setting *Front. Neurol.* **12** 643646

[42] D'Urso G, Magliacano A, Rotbei S, Iasevoli F, de Bartolomeis A and Botta A 2022 Predicting the severity of lockdown-induced psychiatric symptoms with machine learning *Diagnostics (Basel)* **12** 957

[43] Gemelli Against COVID-19 Post-Acute Care Study Group 2020 Post-COVID-19 global health strategies: the need for an interdisciplinary approach *Aging Clin Exp Res.* **32** 1613–1620

[44] Mazza M G, Palladini M, De Lorenzo R, Magnaghi C, Poletti S, Furlan R, Ciceri F, COVID-19 BioB Outpatient Clinic Study group, Rovere-Querini P and Benedetti F 2021 Persistent psychopathology and neurocognitive impairment in COVID-19 survivors: effect of inflammatory biomarkers at three-month follow-up *Brain Behav. Immun.* **94** 138–47

[45] Creamer M, Bell R and Failla S 2003 Psychometric properties of the impact of event scale – revised *Behav. Res. Ther.* **41** 1489–96

[46] Weiss D S 2007 The impact of event scale: revised *Cross-Cultural Assessment of Psychological Trauma and PTSD* (Boston, MA: Springer) pp 219–38

[47] Morina N, Ehring T and Priebe S 2013 Diagnostic utility of the impact of event scale-revised in two samples of survivors of war *PLoS One* **8** e83916

[48] Armour C, Contractor A, Shea T, Elhai J D and Pietrzak R H 2016 Factor structure of the PTSD checklist for DSM-5 *J. Nerv. Ment. Dis.* **204** 108–15

[49] Zung W W 1965 A self-rating depression scale *Arch. Gen. Psychiatry* **12** 63–70

[50] Beck A T and Steer R A 1984 Internal consistencies of the original and revised Beck Depression Inventory *J. Clin. Psychol.* **40** 1365–7

[51] Vigneau F and Cormier S 2008 The factor structure of the State-Trait Anxiety Inventory: an alternative view *J. Pers. Assess.* **90** 280–5

[52] Levine D W, Lewis M A, Bowen D J, Kripke D F, Kaplan R M, Naughton M J and Shumaker S A 2003 Reliability and validity of women's health initiative insomnia rating scale *Psychol. Assess.* **15** 137–48

[53] Foa E B, Huppert J D, Leiberg S, Langner R, Kichic R, Hajcak G and Salkovskis P M 2002 The Obsessive-Compulsive Inventory: development and validation of a short version *Psychol. Assess.* **14** 485–96

[54] Grigoletto F, Zappalà G, Anderson D W and Lebowitz B D 1999 Norms for the Mini-Mental State Examination in a healthy population *Neurology* **53** 315–20

[55] Conti S, Bonazzi S, Laiacona M, Masina M and Coralli M V 2015 Montreal Cognitive Assessment (MoCA)-Italian version: regression based norms and equivalent scores *Neurol. Sci.* **36** 209–14

[56] Goodman W K, Price L H, Rasmussen S A, Mazure C, Fleischmann R L, Hill C L, Heninger G R and Charney D S 1989 The Yale-Brown Obsessive Compulsive Scale. I. Development, use, and reliability *Arch. Gen. Psychiatry* **46** 1006–11

[57] Eisen J L, Phillips K A, Baer L, Beer D A, Atala K D and Rasmussen S A 1998 The Brown Assessment of Beliefs Scale: reliability and validity *Am. J. Psychiatry* **155** 102–8

[58] Steer R A, Beck A T, Riskind J H and Brown G 1987 Relationships between the Beck Depression Inventory and the Hamilton Psychiatric Rating Scale for Depression in depressed outpatients *J. Psychopathol. Behav. Assess.* **9** 327–39

[59] Zung W K 1971 *A rating instrument for anxiety disorders* Psychosomatics **12** 371–9

[60] Patel A S *et al* 2012 The development and validation of the King's Brief Interstitial Lung Disease (K-BILD) health status questionnaire *Thorax* **67** 804–10

[61] Bastien C H, Vallières A and Morin C M 2001 Validation of the Insomnia Severity Index as an outcome measure for insomnia research *Sleep Med.* **2** 297–307
[62] Vagheggini G, Marzetti F, Miniati M, Bernardeschi L, Miccoli M, Boni Brivio G, Meini S, Panait E, Cini E and Gemignani A 2022 Pulmonary function and psychological burden three months after COVID-19: proposal of a comprehensive multidimensional assessment protocol *Healthcare (Basel)* **10** 612
[63] Dunstan D A and Scott N 2020 Norms for Zung's self-rating anxiety scale *BMC Psychiatry* **20** 90
[64] Gartland S, Flynn P, Carneiro M A, Holloway G, Fialho J, de S, Cullen J, Hamilton E, Harris A and Cullen C 2022 The state of web accessibility for people with cognitive disabilities: a rapid evidence assessment *Behav. Sci. (Basel)* **12** 26
[65] Ferrucci R *et al* 2021 Long-lasting cognitive abnormalities after COVID-19 *Brain Sci.* **11** 235
[66] Arica-Polat B S, Gündoğdu A A, Cinar N, Uncu G, Ayas Z O, Iseri P, Karadas O and Adapinar D O 2022 Evaluation of cognitive deficits in patients infected with COVID-19 *Eur. Rev. Med. Pharmacol. Sci.* **26** 678–85
[67] Almeria M, Cejudo J C, Sotoca J, Deus J and Krupinski J 2020 Cognitive profile following COVID-19 infection: clinical predictors leading to neuropsychological impairment *Brain Behav Immun Health* **9** 100163
[68] Jaywant A, Vanderlind W M, Alexopoulos G S, Fridman C B, Perlis R H and Gunning F M 2021 Frequency and profile of objective cognitive deficits in hospitalized patients recovering from COVID-19 *Neuropsychopharmacology* **46** 2235–40
[69] Crispo A *et al* 2021 Strategies to evaluate outcomes in long-COVID-19 and post-COVID survivors *Infect. Agent. Cancer* **16** 62
[70] García-Grimshaw M *et al* 2022 Critical role of acute hypoxemia on the cognitive impairment after severe COVID-19 pneumonia: a multivariate causality model analysis *Neurol. Sci.* **43** 2217–29
[71] Crivelli L, Calandri I, Corvalán N, Carello M A, Keller G, Martínez C, Arruabarrena M and Allegri R 2022 Cognitive consequences of COVID-19: results of a cohort study from South America *Arq. Neuropsiquiatr.* **80** 240–7
[72] Brazier J E, Harper R, Jones N M, O'Cathain A, Thomas K J, Usherwood T and Westlake L 1992 Validating the SF-36 health survey questionnaire: new outcome measure for primary care *Brit. Med. J.* **305** 160–4
[73] Mahoney F I and Barthel D W 1965 Functional evaluation: the Barthel index *Md. State Med. J.* **14** 61–5
[74] Grossi E, Groth N, Mosconi P, Cerutti R, Pace F, Compare A and Apolone G 2006 Development and validation of the short version of the Psychological General Well-Being Index (PGWB-S) *Health Qual. Life Outcomes* **4** 88
[75] Anon 1990 EuroQol - a new facility for the measurement of health-related quality of life *Health Policy* **16** 199–208
[76] Buysse D J, Reynolds C F 3rd, Monk T H, Berman S R and Kupfer D J 1989 The Pittsburgh Sleep Quality Index: a new instrument for psychiatric practice and research *Psychiatry Res.* **28** 193–213
[77] Folstein M F, Folstein S E and McHugh P R 1975 "Mini-mental state": A practical method for grading the cognitive state of patients for the clinician *J. Psychiatr. Res.* **12** 189–98

[78] Caraceni A, Mendoza T R, Mencaglia E, Baratella C, Edwards K, Forjaz M J, Martini C, Serlin R C, de Conno F and Cleeland C S 1996 A validation study of an Italian version of the Brief Pain Inventory (Breve Questionario per la Valutazione del Dolore) *Pain* **65** 87–92

[79] Parker A M, Nikayin S, Bienvenu O J and Needham D M 2017 Validity of the posttraumatic stress symptoms-14 instrument in acute respiratory failure survivors *Ann. Am. Thorac. Soc.* **14** 1047–8

[80] Zigmond A S and Snaith R P 1983 The hospital anxiety and depression scale *Acta Psychiatr. Scand.* **67** 361–70

[81] Guigoz Y 2006 The Mini Nutritional Assessment (MNA) review of the literature – what does it tell us? *J. Nutr. Health Aging* **10** 466–85 discussion 485–7

[82] Haraj N E *et al* 2021 Nutritional status assessment in patients with Covid-19 after discharge from the intensive care unit *Clin. Nutr. ESPEN* **41** 423–8

[83] López M R 2017 Cognitive impairment/dementia: assessment and follow-up. Functional aspects *Rev. Esp. Geriatr. Gerontol.* **52** 28–33

[84] Stricker N H *et al* 2021 Mayo normative studies: regression-based normative data for the Auditory Verbal Learning Test for ages 30–91 years and the importance of adjusting for sex *J. Int. Neuropsychol. Soc.* **27** 211–26

[85] Lunardini F *et al* 2020 Supervised digital neuropsychological tests for cognitive decline in older adults: usability and clinical validity study *JMIR MHealth UHealth* **8** e17963

[86] Górski M, Buczkowska M, Grajek M, Garbicz J, Całyniuk B, Paciorek K, Głuszek A and Polaniak R 2021 Assessment of the risk of depression in residents staying at long-term care institutions in Poland during the COVID-19 pandemic depending on the quality of cognitive functioning *Front. Psychol.* **12** 766675

[87] Keefe R S E, Goldberg T E, Harvey P D, Gold J M, Poe M P and Coughenour L 2004 The Brief Assessment of Cognition in Schizophrenia: reliability, sensitivity, and comparison with a standard neurocognitive battery *Schizophr. Res.* **68** 283–97

[88] Wen D, Xu J, Wu Z, Liu Y, Zhou Y, Li J, Wang S, Dong X, Saripan M I and Song H 2021 The effective cognitive assessment and training methods for COVID-19 patients with cognitive impairment *Front. Aging Neurosci.* **13** 827273

[89] Požar R, Giordani B and Kavcic V 2020 Effective differentiation of mild cognitive impairment by functional brain graph analysis and computerized testing *PLoS One* **15** e0230099

[90] San-Juan D *et al* 2020 Guidance for clinical neurophysiology examination throughout the COVID-19 pandemic. Latin American chapter of the IFCN task force – COVID-19 *Clin. Neurophysiol.* **131** 1589–98

[91] Pinter D, Kober S E, Fruhwirth V, Berger L, Damulina A, Khalil M, Neuper C, Wood G and Enzinger C 2021 MRI correlates of cognitive improvement after home-based EEG neurofeedback training in patients with multiple sclerosis: a pilot study *J. Neurol.* **268** 3808–16

[92] Snider J, Plank M, Lee D and Poizner H 2013 Simultaneous neural and movement recording in large-scale immersive virtual environments *IEEE Trans. Biomed. Circuits Syst.* **7** 713–21

[93] Appel L, Appel E, Bogler O, Wiseman M, Cohen L, Ein N, Abrams H B and Campos J L 2019 Older adults with cognitive and/or physical impairments can benefit from immersive virtual reality experiences: a feasibility study *Front. Med. (Lausanne)* **6** 329

[94] Bauer A C M and Andringa G 2020 The potential of immersive virtual reality for cognitive training in elderly *Gerontology* **66** 614–23

[95] Mancuso V, Stramba-Badiale C, Cavedoni S, Pedroli E, Cipresso P and Riva G 2020 Virtual reality meets non-invasive brain stimulation: integrating two methods for cognitive rehabilitation of mild cognitive impairment *Front. Neurol.* **11** 566731
[96] Thielbar K, Spencer N, Tsoupikova D, Ghassemi M and Kamper D 2020 Utilizing multi-user virtual reality to bring clinical therapy into stroke survivors' homes *J. Hand Ther.* **33** 246–53
[97] Singh S J, Barradell A C, Greening N J, Bolton C, Jenkins G, Preston L and Hurst J R 2020 British Thoracic Society survey of rehabilitation to support recovery of the post-COVID-19 population *BMJ Open* **10** e040213
[98] Woo M S *et al* 2020 Frequent neurocognitive deficits after recovery from mild COVID-19 *Brain Commun* **2** fcaa205
[99] Janiri D *et al* 2020 Psychological distress after Covid-19 recovery: reciprocal effects with temperament and emotional dysregulation. An exploratory study of patients over 60 years of age assessed in a post-acute care service *Front. Psychiatry* **11** 590135
[100] Vance H, Maslach A, Stoneman E, Harmes K, Ransom A, Seagly K and Furst W 2021 Addressing post-COVID symptoms: a guide for primary care physicians *J. Am. Board Fam. Med.* **34** 1229–42
[101] De Lorenzo R *et al* 2020 Residual clinical damage after COVID-19: a retrospective and prospective observational cohort study *PLoS One* **15** e0239570
[102] Korompoki E *et al* 2021 Epidemiology and organ specific sequelae of post-acute COVID19: a narrative review *J. Infect.* **83** 1–16
[103] Cabrera Martimbianco A L, Pacheco R L, Bagattini Â M and Riera R 2021 Frequency, signs and symptoms, and criteria adopted for long COVID-19: a systematic review *Int. J. Clin. Pract.* **75** e14357
[104] Thye A Y-K, Law J W-F, Tan L T-H, Pusparajah P, Ser H-L, Thurairajasingam S, Letchumanan V and Lee L-H 2022 Psychological symptoms in COVID-19 patients: insights into pathophysiology and risk factors of long COVID-19 *Biology (Basel)* **11** 61
[105] Coleman B *et al* 2021 Increased risk of psychiatric sequelae of COVID-19 is highest early in the clinical course https://doi.org/10.1101/2021.11.30.21267071
[106] Akbarialiabad H *et al* 2021 Long COVID, a comprehensive systematic scoping review *Infection* **49** 1163–86
[107] Lopez-Leon S, Wegman-Ostrosky T, Perelman C, Sepulveda R, Rebolledo P A, Cuapio A and Villapol S 2021 More than 50 long-term effects of COVID-19: a systematic review and meta-analysis *Sci. Rep.* https://doi.org/10.1038/s41598-021-95565-8
[108] Leth S, Gunst J D, Mathiasen V, Hansen K, Søgaard O, Østergaard L, Jensen-Fangel S, Storgaard M and Agergaard J 2021 Persistent symptoms in patients recovering from COVID-19 in Denmark *Open Forum Infect Dis* **8** ofab042
[109] Shanbehzadeh S, Tavahomi M, Zanjari N, Ebrahimi-Takamjani I and Amiri-Arimi S 2021 Physical and mental health complications post-COVID-19: scoping review *J. Psychosom. Res.* **147** 110525
[110] Salamanna F, Veronesi F, Martini L, Landini M P and Fini M 2021 Post-COVID-19 syndrome: the persistent symptoms at the post-viral stage of the disease. A systematic review of the current data *Front. Med.* **8** 653516
[111] Venturelli S *et al* 2021 Surviving COVID-19 in Bergamo province: a post-acute outpatient re-evaluation *Epidemiol. Infect.* **149** e32

[112] Moreno-Pérez O *et al* 2021 Post-acute COVID-19 syndrome. Incidence and risk factors: a Mediterranean cohort study *J. Infect.* **82** 378–83

[113] Schou T M, Joca S, Wegener G and Bay-Richter C 2021 Psychiatric and neuropsychiatric sequelae of COVID-19 – a systematic review *Brain Behav. Immun.* **97** 328–48

[114] Poyraz B Ç, Poyraz C A, Olgun Y, Gürel Ö, Alkan S, Özdemir Y E, Balkan İ İ and Karaali R 2021 Psychiatric morbidity and protracted symptoms after COVID-19 *Psychiatry Res.* **295** 113604

[115] Miskowiak K W, Fugledalen L, Jespersen A E, Sattler S M, Podlekareva D, Rungby J, Porsberg C M and Johnsen S 2022 Trajectory of cognitive impairments over 1 year after COVID-19 hospitalisation: pattern, severity, and functional implications *Eur. Neuropsychopharmacol.* **59** 82–92

[116] Reeves R R and Willoughby S G 2022 Significant cognitive impairment likely associated with COVID-19 infection with relatively nonsevere symptoms *J. Osteopath Med.* **122** 119–23

[117] Johanna Josepha Op't Hoog S A, Eskes A M, Johanna van Mersbergen-de Bruin M P, Pelgrim T, van der Hoeven H, Vermeulen H and Maria Vloet L C 2022 The effects of intensive care unit-initiated transitional care interventions on elements of post-intensive care syndrome: a systematic review and meta-analysis *Aust. Crit. Care* **35** 309–20

[118] Puthucheary Z, Brown C, Corner E, Wallace S, Highfield J, Bear D, Rehill N, Montgomery H, Aitken L and Turner-Stokes L 2021 The Post-ICU presentation screen (PICUPS) and rehabilitation prescription (RP) for intensive care survivors part II: clinical engagement and future directions for the national Post-Intensive care Rehabilitation Collaborative *J. Intens. Care Soc.* **23** 223-238

[119] Albu S, Zozaya N R, Murillo N, García-Molina A, Chacón C A F and Kumru H 2021 What's going on following acute COVID-19? Clinical characteristics of patients in an out-patient rehabilitation program *NeuroRehabilitation* **48** 469–80

[120] Schoeps K *et al* 2022 Social and psychological effects of COVID-19 pandemic on adolescents' and young adults' mental health: a cross-cultural mediation study *Psychol. Rep.* **2022** 332941221100451

[121] Hicks B *et al* 2022 A cohort study of the impact of COVID-19 on the quality of life of people newly diagnosed with dementia and their family carers *Alzheimers Dement. (N. Y.)* **8** e12236

[122] Nanwani-Nanwani K *et al* 2022 Prevalence of post-intensive care syndrome in mechanically ventilated patients with COVID-19 *Sci. Rep.* **12** 7977

[123] Latronico N *et al* 2022 Physical, cognitive and mental health outcomes in 1-year survivors of COVID-19-associated ARDS *Thorax* **77** 300–3

[124] Beghi M *et al* 2022 Mid-term psychiatric consequences of the COVID-19 pandemic: a 4 months observational study on emergency room admissions for psychiatric evaluation after the (first) lockdown period in Italy *Soc. Psychiatry Psychiatr. Epidemiol.* **57** 1283–9

[125] Cecchetti G *et al* 2022 Cognitive, EEG, and MRI features of COVID-19 survivors: a 10-month study *J. Neurol.* **269** 3400–12

[126] Vannorsdall T D, Brigham E, Fawzy A, Raju S, Gorgone A, Pletnikova A, Lyketsos C G, Parker A M and Oh E S 2022 Cognitive dysfunction, psychiatric distress, and functional decline after COVID-19 *J. Acad. Consult. Liaison Psychiatry* **63** 133–43

[127] Mattioli F *et al* 2022 Neurologic and cognitive sequelae after SARS-CoV2 infection: different impairment for ICU patients *J. Neurol. Sci.* **432** 120061

[128] Elugbadebo O O and Baiyewu O 2022 Mild anxiety and depression disorders: unusual reactions to COVID-19 lockdown in caregivers of older adults attending a psychogeriatric clinic in Southwest Nigeria *Niger. Postgrad. Med. J.* **29** 13–9

[129] Ceban F *et al* 2022 Fatigue and cognitive impairment in post-COVID-19 syndrome: a systematic review and meta-analysis *Brain Behav. Immun.* **101** 93–135

[130] Nishimi K *et al* 2022 Posttraumatic stress disorder symptoms associated with protective and risky behaviors for coronavirus disease 2019 *Health Psychol.* **41** 104–14

[131] Nishimi K, Neylan T C, Bertenthal D, Seal K H and O'Donovan A 2022 Association of psychiatric disorders with incidence of SARS-CoV-2 breakthrough infection among vaccinated adults *JAMA Netw. Open* **5** e227287

[132] Aaltonen K I, Saarni S, Holi M and Paananen M 2022 The effects of mandatory home quarantine on mental health in a community sample during the COVID-19 pandemic *Nord. J. Psychiatry* **2022** 1–8

[133] Saali A, Stanislawski E R, Kumar V, Chan C, Hurtado A, Pietrzak R H, Charney D S, Ripp J and Katz C L 2022 The psychiatric burden on medical students in New York City entering clinical clerkships during the COVID-19 pandemic *Psychiatr. Q.* **93** 419–34

[134] Cristillo V, Pilotto A, Cotti Piccinelli S, Bonzi G, Canale A, Gipponi S, Bezzi M, Leonardi M, Padovani A and Neuro Covid Next Study group 2022 Premorbid vulnerability and disease severity impact on Long-COVID cognitive impairment *Aging Clin. Exp. Res.* **34** 257–60

[135] Stickley A and Ueda M 2022 Loneliness in Japan during the COVID-19 pandemic: prevalence, correlates and association with mental health *Psychiatry Res.* **307** 114318

[136] Valdes E, Fuchs B, Morrison C, Charvet L, Lewis A, Thawani S, Balcer L, Galetta S L, Wisniewski T and Frontera J A 2022 Demographic and social determinants of cognitive dysfunction following hospitalization for COVID-19 *J. Neurol. Sci.* **438** 120146

[137] Cortés Zamora E B *et al* 2022 Psychological and functional impact of COVID-19 in long-term care facilities: the COVID-A study *Am. J. Geriatr. Psychiatry* **30** 431–43

[138] Costas-Carrera A *et al* 2022 Neuropsychological functioning in post-ICU patients after severe COVID-19 infection: the role of cognitive reserve *Brain Behav. Immun. Health* **21** 100425

[139] Tang S W, Leonard B E and Helmeste D M 2022 Long COVID, neuropsychiatric disorders, psychotropics, present and future *Acta Neuropsychiatr.* **34** 109–26

[140] Tabacof L *et al* 2022 Post-acute COVID-19 syndrome negatively impacts physical function, cognitive function, health-related quality of life, and participation *Am. J. Phys. Med. Rehabil.* **101** 48–52

[141] Liu Y-H *et al* 2022 One-year trajectory of cognitive changes in older survivors of COVID-19 in Wuhan, China: a longitudinal cohort study *JAMA Neurol.* **79** 509–17

[142] Hamlin M, Ymerson T, Carlsen H K, Dellepiane M, Falk Ö, Ioannou M and Steingrimsson S 2022 Changes in psychiatric inpatient service utilization during the first and second waves of the COVID-19 pandemic *Front. Psychiatry* **13** 829374

[143] Crivelli L *et al* 2022 Changes in cognitive functioning after COVID-19: a systematic review and meta-analysis *Alzheimers. Dement.* **18** 1047–66

[144] Henneghan A M, Lewis K A, Gill E and Kesler S R 2022 Cognitive impairment in non-critical, mild-to-moderate COVID-19 survivors *Front. Psychol.* **13** 770459

[145] Badenoch J B *et al* 2022 Persistent neuropsychiatric symptoms after COVID-19: a systematic review and meta-analysis *Brain Commun.* **4** fcab297

[146] Ferrucci R *et al* 2022 One-year cognitive follow-up of COVID-19 hospitalized patients *Eur. J. Neurol.* **29** 2006–14

[147] Na P, Tsai J, Harpaz-Rotem I and Pietrzak R 2022 Mental health and suicidal ideation in US military veterans with histories of COVID-19 infection *BMJ Mil. Health* **168** 15–9

[148] Amanzio M, Palermo S, Prigatano G and Litvan I 2022 Editorial: neuro-covid: neuropsychological implications of the pandemic *Front. Psychol.* **13** 971780

[149] Mansourian M, Khademi S and Marateb H R 2021 A comprehensive review of computer-aided diagnosis of major mental and neurological disorders and suicide: a biostatistical perspective on data mining *Diagnostics (Basel)* **11** 393

[150] Mansourian M, Marateb H R, Mansourian M, Mohebbian M R, Binder H and Mañanas M Á 2020 Rigorous performance assessment of computer-aided medical diagnosis and prognosis systems: a biostatistical perspective on data mining *Modelling and Analysis of Active Biopotential Signals in Healthcare* vol 2 (Bristol: IOP Publishing)

[151] Bajaj V and Sinha G R 2021 *Computer-aided Design and Diagnosis Methods for Biomedical Applications* ed V Bajaj and G R Sinha 1st edn (Boca Raton, FL: CRC Press)

[152] Marateb H R, Mohebbian M R, Shirzadi M, Mirshamsi A, Zamani S, Abrisham chi A, Bafande F and Mañanas M Á 2021 Reliability of machine learning methods for diagnosis and prognosis during the COVID-19 pandemic: a comprehensive critical review *High Performance Computing for Intelligent Medical Systems* (Bristol: IOP Publishing)

[153] Sounderajah V *et al* 2021 Developing a reporting guideline for artificial intelligence-centred diagnostic test accuracy studies: the STARD-AI protocol *BMJ Open* **11** e047709

[154] Sharifi A, Khavarian-Garmsir A R and Kummitha R K R 2021 Contributions of smart city solutions and technologies to resilience against the COVID-19 pandemic: a literature review *Sustain. Sci. Pract. Policy* **13** 8018

Chapter 4

The use of the cognitive Internet of Things for smart sensing applications

Dayarnab Baidya, Ruchee Bhagwat and Mitradip Bhattacharjee

Advancements in information technology have made data management necessary for applications with large data sets. In this context, the Internet of Things (IoT) allows data to be distributed between network objects. This system, however, cannot analyze data and make crucial judgments in real time. To address this issue, the cognitive Internet of Things (CIoT), i.e. the integration of human cognition with the IoT, has recently gained popularity. This approach requires entities to observe behaviors, gain knowledge from the observations, think, and understand the gravity of a decision before implementing it. This is crucial in the case of intelligent sensing applications in order to ensure data security and reliable performance. Moreover, wireless devices connected to IoT networks have inadequate resources to act as the sources and destinations of communication. Various algorithms have been developed to address this problem in order to create CIoT networks that can utilize spectrum and direct data proficiently. This chapter discusses various facets of CIoT in smart sensing, its architectures, and the applications of CIoT in different fields such as healthcare, military, intelligent home applications, and cognitive radio (CR). Finally, this chapter presents the research trends and future scope of this field.

4.1 Introduction

Sensing technologies have become a ubiquitous part of our daily life. Sensors detect physical environmental changes and convert them into measurable electrical outputs that can be displayed or processed further [1]. Sensors can be used in environments that are dangerous for humans. They make more accurate measurements than humans and are particularly useful when they continuously measure large amounts of data. Thus, from small-scale uses such as wearable fitness trackers to large-scale applications in industry, sensors have found applications everywhere.

doi:10.1088/978-0-7503-5326-7ch4

Networking and communication are no longer confined to humans, thanks to the IoT. The IoT has enabled physical objects to sense the environment and communicate among themselves. This networking ability has enabled us to build intelligent systems that improve efficiency and enhance our quality of life, giving rise to the idea of an intelligent world [2]. A smart world is the appealing prospect of a world with an intelligent environment that assists humans and provides ease of living [3]. The IoT has advanced dramatically in recent years. The evolution of IoT technologies has led to numerous applications in healthcare and medicine, industry, the military, business, and consumer applications such as home automation and wearable technologies. But for IoT to truly achieve its full potential, it needs cognition in addition to sensing and communication. Adding a cognitive framework can overcome the existing challenges of IoT networks and boost their performance [4].

The CIoT involves incorporating human reasoning into the design of IoT devices [5]. But what exactly is cognition, and why do we need it in the IoT? Cognition refers to the intellectual activities or processes of learning and acquiring knowledge and understanding through thoughts and experiences [6]. Adding cognition to IoT would make it self-reliant and efficient. The CIoT aims to create intelligent systems that can sense, evaluate, decide, and communicate autonomously with minimal human interaction [7]. This transition of IoT from perception to cognition opens up a world of new possibilities. For instance, consider the COVID-19 pandemic; during the peak of the COVID-19 wave, testing centers were swamped and hospitals ran out of beds. Even operating at maximum capacity, the test facilities were inadequate. A CIoT system for remote testing and health monitoring could have significantly reduced the burden on the healthcare system, allowing non-critical patients to be treated at home and relieving the pressure on overwhelmed hospitals [8]. CIoT should not be misunderstood as being a replacement for humans; as can be seen from this example, it aims to aid humans, not replace them [9].

In this chapter, we discuss the IoT, its limitations, and the concept of Cognitive IoT in detail, followed by CIoT architectures, the use of CIoT in spectrum sensing and routing, and the applications of the CIoT in different fields; the chapter concludes with the research trends and future scope of this field.

4.2 The IoT and the advent of the CIoT

4.2.1 The Internet of Things

Kevin Ashton originated the idea of the 'Internet of Things,' commonly abbreviated as the IoT. The IoT is a network of interconnected gadgets that communicate digitally addressable data via the Internet, allowing systems to share data quickly and intelligently. It is a network of 'things' that can acquire and share information [10]. They communicate through the 'Internet.' These 'things' include but are not limited to physical objects such as sensors, RFID devices, and even humans. The basic idea behind the IoT was to make information accessible to anyone from anywhere and allow us to control systems remotely [11]. One example of such a system is a home automation system that enables the user to monitor and remotely control appliances at home while miles away, say, turning on the heater before

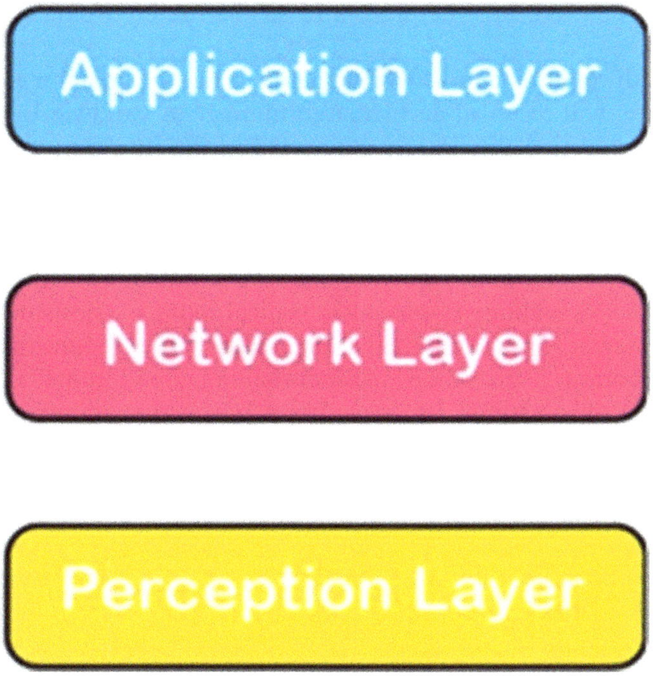

Figure 4.1. Basic IoT architecture.

reaching home [12]. The traditional IoT architecture consists of three core layers: perception, network, and application. Figure 4.1 depicts the basic IoT architecture. The perception layers comprise sensing elements that detect changes in the environment. The network layer connects the device or the system to other systems. The application layer provides user-specific application services such as smart homes [13, 14]. The IoT has found applications in industry, healthcare, home automation, and many more.

4.2.2 Limitations of the IoT

Despite its promise, the IoT has some limitations. Understanding these limitations is crucial for overcoming them.

- Volume—data sources generate a massive amount of data every second. The applications of IoT will proliferate over time, and as more and more IoT devices are deployed, the volume of data collected will explode. The sheer volume of data will lead to data storage issues [15].
- Interoperability—interoperability refers to an IoT system's capacity to send (receive) data to (from) other IoT systems. There are four main interoperability challenges—organizational, technical, semantic, and syntactic [16]. The heterogeneity of IoT systems is the main challenge in interoperability. Various forms of data are produced by different sources. It is difficult to

handle structured, unstructured, and semi-structured data using traditional tools.

- Bandwidth requirements—developments in the IoT, combined with its rising popularity, will result in an increased number of bandwidth-hungry devices. It is predicted that the devices connected to the Internet will outnumber the entire human population of the Earth in the coming years. In such a scenario, bandwidth insufficiency will be a significant future challenge [17].
- Dependency—more often than not, IoT systems need humans to analyze and interpret data. Despite receiving massive volumes of data, they lack the intelligence to evaluate it and make critical real-time decisions.
- Speed, latency, and complexity—the speed of data transmission is crucial when it comes to real-time decision-making. Traditional architectures are lacking in this aspect. Latency arises when the amount of data to be transmitted and the complexity of algorithms increase.
- Security—Data privacy and security are the two of the most significant difficulties with IoT systems. As more and more information is shared by IoT systems, security and privacy breaches become pressing concerns. Thus, there is a need for mechanisms that protect the data in IoT systems and prevent security breaches [18].

4.2.3 The cognitive Internet of Things

The cognitive Internet of Things is a network of interconnected things that can sense, analyze, and interpret data before acting on it by making judgments with minimal human intervention [5]. A context-aware perception–action loop governs CIoT systems. They also follow a learning-by-doing approach; they store the learnt data and constantly adapt to changes and uncertainties. The fundamental goals of the IoT are to bridge the gap between the physical and social worlds, to accomplish smart resource allocation, and to construct intelligent and efficient systems capable of real-time decision-making and enhanced service provisioning. The primary idea is to increase performance by modeling systems using a framework based on human cognitive processes. There are various advantages to such a system. Cognition empowers IoT devices to act independently. They can learn from their surroundings and store the information they gather, as well as continuously update their knowledge and forecast the implications of their actions. In this way, the CIoT reaps the maximum benefits from the acquired information. The perception–action cycle, massive data analytics, semantic derivation and knowledge discovery, intelligent decision-making, and on-demand service provisioning are the essential cognitive tasks of the IoT [19]. The process starts when devices sense data; the amount of data collected is enormous, and its nature is heterogeneous. The next stage is analyzing this massive amount of data; the gathered data does not hold any value unless it is adequately processed. In earlier IoT applications, a significant chunk of the gathered data was never used because traditional tools could not handle such large quantities of data; the CIoT solves this issue. After analysis of the data, the next is interpreting and organizing the data to extract context, meaning, and ontology. The extracted

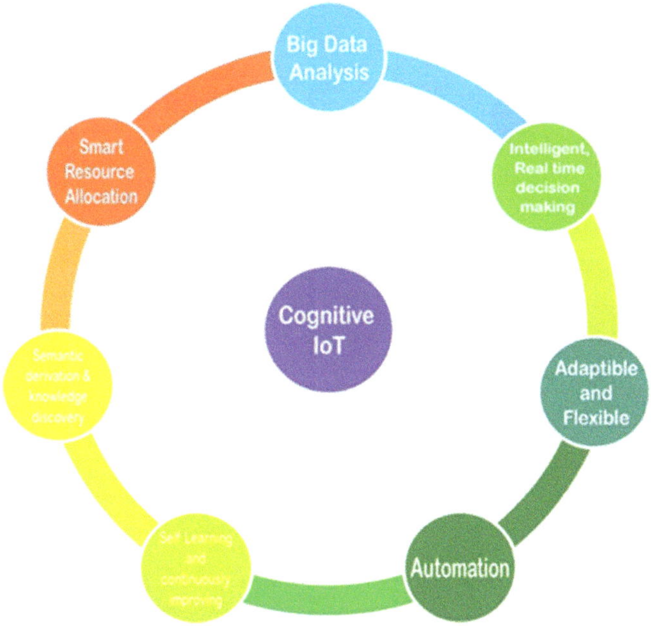

Figure 4.2. Features of the CIoT.

information is stored as knowledge. The subsequent and most crucial step is using the knowledge gained for intelligent decision-making. The final task is on-demand service provisioning. This framework makes systems intelligent, independent, and efficient. It also makes systems flexible and adaptable to various environments. This enhances systems' overall performance and allows for better service provisioning.

There are two aspects to CIoT—networking intelligence and IoT intelligence. The networking aspect focuses on maximizing network performance, minimizing latency, and saving energy, while the latter aims to make IoT systems intelligent. The networking aspect is an extension of CR and cognitive networks. The motive here is to optimize the performance for the network conditions. The IoT intelligence aspect focuses on adding cognitive capabilities to IoT to make it smart [20]. This is done with the help of cognitive computing. So, one way to define the CIoT is that it is the process of implementing cognitive computing techniques for the data generated by IoT devices. Some essential cognitive computing tools and methodologies are natural language processing (NLP), data visualization, machine learning (ML), big data analytics and probabilistic reasoning. These are modeled on human cognitive processes and aid decision-making [21]. Figure 4.2 shows the various features of the CIoT.

4.2.4 The CIoT architecture

The architectural framework of the CIoT aims to fulfill user service demands. Various architectures have been proposed for diverse domains. The perception layer, the networking layer, the information extraction and decision-making layer, and the

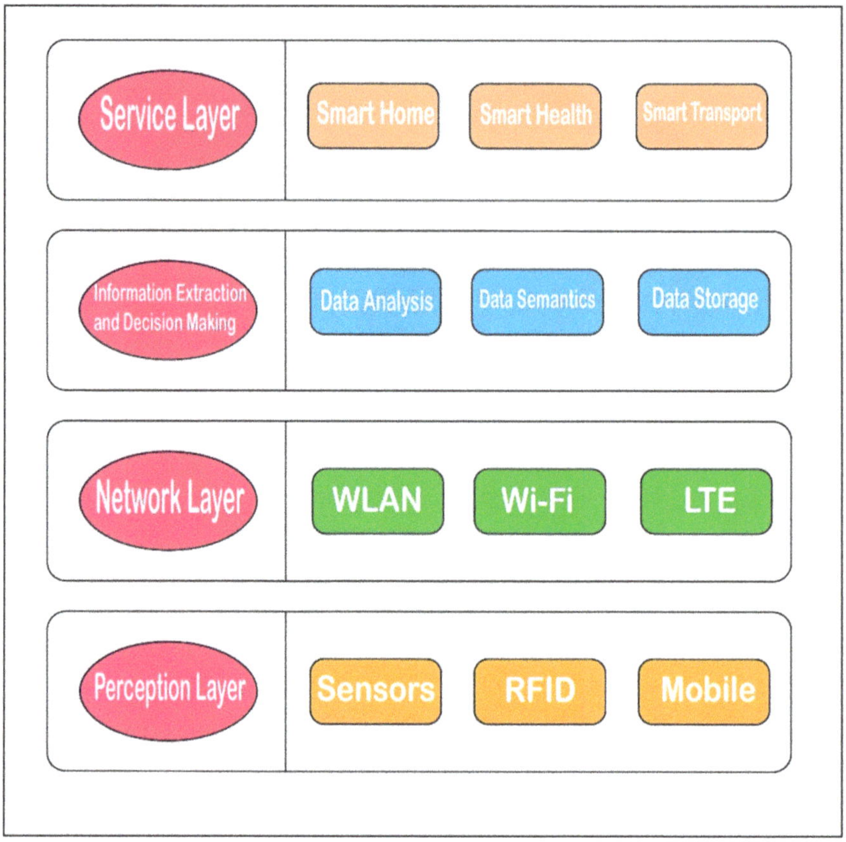

Figure 4.3. CIoT architecture.

service layer are the four critical layers of a generic CIoT architecture [22]. Figure 4.3 depicts the CIoT architecture.

- Perception layer—this layer consists of sensors, actuators, RFID devices, and agents that detect the changes in the physical environment and convert them into electrical signals. This layer is in charge of gathering all of the information required for service provisioning and application development. The collected data can be in various formats, such as video from surveillance cameras, audio input from microphones, readings from different sensors, etc.
- Networking layer—this layer is based on communication technologies and the Internet. The data acquired from the perception layer is delivered to this higher layer for processing via transmission media such as Wi-Fi, Bluetooth, wireless local area networks (WLANs), Ethernet, the Global System for Mobile Communications (GSM) 3G/4G/5G, and so on. Much CIoT-based research is being conducted to increase the networking performance of this layer. Concepts such as CR and cognitive networks are being studied to improve the performance characteristics of existing networks

- Information extraction and decision-making layer—this layer analyzes and interprets the collected data to obtain meaningful information, which is stored as knowledge. Another critical function of this layer is to make intelligent judgments based on the information collected. This is the cognitive part of the CIoT, which differentiates it from the IoT. It is capable of data analytics and independent decision-making. Following further study of cognitive computing approaches, this layer will be able to process vast volumes of data, find patterns, make predictions, and make smarter decisions.
- Service layer—the purpose of the CIoT is to provide intelligent systems such as smart homes, smart healthcare, smart cities, etc. The service layer helps to realize CIoT applications. It interacts with users to provide services based on the processed data. This layer aims to offer consumers pertinent information in a clear, straightforward, and intelligible manner.

4.3 The roles of big data and cognitive computing in the CIoT

Massive data volumes that cannot be processed using typical data processing techniques are called Big Data. This data is complex, voluminous, and continuously increasing over time. Big Data is characterized by three properties—variety, volume, and velocity [23]. A variety of sources generates data, with the result that it can take different forms such as structured, unstructured, or semi-structured; dealing with such diverse forms of data is challenging. The volume of data collected is enormous; it has already reached terabytes and petabytes and continues to increase further. The velocity of big data represents another issue because a large amount of data is accumulated at very high speeds. The practice of analyzing massive amounts of data using advanced data analytics tools is known as big data analytics.

The data generated by the many components of an IoT system grows exponentially as IoT networks continue to expand. Big data refers to a large volume of data. Thus, big data tools are helpful in analyzing such vast amounts of data. Architectures combining big data and the CIoT are more efficient than traditional architectures. Here, we take a Big data and CIoT architecture proposed in [22] as an example. A perceptual layer, an information extraction and context management layer, a knowledge and decision-making layer, and a service layer comprise the four-layer architecture. The perception layer is similar to the one discussed previously for the CIoT architecture. The decision-making layer uses a tool capable of recognizing and extracting data. The data gathered from the tool is fed into a data lake (DL) in the knowledge and decision-making layer. The data is subsequently turned into a data set and saved in a data warehouse (DWH) using extract, load, and transform (ELT). Figure 4.4 depicts the system's decision-making layer.

Combining cognitive computing techniques and big data can extract insights from massive amounts of data. Cognitive computing has the edge over existing big data analytics techniques in terms of scalability, dynamism, and natural interaction. IoT systems can only monitor and communicate; at most, they can analyze [24].

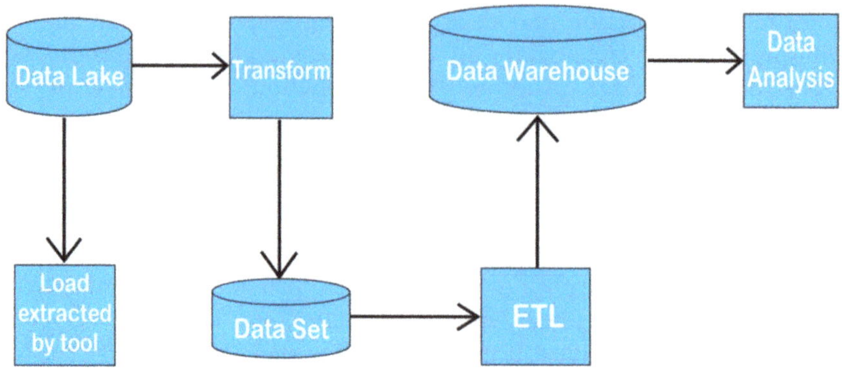

Figure 4.4. The decision-making layer of CIoT and big data.

Cognitive computing allows the system to intervene; it can sense, understand, decide, act, and adapt. In addition, the CIoT system can detect failures and reconfigure itself accordingly. The sensors and IoT devices behave like the senses of the CIoT system, while the cognitive computing methods act like the brain. Some of the essential cognitive computing techniques useful for the CIoT are discussed below:

Computer vision—computer vision is a subfield of artificial intelligence (AI) that trains computers to extract meaningful information from visual inputs. Computer vision, like AI, allows computers to perceive, observe, and comprehend. Computer vision is based on the human vision mechanism. Computer vision algorithms generate meaning from input data such as digital photographs, surveillance camera recordings, and other visual inputs. They can make suggestions based on the information collected [24].

Machine learning—according to Arthur Samuel, ML is the study of how computers or machines can learn without being explicitly programmed. It entails investigating and developing techniques that allow machines to learn and anticipate [25]. ML has emerged as a valuable technique, as it can solve problems at a much greater speed and scale than humans [18]. Through ML, systems can be trained to identify patterns and relationships between the input data. In ML, algorithms are soft-coded to perform tasks, meaning they are designed to alter and adapt automatically through repetition. Thus, these algorithms constantly improve over time. ML is divided into four categories: reinforcement, unsupervised, semi-supervised, and supervised learning [26]. Supervised learning involves the training of models using labeled data sets. Supervised learning deals with two types of problem—regression and classification. Support vector machines, naïve Bayes classifiers, linear regression, and logistic regression are some popular supervised learning algorithms. Supervised learning is used for risk evaluation, sales forecasting, weather forecasting, emotion recognition, etc [27]. Unsupervised learning involves the use of unlabeled data sets to train systems to identify underlying patterns in the data and make predictions. It deals with clustering and association problems [28].

Unsupervised learning is used for customer segregation, anomaly detection, targeted marketing, etc. A semi-supervised learning algorithm is trained using both labeled and unlabeled data. In reinforcement learning, algorithms are trained to react to the environment on their own; it is a feedback-based strategy in which an agent learns by trial and error while attempting to maximize its reward [29, 30]. Reinforcement learning is used in robotics, healthcare, etc. Much research is being done on integrating ML with the IoT. Adding ML to the IoT results in a multifold increase in a system's capabilities and performance [30]. ML, along with other cognitive computing techniques, forms the brain of the various emerging CIoT applications such as smart cities, smart transport, competent healthcare, etc. Another important part of ML is artificial neural networks (ANNs). Intelligent systems modeled on the brain's neurons are called ANNs. In computing, an artificial neural network (ANN) is often a network based on a biological neural network, such as the one found in the brain [31]. The human brain's skills are derived from these biological neural networks. Neurons in ANNs are linked in a similar way to neurons in the human brain. Each layer of an ANN comprises a collection of neurons. These neural connections are also known as nodes. ANNs underlie the deep learning approach to ML. This type of neural network has three or more layers, and it is a subset of ML. Deep learning powers many AI products and services, allowing them to carry out analytical and physical activities without human interaction [32]. Many daily-use products such as credit card fraud detection, digital assistants, and voice-activated TV remotes are powered by deep learning technology. Autonomous vehicles, for example, will rely heavily on deep learning in the future [32]. The number of IoT products on the market has grown steadily during the past few years. These gadgets are made to gather information about their surroundings and then use ML to make sense of it. For instance, Google's Nest Learning Thermostat methodically collects temperature data and then uses algorithms to evaluate the data to learn about its users' temperature preferences and daily routines [33]. Despite this, it cannot comprehend unstructured multimedia data, such as audio and visual signals. More powerful deep learning technologies, such as neural networks, are used by IoT gadgets to understand their surroundings better. The Amazon Echo, for example, can recognize human commands and carry them out. As a result, it can search for relevant information by transforming auditory impulses into words.

Automated speech recognition (ASR) and NLP—computer speech recognition, often known as ASR or simply 'speech-to-text,' is the process by which computer software converts a human voice into written text.

The ability of computer software to comprehend human language in both its spoken and written forms is known as NLP. The term 'natural language' refers to a language that develops spontaneously in its native setting—a part of what we now call AI. NLP allows computers to interpret natural language as humans do. It employs artificial intelligence to convert data from the actual world into a format that computers can understand [34]. This can be done in both written and verbal form.

Cloud computing—The success of the IoT has been partially made possible by cloud computing. Cloud computing enables users to perform computer tasks by accessing services provided through the Internet. Using cloud technologies in conjunction with IoT has become a catalyst for this combination, making the IoT and cloud computing more intertwined. Many new issues have arisen as technology has progressed rapidly, including data storage, processing, and accessibility. Cloud computing not only enables resource sharing but also allows resource optimization. Cloud computing and the IoT will potentially be tremendous agents for change in the future. This combination will enable sophisticated analysis of sensory data streams to take place and the implementation of new monitoring services [31, 35]. For example, cloud computing can be used to store sensor data that is subsequently used by other devices for intelligent monitoring and activation. Preparation for future use is an option here. One goal is to turn data into actionable, productive, cost-effective insights.

4.4 Cognitive radio and its applications in IoT

The typical spectrum management strategy is to issue licenses to operators to operate in specific frequency bands. This makes it incredibly difficult for the system to adapt to changing circumstances. Since most of the usable radio spectrum has already been allotted, it is becoming increasingly challenging to allocate new bands in order to launch new services or improve existing ones [36]. Spectrum management has become problematic because each operator is awarded an exclusive license for a specific frequency band. According to surveys, the allocated spectrum is rarely used continuously in different areas and at different times of the day. Regulatory agencies are currently investigating a new access paradigm that would allow secondary (unlicensed) systems to utilize new primary (licensed) band spectrum [37].

The term CR denotes intelligent radio that can detect environmental variables and change accordingly, thus increasing efficiency [36]. Fundamental bands that primary users are not using are referred to as white spaces; figure 4.5 depicts white spaces in the frequency spectrum. CRs opportunistically make use of these white

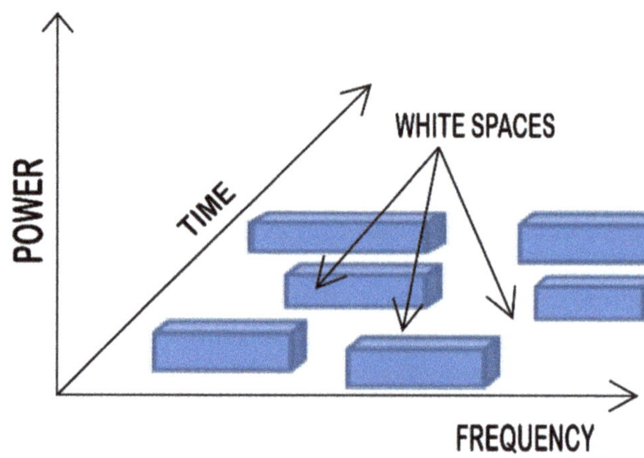

Figure 4.5. White spaces in the frequency spectrum.

spaces in the absence of their primary users. Adaptive changes in operating parameters and efficient use of the radio spectrum are the hallmarks of CR's self-awareness (of its surroundings) and intelligence (of its communication system). The term 'cognitive capacities' refers to all of these attributes taken as a whole [37].

CR offers a solution to the bandwidth issue in IoT systems. Because of the use of CR technology in the IoT, the spectrum can be reused by sensor nodes. In addition, CR's capabilities will improve IoT user device access to a wide range of networks and services, and the system itself will be extendable. It will be easier to use CR-IoT devices due to the changeover between cellular and Wi-Fi networks [38]. CRs give the IoT dynamic spectrum access capability, which boosts performance and helps the IoT to overcome the issue of inefficient spectrum utilization [39].

A typical duty cycle for CR includes detecting unused spectrum, selecting appropriate frequency bands, coordinating spectrum access with other users, and leaving a frequency in the event of the arrival of a primary user. The three responsibilities supporting this conceptual cycle are spectrum sensing and analysis, spectrum management, and spectrum mobility [39]. Using CR's spectrum sensing, it is possible to detect white space, which refers to a region of a frequency band that its primary users are not using; figure 4.6 illustrates a classification of several approaches to spectrum sensing technology and analysis. CR is also able to detect primary users' activity when they return to accessing the licensed spectrum and adapts accordingly. The transmissions of secondary users will thus not cause harmful interference with the primary user signal. In CR, spectrum management and handoff enable secondary users to select the correct frequency band and hop between multiple bands to meet varying Quality of Service (QoS) requirements.

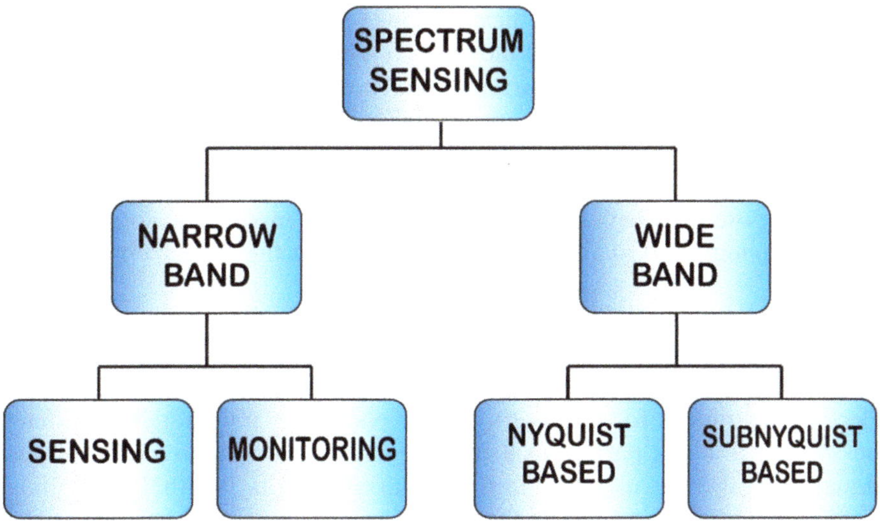

Figure 4.6. Spectrum sensing techniques.

Sensing has already determined the available spectral space; this is now achievable. Secondary users using the licensed band may choose to transmit on other frequencies when a primary user reclaims their frequency band. Secondary users can share spectrum resources with primary users under the dynamic spectrum access model based on channel capacity, which is determined by noise and interference levels, path loss, channel error rate, holding time, and other parameters [40]. As a result, it is critical to develop an effective strategy for distributing and sharing spectrum if we are to achieve high levels of spectrum efficiency [41]. When secondary users share a licensed band with primary users, it is critical to protect the rights of the primary users; thus, the interference level created by secondary spectrum usage should be limited by a predetermined threshold. Secondary users of a frequency band must work together to avoid mishaps and interference. The various spectrum sensing approaches are categorized based on the width of the band of interest, which are narrow band sensing and wide band sensing. For the cognitive cycle to proceed, one of these two sensory operations must be accomplished [42].

4.5 Applications of the CIoT

4.5.1 Smart transport

The purpose of the IoT is to allow intelligent and efficient systems to be built that enhance our quality of life. However, it is the addition of cognition that enables IoT applications to realize their full potential. The CIoT makes systems truly smart. Intelligent transport is a promising application of CIoT. Intelligent transportation has attracted many researchers; applications of the CIoT in traffic assessment, road anomaly detection, smart parking, etc. are some of the emerging areas of research [43]. Topics such as navigation and route optimization have been a key focus of smart transportation. Applications employ data from mobile devices or roadside units deployed in certain spots to assess traffic congestion. As a result, these programs recommend the most efficient routes to save travel time, reducing the amount of energy that cars use and the emissions they cause. It is also recommended that street lights that can assess current traffic conditions and operate accordingly, rather than being on all the time, are installed to reduce energy use. There is a growing demand for intelligent parking solutions that utilize IoT technology. Cameras or other wireless sensors, such as magnetic fields or infrared sensors, have been proposed for new parking reservation systems. These solutions increase parking lot capacity and availability while decreasing the time spent looking for a parking spot.

Parking applications are intended to keep track of parking lot availability, provide users with alternatives for making reservations, and even inform them when a parking space becomes available. These programs include parking detection and alerting techniques for the user's convenience. It is now possible to determine whether or not a car is parked in a particular area using many IoT devices. ML techniques can also be used to find free parking spots by analyzing images of roadways. Such algorithms use image data.

Sensors mounted on vehicles or the driver's phone can also be used to help detect road anomalies, as described by suggestions for new systems that use input data from these sensors. Because road conditions can immediately impact various transportation-related factors, detecting road anomalies is crucial to developing intelligent transportation. The primary purpose of a system designed to identify road irregularities is to detect bumps and potholes and alert vehicles to their presence [44]. Vehicle breakdowns, accidents, and traffic congestion can all be exacerbated by the state of the roadways. Accidents may be averted if poor road conditions are detected and reported promptly. IoT devices have also been utilized to help avoid or detect car accidents. The IoT M2M communication option has enabled the creation of vehicle social networks and vehicle-to-vehicle communication, allowing helpful information to be shared and transmitted. Accident detection and prevention falls under the category of 'smart transportation.' It is an activity that every community should engage in, since it can save lives if done correctly. It is possible to reduce the number of automotive accidents if drivers can focus more on the road. An accident prevention system can warn drivers of potentially hazardous conditions, allowing them to respond quickly. Accident detection can minimize the overall number of accidents and traffic congestion. This is done by identifying accident hotspots and locating incidents that have already occurred in the live traffic network. Evidence confirms that ML can be extremely useful in detecting and alerting drivers to potential road accidents and in identifying patterns that can lead to subsequent incidents. If a driver is looking for specific road conditions, several ML algorithms can help.

The intelligence and informatization of automobiles have emerged as a critical path for the industry's ongoing and future growth due to advancements in intelligent interconnection and artificial intelligence technologies. The concepts of the 'smart car' and 'intelligent traffic' rely heavily on an autopilot system. For practical purposes, however, the requirements for self-driving vehicles are much stricter. Autonomous automobiles that operate at high speeds must be able to handle traffic conditions that are both complex and variable in order to perform sophisticated driving tasks. Ultra-low latency and ultra-high dependability problems can now be solved because of AI technology improvements and rapid breakthroughs in fog computing and 5G communication networks [45].

4.5.2 Industry

Because of the increased usage of wireless sensor networks, cloud computing, industrial robotics, embedded computers, and affordable sensors, there have been considerable developments in industrial IoT technology in recent years. As a result, the range of industries using these tools has expanded (e.g. product lifecycle management). The IoT is a fundamental driver of modernization in the industrial sector. The IoT has made manufacturing intelligent [46]. As one of the essential foundational research topics and key approaches for implementing intelligent manufacturing in the industrial IoT, cognition is becoming increasingly relevant.

Intelligent services and applications are now more readily available as a result. The integration of an intelligent industrial IoT with cognitive computing can therefore deliver services of a suggestive, prescriptive, or instructional nature. The design choices made to make a new class of issues computable could also make it more effective and impactful [47].

The IoT, a new paradigm, connects numerous 'uniquely identified items' via sensors, GPS devices, radio frequency identifiers, actuators, and other wireless and mobile devices. Cisco designed this new paradigm. The term 'smart industry,' which is now known as the industrial IoT (IIoT), arose from a growing interest in the application of IoT technology in industry [48]. Another way to put this is that an intelligent industry uses the IIoT to link all of its devices. There can be no progress in the use of computers in business without this. For example, the automatic monitoring and inventorying of items equipped with GPS or radio tags can be used in manufacturing to improve material tracking and production line control. Sensors and actuators can help to automate other industrial operations, including environmental monitoring and safety and security surveillance [49].

4.5.3 The military

In the near future, the CIoT will find many applications in the military. Research is under way to develop cognitive radio networks to improve military capabilities [50]. Live simulation is another field of research in the military sphere. In real-world situations, units face a tangible enemy in a real-world setting. To strengthen the operational preparedness of the military, live simulation is utilized to simulate these warfare circumstances. These simulations include real people facing real operational challenges, and only the weapons and their effects are simulated [51]. These live simulations train military units for the pressure and challenges of a real battlefield without causing any harm in the process, enabling them to be well prepared whenever an actual battle situation arises.

Another new area of study is the Internet of Battle Things, which is the application of the IoT and, by extension, the CIoT, in battlegrounds. In future, wars will not be fought by humans alone; technology will play a crucial role on the battlefield. These future battlefields will be densely populated by various devices constantly sensing, communicating, interpreting information, and acting on it. Devices such as sensors, drones, weapon systems, and robots will be deployed on the battlefield alongside humans. The information provided by these network devices will provide valuable insights and assistance to human fighters; it will reduce risks, improve operational performance, and aid in the development of better strategies [52].

The vast amount of information obtained from the Internet of Battle Things devices can be analyzed to extract valuable information. Warnings and signals can be sent to the fighters based on the obtained information. Drones can aid fighters by conducting surveillance operations to assess the environment. IoT and CIoT systems developed for military purposes can also be used in times of peace. They can be used in harsh terrains to track the atmospheric conditions of the environment and troop locations and to monitor troops' health. An example of one such scenario could be the Siachen glacier in India, one of the most brutal battlefields in the world. With subzero temperatures that drop below −60 °C, high-speed winds and the constant threat of avalanches, there is no dearth of challenges in this terrain. A system of wearable devices could be used to sense and extract information about the health of the soldiers, relay tactical signals, and constantly monitor the wellbeing of the troops.

4.5.4 Smart cities

An IoT-enabled 'smart city' utilizes this technology to improve citizens' quality of life, ensure their safety, reduce costs, and provide better services and resource management [53]. Sensing devices based on the IoT model deployed in intelligent cities generate massive amounts of data. But when it comes to offering human-centered solutions, traditional approaches have fallen short, which paves the way for CIoT-based smart cities. Academic scholars and corporations alike are interested in cognitive science and AI interfaces. Researchers are implementing cognitive computing methods such as big data and ML in smart cities [54]. Cognitive computing is based on a strategy for teaching computers to think like the human brain, learn from experience, and intelligently utilize the data collected by the IoT. Smart city platforms comprise several sublayers; figure 4.7 depicts the architecture of a CIoT smart city network. Such networks include smart buildings, smart homes, energy, transportation, agriculture, and industry [55, 56]. Big data encompasses both organized and unstructured data, both of which are generated here. A large number of sensors are present in smart buildings. Smart buildings use the data collected from these sensors to provide solutions such as reducing energy consumption, lighting control, air quality control, and other features based on user preferences to maximize users' comfort levels [57].

IoT sensor data can increase efficiency, optimize systems, and save power. For example, it can learn about an occupant's mobility patterns and preferences and control how warm or cool the interior should be. An intelligent house provides this information. The use of cognitive computing can help people to achieve better control of their systems, which can lead to better safety and wellbeing.

IoT layer—the IoT layer refers to data acquired from sensors installed in various devices. Furthermore, these sensors provide a real-time perspective on the performance of the equipment, allowing the cognitive computing algorithm to be built on this data. Transportation, residences, buildings, energy, agriculture, and other electronics are all included in this area. The smart city platform uses various data collected from these sensors to better manage the multiple smart city services.

Data layer—The data layer classifies the many sorts of information acquired from the sensors.

Cognitive computing layer—this layer describes how cognitive computing algorithms are created. It includes data preparation and analysis, the extraction of cognitive characteristics, and ML. Algorithms developed for this layer, which employ user-specified feature selections to build algorithms based on individual

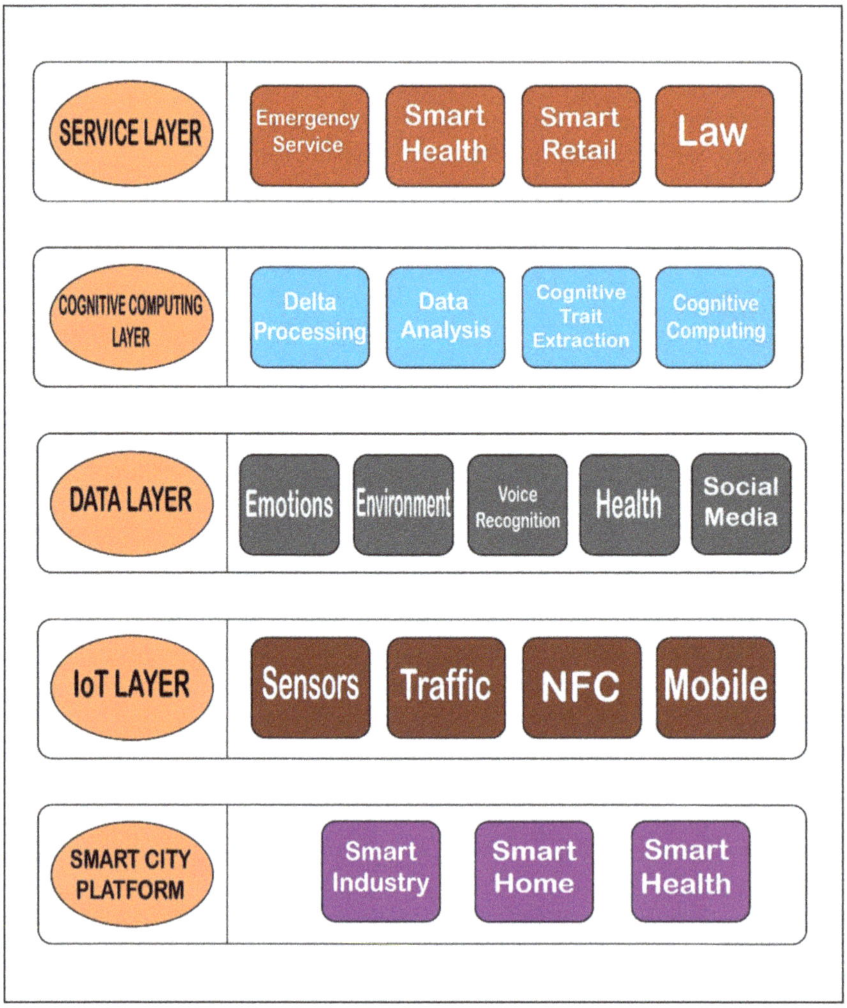

Figure 4.7. CIoT-based smart city architecture.

user specifications, can help smart cities. Because the raw data obtained from data sources may contain noisy or missing data, preprocessing is required. Including noisy or missing data leads to less effective cognitive computing and artificial intelligence models. Depending on the situation, the missing data can either be restored; alternatively, it can be removed if it is decided that it does not impact the outcome.

Analyzing data means identifying cognitive features crucial to the model's development. Several features are chosen because they can be utilized for many purposes, avoiding the need to construct a different cognitive model for each smart city use case. For instance, the creation of solutions for intelligent energy systems and agricultural management systems can use brain activity, including cognitive and perceptual processing. Optimization in both the smart home and transportation industries can benefit from the user's emotional state and the context of their surroundings.

Service layer—the smart city aims to enhance the living conditions of citizens and reduce the cost of living [58]. The service layer investigates the various cognitive computing applications in a smart city context. It is the most crucial layer of smart city architecture; it allows smart cities to achieve their goal of delivering sustainable development while also improving the quality of life. Energy harvesting, emergency services, medical care, self-driving cars, retail, and the media are prominent applications of smart cities. Cognitive computing's analytical capabilities are vastly enhanced by using unstructured data, as opposed to previous ML models and databases that relied on structured data; thus, cognitive computing gives CIoT smart city platforms an edge over traditional IoT systems. Some of the services that smart cities can provide are discussed below.

- Emergency services—smart cities can facilitate quicker and more efficient emergency services in times of need. Fighting fires is a dangerous job that necessitates constant vigilance on the part of firefighters. Cognitive-based apps can now assist firefighters by providing real-time analysis of an emergency based on a system that has been trained using image analysis, primary missions, cognitive-based contextual settings, gesture recognition, and brain activity. Such systems can detect the level of risk involved, warn firefighters of the risk and the circumstances that they will encounter once they arrive at the scene, and provide real-time analysis if the risk factor increases. In addition to saving the lives of first responders, cognitive-based applications could also save those directly or indirectly involved in the emergency. Real-time traffic conditions can be monitored and altered to provide fire engines and ambulances with the fastest possible route to their destinations.
- Security and surveillance—video data from surveillance cameras across the city can be processed, analyzed, and monitored in real time using CIoT-based systems. This helps in the detection of suspicious activity and threats. Researchers have analyzed the behavior of crowds using obtained video data [59].
- Medical care—healthcare providers can use cognitive applications to improve patient care by identifying symptoms and offering correct diagnoses, among other things. To provide an accurate diagnosis, medical history, symptoms, and cognitive data such as brain activity, emotional state, gesture recognition, and voice recognition are all utilized.

One challenge faced by smart city platforms is energy efficiency. Researchers are investigating resource management and energy harvesting for smart city CIoT platforms [58]. Future work needs to be done on building energy-efficient smart cities that promote sustainable development and provide better real-time, on-demand services.

4.5.5 Smart healthcare

A significant amount of recent research has concentrated on the IoT, which has earned global recognition as a viable option for relieving pressure on healthcare systems; figure 4.8 depicts smart healthcare applications. One emerging area of smart healthcare is telemedicine. Telemedicine involves providing healthcare at a distance [59]. The significance of telemedicine has become evident in the current COVID-19 pandemic situation. Providing healthcare at a distance through rapid diagnosis, tracing, clustering, and remote patient monitoring to reduce the load on hospitals is the need of the hour.

Medical personnel, patients, hospitals, and university research organizations all provide intelligent healthcare. It encompasses illness prevention and surveillance, disease diagnosis and treatment, hospital administration, healthcare policymaking, and medical research. Smart healthcare's target groups include clinical and scientific research institutes, regional health decision-making bodies, and individual or family clients. Smart healthcare can be split into the following categories based on its many requirements: diagnosis and treatment; health management; disease mapping,

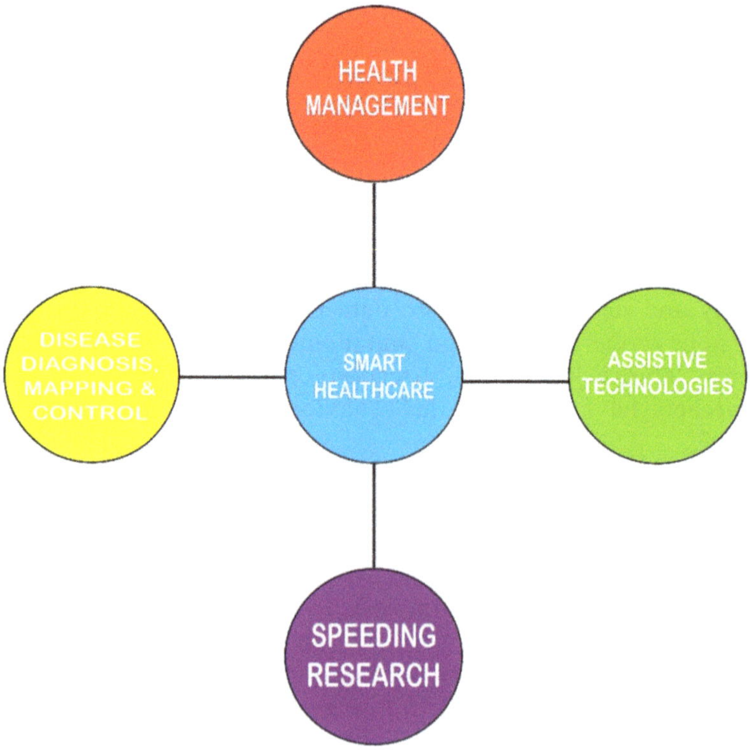

Figure 4.8. Applications of smart healthcare.

control, and risk monitoring; smart hospitals; supporting drug development; and assistive technologies [62].

Diagnosis and treatment: any form of treatment begins with diagnosis. Disease detection, for example, the detection and monitoring of epileptic seizures and heart diseases using CIoT systems, is an emerging area of research [60, 61]. Several studies have been devoted to determining how to keep track of patients with a variety of illnesses. A precise assessment of the patient's condition and sickness state can aid in developing a personalized treatment plan. This aids both the diagnosis and the management of the condition. The treatment is carried out more precisely due to this improvement. Smart radionics, for example, can be used to dynamically monitor the radiotherapy process being provided to a patient suffering from malignancies. It is possible to improve the radiotherapy program, monitor the evolution of the sickness, and prevent the unpredictability of manual operation. The development of surgical robots has made it easier for doctors to perform surgeries at a higher level than ever before. Additionally, research efforts are trying to aid rehabilitation by constantly monitoring a patient's progress.

Health Management: chronic diseases have become a new epidemic in health management. Chronic diseases are long-lasting, incurable, and costly; thus, efficient disease management is crucial. The standard healthcare management paradigm, which focuses on hospitals and physicians, appears inadequate to handle the expanding number of patients and diseases. Smart healthcare's new health management paradigm emphasizes patient autonomy in managing care. It focuses on real-time patient self-monitoring, immediate health data feedback, and medical behaviour change. Smart gadgets that can be implanted or worn, smart homes, and health information systems connected to the IoT provide a solution. Recently, much emphasis has been placed on developing multifunctional wearable gadgets based on flexible sensors. Because of their flexibility and cost-effectiveness, flexible sensors have piqued the interest of researchers due to their vast applications in healthcare and biomedical disciplines [63–66]. Modern sensors, microprocessors, and wireless modules can be integrated into third-generation wearable and implantable devices to reduce power consumption and improve patient comfort. This allows devices to monitor patients' physiological markers intelligently. Medical facilities must transition from scenario monitoring to continuous perception and integrated care to track sickness progression. Smartphones, smartwatches, and other network-connected devices have created a new monitoring platform. Mobile phones have been tested with biosensors. High-performance smartphones are becoming more mobile, allowing users to monitor their surroundings and bodies. Smart housing operations improve life quality. Smart homes help the healthcare business through home automation and health monitoring, and they monitor inhabitants' physical signs and environment using integrated sensors and actuators. Thanks to this technology, people who need care can become less reliant on healthcare providers while enhancing their home lives. Apps and a health information portal enable patients to self-manage their ailments. A wearable medical sensor is used in stress detection and alleviation to assess body pressure and alleviate stress. Integrating health data from many portable devices into a clinical decision support system creates a hierarchical system capable of diagnosing disease [67]. The use of a clinical decision support system aids healthcare decision-making and predicts patient hazards using cloud calculators and big data. Establishing an open framework for mobile healthcare will lower entry barriers for medical professionals, patients, and researchers, allowing them to communicate. Physicians can dynamically monitor patient status and provide telemedicine advice and services to patients. Specialist peers and researchers can help clinicians. Such systems can also provide emergency medical services in case of health emergencies such as heart attacks, strokes, and seizures by timely detection and relaying information to healthcare providers; however, this still needs more research.

Monitoring potential dangers and disease prevention: disease mapping can be crucial in preventing and controlling the spread of diseases. It can detect and monitor clusters and thus help improve strategies to tackle disease outbreaks. Health officials gather patient data, compare it to the organization's criteria, and disseminate the forecast. This method causes delays and is ineffective. Smart healthcare allows illness risk to be forecasted dynamically and individually. It enables patients and doctors to proactively estimate their disease risk and engage in targeted prevention activities. The latest methodology for forecasting disease risk uses wearable gadgets and smart apps to collect data, upload it to the cloud, and evaluate it using big data-based algorithms. The model sends real-time SMS predictions to subscribers. These safeguards have already worked. They help medical professionals and the public to improve their medical practices and lifestyles and enable decision-makers to establish regional health initiatives that minimize the risk of illness.

Assistive technologies

Virtual assistants are algorithms, not people. Virtual assistants communicate with consumers via speech recognition, massive volumes of data, and user preferences or requirements. They include Apple Siri (Cupertino, CA), Microsoft Cortana (Redmond, WA), and Google Assistant (Mountain View, CA). Users can generate reminders and automate their houses with virtual assistants, leveraging session history and NLP. Virtual assistants connect patients, professionals, and medical organizations in linked healthcare. They enhance medical care. The virtual assistant may transform normal language into medical terminology using the patient's smart device. This helps patients to find the proper healthcare. Based on patient information, the virtual assistant can provide appropriate information to doctors. This helps doctors to manage patients and coordinate medical operations, saving time. The use of virtual assistants in medical facilities can save labor and material resources and also improve responses. To improve the medical experience, nuance technology can also be utilized for virtual assistant interactions, for both general and highly specialized assistants. Using virtual assistants to boost mental health could help to mitigate the lack of psychotherapists while providing spiritual wellness to more patients. Virtual assistants can also be utilized to improve patients' physical health. Another critical application area is assistive technologies for the elderly and the disabled, for example, IoT assistance systems for the visually impaired [68, 69].

CIoT systems can be used to provide assistance and support in smart homes to help the elderly and disabled [70].

Smart hospital: intelligent healthcare is divided into three categories: regional, hospital, and family. Smart hospitals integrate IoT optimization and automated processes to improve patient care and establish new capabilities. Smart hospitals provide three services. Medical staff, patients, and administrators can all benefit from these services. Their needs must be considered when making healthcare decisions. In hospital management, IoT-based information platforms connect digital devices, intelligent buildings, and staff. This technology can be used to identify and monitor patients, manage medical staff, track tools, and handle biological specimens. A management platform can be used to distribute resources, assess quality, and track performance. It can reduce medical costs, maximize resource utilization, and assist hospitals in facility growth planning. Physical examination methods, online appointment scheduling, and doctor–patient communications can all be made available to patients. Automated methods can improve patient care while saving time. Patients wait less and receive more tailored care. The future of smart hospitals is one of integration, refinement, and automation. Pharmaceutical businesses can use smart healthcare for medicine manufacture and distribution, inventory management, and anti-counterfeiting.

Drug research: drug target screening has traditionally been done manually, but it is time-consuming and labor-intensive. Autonomous AI-based drug screening has boosted the target screening speed in drug trials. Drug discovery uses high-throughput screening. Computer pre-screening reduces the number of medication molecules to be physically screened. It can increase compound discovery, assess medicinal molecule activity, locates prospective compounds, and assemble a group of compounds with suitable properties. Clinical tests of new drugs use IoT, big data, and AI. AI is used to screen and assess many cases to select the best target subjects based on the trial criteria. This can save recruitment time as well as improve subject targeting. All the data is available to the researchers on the right platform in the right format.

Epidemic prevention, control, and management—one crucial application of smart healthcare is the prevention and management of disease outbreaks. CIoT-based systems can remotely and rapidly diagnose diseases, reducing exposure and decreasing the risk of transmission. For example, a cognitive IoT-based framework for the remote diagnosis of COVID-19 was proposed in [71]. Self-care and remote health monitoring can relieve the pressure on hospitals. CIoT systems can also integrate patient data in order to trace patients and map clusters.

4.5.6 Smart homes

A smart home is an environment where computing and communication technologies are used to control home appliances remotely. These smart home systems can be remotely controlled and optimized according to user preferences. Popular applications in smart homes include thermostat control, security and surveillance, and automated lighting [72]. Computing methods such as neural networks can be used to make suggestions and predictions and for anomaly detection. These systems can continuously learn and adapt by observing user behaviour and can thus improve over time.

4.6 Conclusions

There is no doubt that the concept of the IoT is amazing, but putting it into practice will be difficult. A number of technological issues must be tackled before the CIoT concept can come to fruition. Some of these difficulties are listed below.

A lack of battery power is a major concern for any wireless gadget. The same is true for CIoT devices. CIoT devices need more power than typical IoT devices because, in addition to standard sensing and data transport, they also perform cognitive functions. Human intervention is required to recharge and replace these gadgets' batteries. This inconvenience may outweigh the benefits of CIoT.

The CIoT deals with a wide range of different kinds of information. Because of the sheer volume of IoT data and the fact that it is primarily unstructured, analyzing IoT data to determine its relevance is difficult. Data is collected via various sensors, including visual, audio, gesture, text, and other inputs. Most data inputs can represent many meanings and sources of information, making them entirely heterogeneous; tackling such heterogeneous data is a critical challenge in CIoT systems.

At many network layers, software and algorithms are incorporated into 'things,' which transmit and evaluate network data. Various objects communicating over various protocols make up the contexts in which this software can be used. Integrating multiple software modules running in diverse environments is a significant problem in developing CIoT systems. Distributed software must be knit together to address service-oriented issues and allow machine-to-machine and thing-to-thing network interactions. This is a significant challenge.

The necessity to process and assess data in real time is a significant problem for IoT systems due to the nature of the continuous time series generated by the devices. It is possible for a device to feed a result back to itself for actuation or feed a result back to another device. Thus, data must be transferred from one device to another with high speed and efficiency. Managing and routing data flows between different devices is a challenge. IoT networks with many connected devices confront challenges in identifying devices and establishing protocols for sharing various data types.

Privacy and security are two of the most urgent challenges in networked systems. As the popularity of the IoT grows, so does the risk. Hackers can gain access to any device they may find. Using cognition as opposed to transmitting raw data may yield significant insights, but it may also expose sensitive information, posing privacy and security concerns. One of the most pressing issues is protecting the privacy of personal and company data and information. Data can be protected using a variety of encryption algorithms. These algorithms are complex, time-consuming, and resource-constrained. Encryption algorithms that enable distributed vital mechanisms and are fast and energy efficient are among the most difficult to find. Google Home and Amazon Alexa are smart home systems [73].

References

[1] Javaid M, Haleem A, Rab S, Singh R P and Suman R 2021 Sensors for daily life: a review *Sensors Int.* **2** 100121

[2] Paulovich F V, De Oliveira M C F and Oliveira O N Jr 2018 A future with ubiquitous sensing and intelligent systems *ACS Sens.* **3** 1433–38

[3] Abdoullaev A 2011 *Keynote: A smart world: A development model for intelligent cities The 11th IEEE Int. Conf. on Computer and Information Technology (CIT)* http://www.cs.ucy.ac.cy/CIT2011/files/SMARTWORLD.pdf

[4] Aggarwal P K, Jain P, Mehta J, Garg R, Makar K and Chaudhary P 2021 Machine learning, data mining, and big data analytics for 5G-enabled IoT *Blockchain for 5G-Enabled IoT* (Berlin: Springer) pp 351–75

[5] Pramanik P K D, Pal S and Choudhury P 2018 Beyond Automation: The Cognitive IoT. Artificial Intelligence Brings Sense to the Internet of Things *Cognitive Computing for Big Data Systems Over IoT* (*Lecture Notes on Data Engineering and Communications Technologies* vol 14) (Berlin: Springer) pp 1–37

[6] Wolff C E, Jarodzka H and Boshuizen H 2021 Classroom management scripts: a theoretical model contrasting expert and novice teachers' knowledge and awareness of classroom events *Educational Psychology Review* **33** 131–48

[7] Wu Q, Ding G, Xu Y, Feng S, Du Z, Wang J and Long K 2014 Cognitive Internet of Things: A new paradigm beyond connection *IEEE Internet of Things J.* **1** 129–43

[8] Swayamsiddha S and Mohanty C 2020 Application of cognitive internet of medical things for COVID-19 pandemic *Diabetes & Metabolic Syndrome* **14** 911–15

[9] Zhang C and Chen Y 2020 A review of research relevant to the emerging industry trends: industry 4.0, IoT, blockchain, and business analytics *J. Industrial Integration Management* **5** 165–80

[10] Mishra J, Choubey S, Bagga J and Choubey A 2020 Surveying various effective modes and research trends on cognitive Internet of Things over wireless sensor network *Cognitive Informatics, Computer Modelling, and Cognitive Science* (Amsterdam: Elsevier) pp 233–54

[11] Shaukat K, Alam T M, Hameed I A, Khan W A, Abbas N and Luo S 2021 A review on security challenges in internet of things (IoT) *2021 26th Int. Conf. on Automation and Computing (ICAC)* (Piscataway, NJ: IEEE) pp 1–6

[12] Wood G and Newborough M 2007 Energy-use information transfer for intelligent homes: enabling energy conservation with central and local displays *Energy Build.* **39** 495–503

[13] Jabraeil Jamali M A, Bahrami B, Heidari A, Allahverdizadeh P and Norouzi F 2020 IoT architecture *Towards the Internet of Things* (Berlin: Springer) pp 9–31

[14] Ande R, Adebisi B, Hammoudeh M and Saleem J 2020 Internet of Things: evolution and technologies from a security perspective *Sustainable Cities Soc.* **54** 101728

[15] Kar H and Rather G 2020 Multilayer software defined networking architecture for the internet of things *Int. J. Comput. Digital Syst.* **9** 735–45

[16] Van Der Veer H and Wiles A 2008 Achieving technical interoperability *European Telecommunications Standards Institute* https://www.etsi.org/images/files/ETSIWhitePapers/IOP%20whitepaper%20Edition%203%20final.pdf

[17] Samuel S S I 2016 A review of connectivity challenges in IoT-smart home *2016 3rd MEC Int. Conf. on Big Data and Smart City (ICBDSC)* (Piscataway, NJ: IEEE) pp 1–4
[18] Hwang Y 2015 IoT security and privacy: threats and challenges *Proc. 1st ACM Workshop on IoT Privacy, Trust, and Security* (New York: ACM) pp 1–1
[19] Hurwitz J S, Kaufman M and Bowles A 2015 *Cognitive Computing and Big Data Analytics* (New York: Wiley)
[20] Park J-h *et al* 2019 CIoT-Net: a scalable cognitive IoT based smart city network architecture *Hum. Cent. Comput. Inf. Sci.* **9** 1–20
[21] Sagiroglu S and Sinanc D 2013 Big data: a review *2013 Int. Conf. on Collaboration Technologies and Systems (CTS)* (Piscataway, NJ: IEEE) pp 42–7
[22] Sassi M S H, Jedidi F G and Fourati L C 2019 A new architecture for cognitive internet of things and big data *Procedia Computer Sci.* **159** 534–43
[23] Gupta S, Kar A K, Baabdullah A and Al-Khowaiter W A 2018 Big data with cognitive computing: a review for the future *Int. J. Inf. Manage.* **42** 78–89
[24] Rosenfeld A 1988 Computer vision: basic principles *Proc. IEEE* **76** 863–68
[25] Mahesh B 2020 Machine learning algorithms-a review *International Journal of Science and Research (IJSR)* **9** 381–86 https://www.ijsr.net/archive/v9i1/ART20203995.pdf
[26] El Naqa I and Murphy M J 2015 What is machine learning? *Machine Learning in Radiation Oncology* (Cham: Springer) pp 3–11
[27] Shanthamallu U S, Spanias A, Tepedelenlioglu C and Stanley M 2017 A brief survey of machine learning methods and their sensor and IoT applications *2017 8th Int. Conf. on Information, Intelligence, Systems & Applications (IISA)* (Piscataway, NJ: IEEE) pp 1–8
[28] Kang M and Jameson N J 2018 Machine learning: fundamentals *Prognostics and Health Management of Electronics: Fundamentals, Machine Learning, and the Internet of Things* (New York: Wiley) pp 85–109
[29] Kaelbling L P, Littman M L and Moore A W 1996 Reinforcement learning: a survey *J. Artif. Intell. Res.* **4** 237–85
[30] Khattab A and Youssry N 2020 Machine learning for IoT systems *Internet of Things (IoT)* (Berlin: Springer) pp 105–27
[31] Iqbal U, Barthelemy J, Li W and Perez P 2021 Regression on deep visual features using artificial neural networks (ANNs) to predict hydraulic blockage at culverts arXiv:2105.03233
[32] LeCun Y, Bengio Y and Hinton G 2015 Deep learning *Nature* **521** 436–44
[33] Verma R, Mishra P K, Nagar V and Mahapatra S 2021 Internet of Things and smart homes: a review *Wireless Sensor Networks and the Internet of Things* (Boca Raton, FL: CRC Press) pp 111–28
[34] Hirschberg J and Manning C D 2015 Advances in natural language processing *Science* **349** 261–6
[35] Sadeeq M M, Abdulkareem N M, Zeebaree S R, Ahmed D M, Sami A S and Zebari R R 2021 IoT and cloud computing issues, challenges and opportunities: a review *Qubahan Academic J.* **1** 1–7
[36] Sansoy M and Buttar A S 2015 Spectrum sensing algorithms in cognitive radio: a survey *2015 IEEE Int. Conf. on Electrical, Computer and Communication Technologies (ICECCT)* (Piscataway, NJ: IEEE) pp 1–5
[37] Ivanov A, Dandanov N, Christoff N and Poulkov V 2018 Modern spectrum sensing techniques for cognitive radio networks: practical implementation and performance evaluation *Int. J. Comput. Info. Eng.* **12** 572–77

[38] Khan A A, Rehmani M H and Rachedi A 2016 When cognitive radio meets the Internet of Things? *2016 Int. Wireless Communications and Mobile Computing Conf. (IWCMC)* (Piscataway, NJ: IEEE) pp 469–74
[39] Kumar S, Sahay J, Mishra G K and Kumar S 2011 Cognitive radio concept and challenges in dynamic spectrum access for the future generation wireless communication systems *Wirel. Pers. Commun.* **59** 525–35
[40] Bouras M A, Farha F and Ning H 2020 Convergence of computing, communication, and caching in internet of things *Intelligent and Converged Networks* **1** 18–36
[41] Wu P, Cui Y, Wu J, Liu J and Metz C 2012 Transition from IPv4 to IPv6: a state-of-the-art survey *IEEE Communications Surveys & Tutorials* **15** 1407–24
[42] Ali A and Hamouda W 2016 Advances on spectrum sensing for cognitive radio networks: theory and applications *IEEE Communications Surveys & Tutorials* **19** 1277–304
[43] Zantalis F, Koulouras G, Karabetsos S and Kandris D 2019 A review of machine learning and IoT in smart transportation *Future Internet* **11** 94
[44] Ng J R, Wong J S, Goh V T, Yap W J, Yap T T V and Ng H 2018 Identification of road surface conditions using IoT sensors and machine learning *Computational Science and Technology* (Berlin: Springer) pp 259–68
[45] Lu H, Liu Q, Tian D, Li Y, Kim H and Serikawa S 2019 The cognitive internet of vehicles for autonomous driving *IEEE Network* **33** 65–73
[46] Haleem A, Javaid M, Singh R P, Rab S and Suman R 2021 Hyperautomation for the enhancement of automation in industries *Sensors Int.* **2** 100124
[47] Rathee G, Ahmad F, Iqbal R and Mukherjee M 2020 Cognitive automation for smart decision-making in industrial internet of things *IEEE Trans. Ind. Inf.* **17** 2152–9
[48] Chowdhury A and Raut S A 2019 Benefits, challenges, and opportunities in adoption of industrial IoT *Int. J. Comput. Int. IoT* **2** https://ssrn.com/abstract=3361586
[49] Kaur N and Sood S K 2015 Cognitive decision making in smart industry *Comput. Ind.* **74** 151–61
[50] Vogel D, Rauschen D and Gläsel D 2019 Susceptibility of LTE to downlink jamming *2019 Int. Conf. on Military Communications and Information Systems (ICMCIS)* (Piscataway, NJ: IEEE) pp 1–6
[51] Sfar A R, Chtourou Z and Challal Y 2017 A systemic and cognitive vision for IoT security: a case study of military live simulation and security challenges *2017 Int. Conf. on Smart, Monitored and Controlled Cities (SM2C)* (Piscataway, NJ: IEEE) pp 101–5
[52] Gotarane V and Raskar S 2019 IoT practices in military applications *2019 3rd Int. Conf. on Trends in Electronics and Informatics (ICOEI)* (Piscataway, NJ: IEEE) pp 891–94
[53] Silva B, Khan M and Han K 2018 Towards sustainable smart cities: a review of trends, architectures, components, and open challenges in smart cities *Sustainable Cities and Society* **38** 697–713
[54] Vlacheas P, Giaffreda R, Stavroulaki V, Kelaidonis D, Foteinos V, Poulios G, Demestichas P, Somov A, Biswas A R and Moessner K 2013 Enabling smart cities through a cognitive management framework for the internet of things *IEEE Commun. Mag.* **51** 102–11
[55] Mohammadi M and Al-Fuqaha A 2018 Enabling cognitive smart cities using big data and machine learning: approaches and challenges *IEEE Commun. Mag.* **56** 94–101
[56] Ploennigs J, Ba A and Barry M 2017 Materialising the promises of cognitive IoT: How cognitive buildings are shaping the way *IEEE Internet of Things J.* **5** 2367–74
[57] Varghese E B and Thampi S M 2021 A cognitive IoT smart surveillance framework for crowd behavior analysis *2021 Int. Conf. on COMmunication Systems & NETworkS (COMSNETS)* (Piscataway, NJ: IEEE) pp 360–62

[58] Alzahrani B and Ejaz W 2018 Resource management for cognitive IoT systems with RF energy harvesting in smart cities *IEEE Access* **6** 62717–27

[59] Garai Á, Péntek I and Adamkó A 2019 Revolutionising healthcare with IoT and cognitive, cloud-based telemedicine *Acta Polytechnica Hungarica* **16** 163–81 http://acta.uni-obuda.hu/Garai_Pentek_Adamko_89.pdf

[60] Alhussein M, Muhammad G, Hossain M S and Amin S U 2018 Cognitive IoT-cloud integration for smart healthcare: case study for epileptic seizure detection and monitoring *Mobile Networks and Applications* **23** 1624–35

[61] Devi K R R, Chozhan R M and Murugesan R 2019 Cognitive IoT integration for smart healthcare: case study for heart disease detection and monitoring *2019 Int. Conf. on Recent Advances in Energy-efficient Computing and Communication (ICRAECC)* (Piscataway, NJ: IEEE) pp 1–6

[62] Diwakar M, Singh P, Shankar A, Nayak S R, Nayak J, Vimal S, Singh R and Sisodia D 2022 Directive clustering contrast-based multi-modality medical image fusion for smart healthcare system *Network Modeling Analysis in Health Informatics Bioinformatics* **11** 15

[63] Escobedo P, Bhattacharjee M, Nikbakhtnasrabadi F and Dahiya R 2020 Smart bandage with wireless strain and temperature sensors and batteryless NFC tag *IEEE Internet of Things J.* **8** 5093–100

[64] Bhattacharjee M, Nikbakhtnasrabadi F and Dahiya R 2021 Printed chipless antenna as flexible temperature sensor *IEEE Internet of Things J.* **8** 5101–10

[65] Baidya D and Bhattacharjee M 2022 Substrate optimisation of flexible temperature sensor for wearable applications *IEEE Sens. J.* **22** 15393–401

[66] Singh L, Baidya D and Bhattacharjee M 2022 Structurally modified PDMS-based capacitive pressure sensor *2022 IEEE Int. Conf. on Flexible and Printable Sensors and Systems (FLEPS)* (Piscataway, NJ: IEEE) pp 1–4

[67] Tian S, Yang W, Le Grange J M, Wang P, Huang W and Ye Z 2019 Smart healthcare: making medical care more intelligent *Global Health J.* **3** 62–5

[68] Darvishy A 2022 Internet of things–services and applications for people with disabilities and elderly persons *Int. Conf. on Computers Helping People with Special Needs* (Berlin: Springer) pp 101–4

[69] Mallikarjuna G C, Hajare R and Pavan P 2022 Cognitive IoT system for visually impaired: machine learning approach *Mater. Today Proc.* **49** 529–35

[70] Prakash R and Balaji Ganesh A 2019 Cognitive wireless sensor network for elderly home healthcare *Wirel. Pers. Commun.* **107** 1815–22

[71] Jayachitra V, Nivetha S, Nivetha R and Harini R 2021 A cognitive IoT-based framework for effective diagnosis of COVID-19 using multimodal data *Biomed. Signal Process. Control* **70** 102960

[72] Feng S, Setoodeh P and Haykin S 2017 Smart home: cognitive interactive people-centric internet of things *IEEE Commun. Mag.* **55** 34–9

[73] Serianni A, De R F and Raimondo P 2021 *Cognitive IoT enabled by Layered Architecture and Neural Networks in a Smart Home Environment 2021 Wireless Days (WD)* (Piscataway, NJ: IEEE) pp 1–7

Chapter 5

Design challenges and issues in cognitive sensor networks: a mitigation and deployment perspective

Chandramouleeswaran Sankaran (Mouli)

Wireless sensor networks (WSNs) are the most popular way of interconnecting various Internet of Things (IoT) sensors and actuators for various application areas, including smart buildings, healthcare, transportation, smart cities, etc. Normally, data with different rates and importance are shared over WSNs. Although IoT devices are low-cost and constrained in terms of both energy and computing power, they have complementary requirements, such as good coverage, enhanced security, and energy efficiency. When the cognitive IoT is to be implemented over a WSN, it creates additional complexity in terms of its implementation and providing service guarantees as well as meeting the stringent quality of service (QoS) requirements. This chapter provides an overview of the WSN design challenges and additional design issues that need to be addressed in order to convert WSNs into cognitive WSNs. It also reviews different methodologies and protocol enhancements that are used to realize an efficient implementation of cognitive WSNs.

5.1 Introduction

WSNs are becoming very popular and are the most preferred communication protocol for interconnecting IoT devices in variety of usage scenarios because of their features and ease of use. Various kinds of data that have different data rates and QoS requirements can be sent over WSN networks. The IoT domain needs connectivity solutions that are secure and energy efficient and that can ensure reliable connections are established, which at the same time are low-cost and miniaturized. The cognitive IoT introduces additional requirements in terms of cognitive communications, which is a new concept and has been an emerging field in the last few years, especially in the field of cognitive radio (CR).

doi:10.1088/978-0-7503-5326-7ch5

WSNs are the most suitable communication systems for interconnecting IoT devices, and they can be enhanced by the addition of cognition and intelligence. They are also easy to develop, and those enhancements can also be used to improve overall network efficiency and meet user requirements. Because the basic nature of WSNs is that they are distributed in nature, their network properties and configuration parameters are well suited to the addition of cognitive capabilities. The distributed information available about a WSN includes network conditions, environmental properties such as channel status and power levels, device battery levels, the health of the nodes in the network, etc.

Security features are also most important; this is an area in which the cognitive features can have a greater impact. In this chapter, we review different cognitive methodologies and approaches used to realize WSNs with cognitive capabilities. Since many WSN-based applications need real-time node position information, several cognitive approaches expect the node position information to be available. WSN networks can be equipped with real-time positioning devices, such as Global Positioning System (GPS) receivers; other technologies and algorithmic approaches are also available, which can provide the node position information if the GPS system is down or unreachable.

In this chapter, after reviewing the architecture and features of WSNs, we cover specific issues related to CR and spectrum access that need to be addressed by WSN designers. Such issues are associated with the physical layer of the network protocol. We also look in detail at various applications that can benefit from CR-based WSNs.

We also take a a look at the routing protocol, which needs changes to make the cognitive IoT possible on WSNs. Since the hardware implementation of cognitive WSNs is not yet mature, this chapter addresses important issues and challenges that need to be faced in implementing CR WSNs and designing location-aware nodes by adding miscellaneous units such as GPS/inertial navigation system (INS) devices to WSNs.

5.2 Wireless sensor networks

WSNs are built using low-power, small, lightweight nodes, which are normally deployed in large numbers [1]. Recent advancements in very large-scale integration (VLSI) and embedded systems make it possible to build tiny nodes that are power efficient and have advanced communication and computing capabilities. Such nodes, which have sensing and processing capabilities, which are miniaturized, and that cater to various application areas, make it possible to realize WSN networks in the IoT domain. These WSN nodes are environment-aware; they can be personalized, they are capable of performing computation and meeting real-time system requirements, and they are adaptable to various conditions.

A WSN is normally a self-configured system that has reduced infrastructure needs and can monitor the physical and environmental conditions by itself. Such conditions include sound, temperature, vibration, motion, the concentrations of pollutants, etc. The nodes in a WSN work cooperatively and exchange data among themselves without much external intervention from the user or a central control unit. A WSN can effortlessly collect and exchange the data collected by the sensors in its nodes. Such data is forwarded to the data sink over the Internet. This makes it an ideal choice for various IoT applications that are deployed over large areas.

Extracting or querying a particular data type or elements from a WSN is also made easier by the presence of a centralized sink that holds the data collected from the various nodes in a WSN, even though a WSN is itself a distributed and ad-hoc network with intelligence and sensing activities spread over a wide area [2].

Typically, a WSN is built from thousands of nodes that gather various parameters about the environment in which the network is deployed [3]. These sensor nodes communicate among themselves using radio signals. Once the sensor nodes are deployed in physical proximity to other nodes, they self-configure and self-organize by automatically exchanging information using multihop communication, without much human interference.

Once deployed, the sensor nodes start periodically collecting data they are supposed to sense about the environment based on their initial configuration or system requirements, depending on the application they are expected to support. A sensor or set of sensors in a WSN typically supports more than one application and feeds its data to the sink. The sink, in turn, is interfaced with different application interfaces that query the data they are interested in, or alternatively, the sink may collate multiple data items using postprocessing so that they are suitable for the end application to consume effectively. The sensor nodes in a WSN are also capable of responding to the queries generated by the control center based on either user requirements or the application.

The working mode of the sensor nodes can be configured to be either continuous or event driven. To make WSN nodes position aware, either GPS- or IPS-based systems are normally added, or local positioning algorithms can also be used for this purpose. Since WSN nodes are always power constrained, the chosen algorithms or protocols have to be power efficient, and the algorithms used need to be efficient and optimal. To optimize power consumption and extend the battery lives of the sensors, the application domains considered for WSN-based networks are sometimes only data-oriented monitoring or status-reporting applications, which do not need very complex compute-intensive operations to be performed by the sensor nodes.

Figure 5.1 shows a typical WSN network that connects to the user through a base station. All the nodes which are part of the WSN have the capability to interact with their adjacent nodes and communicate wirelessly, so that the data sensed by a node is sent via the nearby nodes, finally reaching the base station though intermediate nodes which are in the direction of the base station and also closer to the base station.

The base station is, in turn, connected to the Internet. The base station is not normally powered by a battery, as it has better accessibility to a constant and large power source, which helps in collecting data from remote WSN nodes and sending it across to the Internet to the user or to the central sink or control unit. The commands or any other communication sent back to the nodes requesting them to take action in the field again travel back from the user or central control center

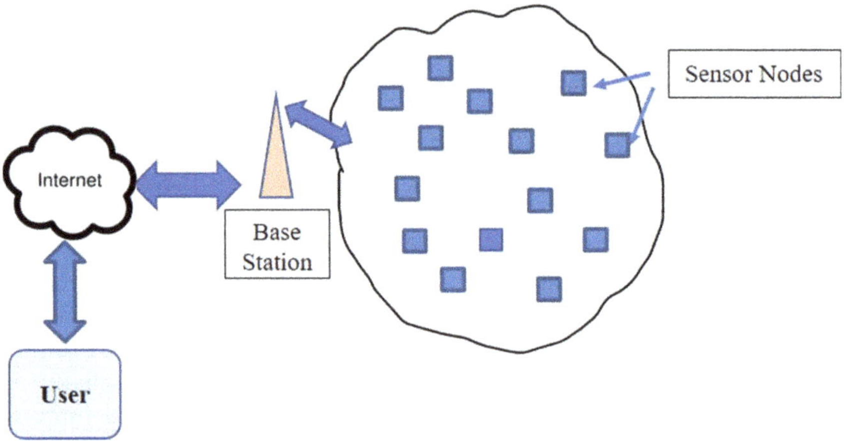

Figure 5.1. Wireless sensor networks.

through the Internet. After passing through the base station, the commands reach the individual nodes, which take appropriate action in the field.

Since WSNs are datacentric, the commands that are delivered from the central control unit to the WSN are for the entire network as a whole rather than addressed to individual nodes. Any action to be taken by the WSN is always sent to all the nodes in the WSN for a collective action to make sure that the large area they control effectively reaches the expected condition.

For example, in an agriculture-related application, watering a field or applying a pesticide based on the conditions sensed in the field needs to be completed for the entire field, and the action to be taken is common to all the nodes in the area; therefore, sending a command that does not address an individual node is fully understandable and also logical.

5.2.1 Sensor network architecture

The WSN follows the layered architecture of the Open System Interconnect (OSI) model. A sensor network has five layers, which are the application layer, the transport layer, the network layer, the data link layer, and the physical layer.

There are three common planes to these five layers, which are the power management plane, the mobility management plane, and the task management plane. The architecture used for sensor networks is shown in figure 5.2.

The three cross planes are layers that control and coordinate the five horizontal layers, ensure that the sensors work together in the network in a coordinated manner, and contribute to the overall efficiency of the network.

The mobility management plane tracks the movement of the nodes with the help of sensors installed to detect the current position of the nodes and determines the neighboring nodes and their relative power levels, so that each node is aware of the nearby nodes that it can communicate with. This tracking data is used to exchange

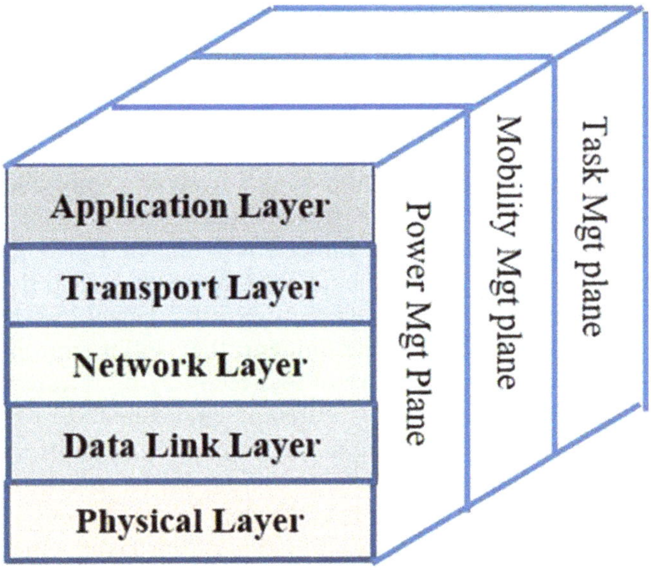

Figure 5.2. Sensor network architecture.

information as well as to pass information from other nodes across the network so that it finally reaches the base station and eventually the end destination, the data sink on the Internet.

The power management plane controls the power levels and conserves battery power, thus increasing the lifetime of the nodes by selectively choosing the nodes for the internode communication tasks, thus balancing the energy levels across thousands of nodes in the network. This keeps the network alive and improves the overall efficiency and lifetime of the network.

The task management plane takes care of scheduling the sensing tasks in a given area by selectively choosing the nodes used to perform the sensing job from the nodes which are close by. Since the data being sensed (such as the pressure, temperature, or humidity) in a given small area do not vary much in value, using multiple sensors to sense the same area would be a waste of energy, both in terms of collecting the data as well as transporting it over multiple nodes across the WSN network to the data sink. The data aggregation and proper scheduling of sensing jobs by the task management plane ensures that the entire WSN network is efficient and optimal without compromising the data quality. In addition, the frequency of the sensing job is chosen in such a way that minute changes in the sensing data are not missed as a result of the optimization performed while sensing the data.

The other layers are same as those normally found in the layers of the OSI network protocol; they perform similar jobs in WSNs as well.

- **Application layer**: this interfaces with the applications that run on the WSN network. It passes control information from this layer to the layers below for further action and processing. It also receives responses to commands previously issued by this layer from the same node or from other nodes on the network. This layer is also responsible for receiving the health and status of the network and the overall health and status of the application running on it.

- **Transport layer:** the main function of this layer is to provide a reliable and congestion-free path for the data packets over the network. It ensures that the flow of data takes place in both directions, from the top layer to the bottom and for the data coming from the network to be passed up to the application layer. This layer is also responsible for detecting packet loss. Since WSNs are wireless and work in a noisy environment, they suffer from increased packet loss [4]. This layer also takes care of loss recovery by retransmitting the data packets which were lost due to different environmental conditions or the misbehavior of any intermediate nodes. Packet loss can be caused by power loss; noisy surroundings; packet corruption; buffer overflow in any of the intermediate nodes, which thus lose the data in transit; etc.
- **Network layer**: the most important function of this layer is the routing of data packets. The major challenge and responsibility of this layer is to save the power of the network nodes involved in routing the packets through the network as well as managing the memory and buffer availability of the nodes, which is at premium because they are resource-constrained nodes. This reduces packet loss and improves the reliability of the network. The IPv6 based protocol is used on WSNs, since it does not suffer from a shortage of Internet Protocol (IP) addresses for the thousands of nodes in a WSN with the plethora of sensors that are needed to sense and collect data over a large area, such as soil quality or the water content of agricultural fields, etc.
- **Data link layer**: this takes care of multiplexing data units belonging to different streams of data received through different nodes and aggregating them into a single data unit. Similarly, when larger data items are transmitted, although it is quite rare, depending on the maximum transmission unit (MTU), data packets could also be segmented before they are sent over the network. The most important goal here is that each layer needs to take care of its responsibilities by consuming the minimum amount of memory, buffer space, and air time when data is transmitted over the wireless interface either with the nearby nodes or with the base station. Although this is the job of the physical layer, the upper layers, including the data link layer, need to follow this principle while performing their tasks in order to optimize the use of the available resources by taking the appropriate actions during data transmission and reception.
- **Physical layer**: This layer takes care of converting data bits into physical signals by converting them into analog signals based on the physical protocol being used to send the data. For example, carrier frequency generation, signal detection, modulation, and data encryption/decryption activities are performed by hardware units.

5.2.2 Congestion management with common planes

Apart from the responsibilities of the individual layers, the layers need to take care of interlayer communication requirements as well; these allow information sharing between layers to enable joint optimization across different layers [5]. Information sharing between layers reduces modularity due to the increased interactions and makes the overall system complex both in terms of adding new features and maintaining it.

Congestion can happen at a node when every nearby node sends all the data it receives to a particular node. This causes buffer overflows and packets get dropped by the node which is receiving data from most of its neighbors. To avoid this, every node should have a better view of the loads of the nearby nodes and the past history of the data packets forwarded to them so that every node can be cautious in choosing the next-hop node to send data packets to. In this way, load balancing is performed and no congestion occurs in any of the network paths or at the node level. Implementing congestion avoidance requires the correct future operational decisions to be chosen based on the past actions taken by a node as well as by other nodes in the vicinity.

Thus, the ability to learn from the past actions is an ability of cognitive networks and the major differentiator between cognitive networks and those without cognitive abilities. Figure 5.3 show how congestion avoidance can be achieved by nodes when they start learning about their surroundings and learn from their past actions to control their future routing decisions.

In figure 5.3, all four nodes in the vicinity of the node at the center choose to send their data packets to be forwarded by the same node at the center [6]. This causes congestion at the central node because of the frequent generation of data packets by those nodes, which are configured to provide data at a rate which can make the data buffers in the central node overflow, thus causing congestion.

5.2.3 The need for cognitive abilities

If these nodes have the cognitive abilities and each node exchanges information about its current load, its additional load-bearing capabilities, and its health and power levels with all the nearby nodes which forward their data packets to it, then

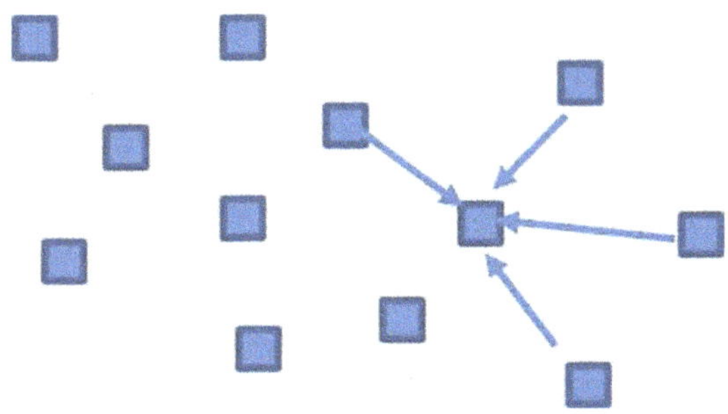

Figure 5.3. Congestion avoidance due to cognitive abilities.

the nodes which forward the data packets can learn from their own past actions as well as from the load at the node to which they are forwarding data. Thus, they can choose an alternative route, preventing any particular node from overflowing by avoiding congestion.

This ensures that each individual node is not overloaded by more than it can handle and also ensures that no data is lost due to wrong choices made by a next-hop routing node, thus triggering a sequence of retransmission actions. Retransmission causes more such events in the future, resulting in the entire network becoming overloaded. This causes a domino effect; the network thus becomes inefficient and finally runs out of power, and the entire network breaks down completely.

Therefore even though the cognitive abilities built into each node consume energy in running suitable optimization algorithms, this actually saves the nodes from wasting their energy on inefficient routing and forwarding decisions, thus improving the overall efficiency of the entire system, reducing packet loss, and avoiding congestion before it happens.

If by any chance congestion does happen, the event is captured and shared with the nodes which are closest to the affected nodes. This means that traffic can be diverted in such a way that the congestion problem does not become worse. The system recovers from the failure and returns to normality quickly, without spreading the impact beyond a couple of nodes in close proximity in the system.

5.2.4 Summary

We can summarize the need to incorporate cognitive capabilities into sensor networks as follows:

- To make sure that the network is made aware of the environmental and traffic conditions as well as the health of the nodes, so that the application requirements are always met.
- Gather information on the network channel conditions and environmental conditions in terms of error rates and signal-to-noise ratio (SNR) and take appropriate actions to ensure that the network meets the overall application requirements without any compromise.
- Make the sensors behave intelligently by being more sensitive and responsive to the current conditions of each node and the environment in such a way that the overall network performance never drops below an acceptable level.
- Detect and take action proactively, i.e. before problems such as congestion or packet loss happen, so that the network always maintains the required QoS.

By gathering information on the status and error conditions (if any), all the layers can be made aware of it by making such information available to all the other layers through the common management plane. This makes it possible to obtain a holistic view of the entire network from the perspective of the individual layers of the network. A great deal of information related to the application layer, channel status, the quality of connectivity, etc. has to be gathered and shared across the layers to make them cognitive.

Although having common planes that interact with multiple layers is effective in terms of making the network more reliable and less error prone, it is possible to introduce a new layer called the knowledge plane (KP), which is different from the control or data planes that exist in current protocol stacks. In the next section, we describe the effect of introducing the KP and the functionalities that it can support in wireless networks, thereby applying a concept that was successfully implemented in wired networks, in order to realize cognitive sensor networks (CSNs).

5.3 The knowledge plane in cognitive networks

The concept of the knowledge plane (KP) was proposed by Clark *et al* [7] in response to the limitations of the cross-layered design approach of the existing WSN architecture. Its aim is to remove the barriers that are present in the communication between layers and make the exchange of information between layers seamless, so that data is exchanged between them effectively.

The KP is designed to be based on knowledge instead of depending on individual data items, so that the information and observations from different parts of the network are properly correlated in order to extract meaningful information from the conflicting data received from the dynamically changing environment [8]. The intention for the KP is that this plane should be able to take meaningful and dependable decisions, even in the presence of incomplete, inconsistent, and conflicting information obtained from the dynamically changing environment.

The aim was to build a network which can adapt itself and take decisions on its own based on the network conditions, and reassemble itself if required based only on high-level instructions and requirements, without any user-level or human intervention in the continuously changing network conditions. In this way, the network would be able to meet the performance goals for which it was built by automatically adjusting itself, and, if required, by modifying and changing its behavior to meet its end goals successfully.

What is needed is that the network should have sufficient knowledge and information about itself and that it should be able to adapt its actions. It should also have sufficient learning abilities and mechanisms to take decisions based on its current status as well as its past actions and statuses.

The artificial intelligence (AI) and cognitive techniques that are capable of representing knowledge, learning, and reasoning abilities can make the KP most suited to meeting the complex objectives of making the network more efficient than traditional algorithm-based approaches. This also adds a lot of flexibility and scalability; even if the knowledge used to make the decisions taken by the network grows exponentially, KP-based decision making simplifies the implementation and enhances its adaptability to changing conditions.

Cognitive networks (CNs) aim to have end-to-end network capabilities and goals, which requires all the components and data elements of every layer to contribute to the overall performance and goals of the entire network. This is a holistic approach, wherein knowledge of the entire network status is captured to allow every node in the system to understand the overall behavior of the network and adapt itself to meet the overall goals.

5.4 Cognitive networks

CNs are designed to be self-aware, self-correcting, and adaptive so that they can make intelligent decisions based on the following:

- An individual node's observations about the state of the network
- Effective information sharing between layers, thus avoiding the limitations of the cross-layer sharing of data
- The ability of the individual nodes to learn and have the capability to reason effectively based on different conditions in order to generally make the network's performance better.

To implement a CN [9], the following three-layer approach was proposed by Thomas [9]. The three layers are the requirements layer, the cognition layer, and the software adaptable network (SAN) layer.

Figure 5.4 shows the three layers of the CN architecture with its properties and features.

The SAN is the interface that connects the system to the outside world. In the case of CR, the transmission power of a directional antenna can be used as one of the configurable elements of the network, and is thus called a modifiable element. The second layer, which is the cognitive layer, has a one-to-one mapping between itself and each modifiable element of the network [10]. These elements, which are part of the cognition layer, help the operation and functionality of the network to be spatially distributed within the system. Network status sensors which collect information about the status and error conditions of the network channels provide partial knowledge of the network status to the system. The cognitive layer has the

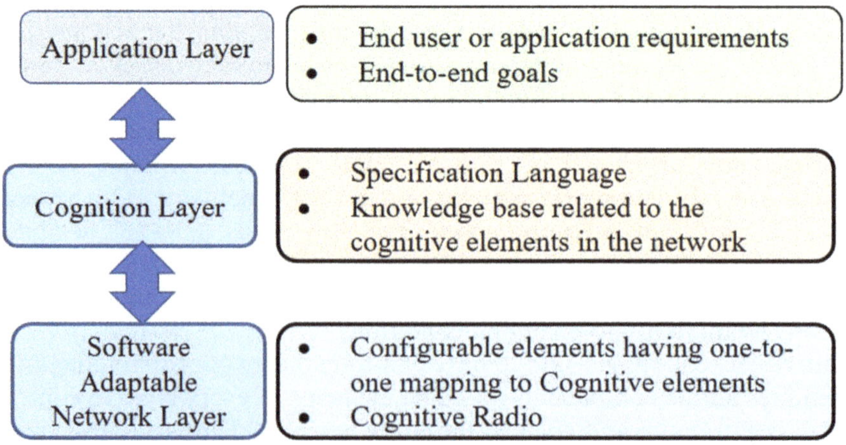

Figure 5.4. Example cognitive network architecture.

information and data structures required to hold the information gathered from different sources [11].

The highest layer, which provides the most hierarchical abstraction, is the requirements layer, which abstracts out the network status information and the properties and states of the configurable parameters of the network elements. These are then transformed to meet the end-to-end goals of the cognitive elements in the cognition layer, each of which is represented using a cognitive specification language (CSL).

The cognitive layer is the most important and central part of this system architecture and one that has a holistic knowledge of the entire network [12]. This layer is the brain behind the actions taken by the cognitive node. The future actions to be taken by the node are based on the current state of the network and the past actions.

This follows the observe, orient, decide, and act (OODA) control feedback loop that we studied in the previous chapter on the cognitive IoT.

The above features help the network to learn, achieve good network coverage, and become an efficient network that has good QoS based on the network status and traffic conditions.

5.5 Cognitive radio in wireless sensor networks

The cognitive radio wireless sensor network (CR WSN) is a specialized version of the WSN which is equipped with the advanced capabilities of CR. A CR WSN is different from and more efficient than a conventional WSN or a conventional distributed cognitive radio network (CRN). This requires cognition capabilities to support a higher degree of cooperation and adaptability to perform better and achieve the desired performance goals through a higher degree of task coordination between thousands of nodes in a WSN and their radio transmitters.

These CR nodes communicate collaboratively among themselves in the dynamically changing available spectrum bands in a multihop manner [13] with all the nodes or with the nearby ones in a WSN to meet application-specific goals and requirements.

In a CR WSN, each sensor continuously monitors the radio spectrum looking for an idle channel which is the most appropriate for communication and chooses it. It also frees up the currently used channel when the licensed user wants to use that channel for its communication. This CR technique [6] is the most unique and efficient way of improving the efficiency of spectrum usage. It achieves a higher degree of spectral efficiency, increases the overall effectiveness of the network, and also extends the overall lifetime of the network by improving the battery life of all the nodes in the network.

The advantage of CR WSNs is that they utilize the spectrum [14] even during bursty traffic conditions. They also have the ability to reduce packet loss and reduce power wastage. They also control the buffer management of individual nodes by optimizing the overall utilization of memory in each node, thus consuming the smallest amount of buffer space and improving the performance-per-watt metric of the nodes.

Given increasing demands and the exponential costs associated with the acquisition and use of spectrum in a highly competitive and dense environment in which many competing vendors provide services, spectrum optimization and utilization are critical. Since the total amount of spectrum available cannot be increased beyond what is available in the band, it is imperative that the available spectrum is used at the maximum efficiency. With the exception of the industrial, scientific, and medical (ISM) radio bands, the vendors offering services need to acquire the bands through very expensive auctions hosted by the government [15]. Once these bands are acquired by paying a huge upfront cost, each of the service providers is forced to utilize the allocated bandwidth most efficiently in order to earn the highest return on investment (RoI). This goal can be achieved only by a full implementation of CR that helps the service provider to achieve the best possible network performance with the allocated spectrum available at its disposal [16].

CR can make effective use of the white spaces, i.e. the unutilized spectrum, without disturbing the users of the licensed spectrum [17]. Unlicensed bands such as the ISM bands can also be used most effectively using CR techniques. More services, applications, and new technologies can be developed using unlicensed spectrum for the benefit of humanity if CR support is implemented by sensors in WSNs.

The CR wireless sensor has the following six basic units: sensing, a processing and storage unit, a CR unit, a transceiver, a power unit, and a miscellaneous unit. The sensor units contain analog-to-digital converters (ADCs). The sensed analog signal is converted into a digital signal which is passed on to the processing unit to remove any noise and transmit the processed signal.

Multiple channels are not used by traditional WSNs, since they operate on a single channel. In WSN scenarios in which event-based transmission is used, all the nodes can start to communicate over a single channel. In an event-driven implementation of WSNs, when all the nearby nodes sense changes in the status of the environment, all the nodes are triggered by the same kind of event. This causes serious network congestion issues, because all the nodes try to transmit over the same frequency band, causing all the nodes to encounter collisions and revert to exponential backoff. Exponential backoff in turn creates serious network performance issues, especially when a critical event triggers transmission by all the nodes in the vicinity. CR-based WSNs can access multiple channels, thus eliminating the above issue, in contrast to the single-channel operation of traditional WSN solutions [18].

This makes CR WSNs more power efficient; it reduces packet loss, improves network efficiency, and increases reliability. A large amount of energy is normally wasted due to collisions and multiple retransmission attempts made by the transmitters in traditional WSNs. This problem is avoided by CR WSNs, since they can change their operating parameters by selecting the channels they transmit on from the multiple channels available to each node based on the network conditions and the current transmission and utilization of the channels by the neighboring nodes [19]. This optimizes the power consumed by individual nodes, thus making the overall lifetime of the WSN longer.

Systems built with CR capabilities are easier to deploy in various countries, because each country has different frequency channels and bands which are usable and commercial equipment is prohibited from using some channels due to their potential use by the military, sensitive security, or police departments in the country or specific regions in a country.

The CR ability makes it easier to configure the system not to use specific frequency bands and channels based on the local regulations and to only use a subset of channels based on the local country's telecommunication rules and regulations [20]. Thus, service providers can build a system and use it with software-based configuration in a wide range of countries across the whole world. This makes CR wireless sensors usable and potentially deployable without much in the way of administrative or technical hurdles almost anywhere in the world.

This CR WSN capability also makes the system more resilient to attacks by cybercriminals and people who aim to break the system by transmitting signals on the fixed frequency used by the WSN nodes. With the invention of CR WSNs this issue is easily mitigated by selectively avoiding the frequencies which are subjected to cyber attacks or by dynamically changing the frequency bands, thus avoiding such attacks by protecting the system and making it more reliable [21].

This is an important feature to support, especially if the system is to be used in military applications where border areas are to be controlled using WSN-based systems, i.e. close to unfriendly or enemy countries when a nation is trying to protect its sovereignty with the help of a CR WSN.

5.6 Areas of application of cognitive radio wireless sensor networks

CR WSNs have potential uses in the following areas because of their inherent advantages and benefits and their adaptability to continuously changing radio spectrum usage in the environment: intelligent border security systems, military applications, object tracking, precision agriculture, medicine, healthcare, logistics, the maintenance of complex systems, and also the monitoring of indoor and outdoor environments.

CR WSNs have very wide range of application areas which can exploit and make use of their features. These can replace the existing systems which are based on conventional WSNs.

In the following, we review some of the areas in which CR WSNs can be deployed, including facility management, preventive and machine maintenance, precision agriculture, logistics, object tracking, etc. to name a few. The following sections discuss some of these potential areas where CR WSNs can be deployed by giving example usage scenarios.

5.6.1 Military and public security applications

Conventional WSNs are already being used in many military and public security applications. Some of these are as follows: chemical, biological, radiological, and nuclear (CBRN) attack detection and investigation, command control, battle damage evaluation and data collection, battlefield surveillance, intelligent assistance, etc [22].

In the battlefield or border regions disputed with an enemy, countries or adversaries may try to jam CR WSN systems' transmitters by broadcasting high-energy signals at the same operational frequency as that of the transmitters, which can make the most sensitive wireless communication systems blind [23]. This makes them ineffective and prevents them from communicating with their own users and command center because their receivers are saturated by high-energy jamming signals transmitted by their adversaries.

Jamming is the most popular method used by enemy countries. The jamming process starts by sensing the frequency of military communications used by a country at the time of the jamming attack; the adversary then sends powerful, higher-amplitude signals at the same frequency to make the operational communication systems incapable of performing their normal operations. This works because the country's own receivers become blinded to the actual communication signals between their own friendly units because they are overwhelmed by high-energy signals at the same frequency sent by the adversary [24].

In such situations, due to the inherent property of CR WSN whereby the operating frequency of CR is dynamic in nature, CR WSN-based systems can switch over to another adjacent or different frequencies which are also known to other friendly systems in the field, thus rendering the adversary's jamming ineffective. In this scenario, the system still continues to work effectively and can communicate with friendly units and sensors without any disturbance or hindrance.

Since the adversary requires time to sense a wide range of frequencies in order to locate the new frequency band the CR WSN has switched over to, it takes time for the adversary to keep track of the changes in operating frequency, which makes it hard for them to jam the transmitters [25]. By the time the adversary finds the new frequency of operation and tries to jam it, the CR WSN has system switched over to a different frequency, thus the adversary's attempts to jam them fail miserably [26].

The availability of a wider range of frequencies for military systems and the implementation of a CR WSN make it possible for military systems to use a set of frequency bands in a predefined pattern using a sophisticated algorithm which is very hard for the adversary to predict, while it is known to the participating systems.

This makes it difficult for the adversary to find the sequence of operational frequencies, thus making CR WSNs more reliable and invulnerable to jamming. This allows communication to continue on the battlefield or in border areas, which is crucial in highly sensitive and critical military communication systems.

5.6.2 Security threats in wireless sensor networks

Security issues are very critical and important for WSNs because the wireless communication between nodes can easily be broken if the wireless signals are not well protected from attacks. In the following, we review the different kinds of attack which can be used against WSNs and how CR WSNs protects their systems from such attacks.

- **Passive attacks**: these are an unintrusive kind of attack in which eavesdroppers monitor the communication channel between WSN nodes and try to collect and gather useful data which is being exchanged over the wireless medium.
- **Active attacks**: active attacks can be further subdivided into internal and external attacks, in which the active nodes operating in the field are replicated and their identity is stolen. By mimicking an active node, the copied nodes mislead other nodes into sharing sensitive information with an adversary without knowing that they are communicating with an adversary and not with the trusted nodes which are part of the network. There are different types of attacks in this category, including replication attacks, Sybil attacks (in which a major influence is illegally exerted on the decisions taken by the node under attack), and wormhole attacks, which are very dangerous, wherein the data packets exchanged between any nodes are tunneled to another malicious node without the knowledge of the genuine nodes which are communicating [27].

In wormhole attacks, the malicious node is kept within the transmission range of the legitimate node and mimics another legitimate node which is not even within the transmission range of the sending node. By tunneling data from the sender to the malicious node, the malicious node can disrupt the communication mechanism of the entire network by attacking the routers in the system.

Internal attacks work by compromising one of the legitimate nodes in the system, which is very difficult to detect and defeat once one of the legitimate nodes is compromised and comes under the control of an adversary.

Another novel method of attack is node replication, wherein the compromised node is replicated by the adversary into many nodes in different areas by populating the network with many replicated copies of one legitimate node. This type of attack is more dangerous and difficult to detect and remove after an adversary has sabotaged the entire network by injecting false sensor data and making the entire WSN system unusable [28].

Many of these attacks succeed due to the vulnerability of traditional WSN networks. Most of these issues are prevented and made close to impossible by CR WSNs, because CR support makes it very hard or nearly impossible for adversaries to attack such a system.

Apart from CR-based protection, other protective and preventive methods are employed by WSNs to protect systems from these potential threats, which we briefly list here.

- **Node authentication**: every node in the WSN need to prove its validity to all its neighboring nodes as well as the management or control node. This prevents any malicious outside node from replicating any of the legitimate nodes because of the node authentication step involved. This ensures that any

node which is part of the network is properly authenticated and that every node has right to access the sensor network and exchange information over it.
- **Availability**: although the entire WSN is power constrained, at any point in time, the network should be active and available for the outside world to communicate with it and control the operations within it. Even though individual nodes can enter sleep modes during power-saving cycles, the entire network does not go down at any point in time. This is achieved by carefully ensuring that some subsets of nodes are active at any point in time, though with limited capabilities, making the entire system active and functional while at the same time extending the lifetime of the network by properly planning and orchestrating the distributed sleep cycles.
- **Location awareness**: as previously mentioned in this chapter, location awareness can be built into the system either by applying algorithm-based localization or with the support of GPS/INS systems added to the WSN nodes. This means that any security threats or compromised nodes in a given area are not widespread and that the attack is localized by restricting the spread of the attack, thus providing improved performance and safety.

5.6.3 Healthcare

Healthcare systems such as wearable body sensors and telemedicine are becoming increasing popular and their widespread usage can be seen due to their relevance and importance for healthcare, which is a most important field and critical for the wellbeing of a nation.

Patients who are critically ill and who need continuous remote monitoring by doctors have many wireless sensor nodes installed to monitor their vital signs so that healthcare providers can take preventive actions and ensure that medical attention is provided if the condition of the patient becomes critical.

In 2011, IEEE 802.15 Task Group 6 approved a draft proposal for body area network (BAN) technology. In various countries, BAN-based WSN networks are already used to remotely monitor the health of critically ill patients with the help of sensors installed on their bodies in order to track their vital signs [29].

Medical data is very sensitive and very critical; any security breaches have very serious consequences and create legal hurdles for the service providers. Moreover, any medical report or information is also very time-sensitive, and any unexpected delay could be life-threatening; therefore, the network should be reliable and also needs to be responsive without introducing much of a delay into the exchange of time-sensitive information with doctors or hospitals. Because of these requirements, traditional WSN networks cannot meet these challenging needs, which limits their potential to be used for telemedicine applications [30]. Traditional WSNs cannot achieve the required QoS due to traffic and conflicts caused by BANs, making traditional WSNs unusable for medical applications.

In contrast, CR WSN wearable devices can mitigate all these issues due to the ease of bandwidth availability for these wearable body sensors, thus making CS WSN wearable devices more suitable for medical and telemedicine applications.

Since CR WSNs are globally operable and jam proof and have improved reliability, they are a practical choice for medical applications. The use of CR implementations in WSNs is a boon and the right solution for the most important and critical field of medicine.

5.6.4 Home appliances and indoor applications

Many of home applications and appliances have a need for dense sensors to be deployed with better QoS and improved reliability, for which CR WSNs are most suitable. Since most home appliances operate in the free ISM band, the use of traditional WSNs makes it hard to achieve the desired QoS, because the spectrum is already crowded.

A few examples of indoor applications are intelligent buildings, home automation and monitoring systems, personal entertainment, etc.

CR WSNs can mitigate the difficulties faced by traditional WSN systems due to flexible and dynamic allocation of spectrum bands by the CR, which learns from the environment, thus improving the reliability and operational efficiency of the system as a whole.

Multimedia home entertainment applications are very data sensitive as well as latency sensitive. In particular, if a system is used for home automation, home security, or surveillance based on camera feeds from a home or factory [31], then any delay or lack of quality in the video delivery has a larger impact on the safety of property and human life [32].

Other WSN applications in hospital environments, vehicular WSNs, or the tracking and surveillance of a home or factory, have very large temporal and spatial variations in the data density which are linked to the number of nodes in a given area or the node density of the WSN system. Since these applications are bandwidth-hungry and delay or latency sensitive and have occasional bursty traffic, traditional WSNs fail to meet the QoS requirements of these systems. In contrast, CR WSNs fit the requirements pretty well and can cater to these kinds of varied performance expectation.

Channel banding is another technique which is widely used in CR WSNs, in which two adjacent channels are allotted for a particular communication between nodes. This improves the bandwidth by combing the frequency bands and doubling the data rate in a given scenario, thereby improving the overall throughput between two or more wireless nodes. Channel bonding is also called Ethernet bonding, but it is mostly used in the wireless domain and in particular in CR WSN systems.

5.6.5 Real-time surveillance applications

Applications such as traffic monitoring, vehicle and inventory tracking, biodiversity mapping, the environmental monitoring of crops or livestock, irrigation, underwater sensing, disaster relief operations, and bridge and tunnel monitoring require the most efficient channel access and reduced communication delay or latency.

These real-time surveillance applications are delay sensitive and require very high reliability. Delays due to channel failure are common in traditional WSNs, whereas in CR WSNs, if a channel fails, the wireless sensor nodes in a CR WSN can hop to another channel which is idle and keep the delay and latency within the expected limits without sacrificing the QoS of the network.

Channel aggregation and the concurrent use of multiple channels are other ways in which CR WSNs can meet QoS requirements by dynamically increasing the available channel bandwidth based on the need and the channel conditions.

5.6.6 Transportation and vehicular networks

The IEEE 1609.4 standard defines the multichannel operations used in wireless access for vehicular environments (WAVE). WAVE systems operate in the 5.9 GHz band with a bandwidth of 75 MHz that is divided into one control channel and six service channels. However, these systems need much higher bandwidth and suffer from spectrum inefficiency [33]. Today, there is active research into spectral inefficiency in CR WSN systems.

Vehicular WSNs are a new emerging area and this paradigm is also gathering traction in urban environment monitoring systems. CR WSNs are likely to be more suitable and relevant in this field because of their reliability in meeting the QoS requirements of the system. There is also opportunity for CR WSNs to be used in highway safety systems due to their features, scalability, and flexibility.

5.6.7 Multipurpose sensing

It is becoming more common to use wireless sensors located in an area for different purposes simultaneously. As we are aware, in a conventional WSN the wireless sensors in the network access the channel in a non-cooperative manner.

The efficient implementation of the media access control (MAC) protocol in CR WSNs, which is capable of selecting different channels for different applications, helps to balance the load on the network. In addition, it can also choose channels with differing bandwidth, which helps to cater to different data rates or bandwidth requirements that different applications need the WSN to provide.

5.7 Challenges in cognitive radio wireless sensor networks

Communication in sensor networks normally has to be energy efficient because the sensors are battery operated and power constrained, which makes it hard to implement features such as cognitive radio and intelligent nodes in WSNs. Moreover, the interactions between different layers of WSN are also restricted. It is therefore a challenging task to cater to the goals of various network elements [34].

The flexibility of CR WSNs in terms of bandwidth and channel allocation also causes many challenges that the designers of CR WSNs need to address [35]. The reminder of this chapter provides an in-depth discussion of some of the challenges faced in implementing CR WSNs.

5.7.1 False alarms and misdetection in cognitive radio wireless sensor networks

An important feature of WSNs is to measure channel quality while a sensor is looking for a free channel for the node to acquire and use. A metric known as the detection probability is a measure of the correct detection of the presence of a primary signal on the channel. Misprediction results in wasted energy for the node and increases the time taken to detect a channel and establish a connection.

A missed detection of the primary signal results in spectrum underutilization, resulting in reduced performance and power wastage. This also increases the delay involved in acquiring the channel, thereby causing a further loss of data or missed commitments from a system perspective. Misdetection and false alarms cause long delays, frequent channel switching, and significant degradation of the overall throughput. However, much research is ongoing in this area in order to study the behavior of CR WSNs in terms of their impact on network performance.

5.7.2 Complex hardware design

Compared to conventional WSNs, CR WSNs have additional computational power and storage requirements, because the physical layer of a CR WSN has additional responsibilities, e.g. to sense channels, analyze, decide, and act according to the signals sensed in the wireless channels by the sensors. These sensors are expected to change and adapt to the dynamic conditions of the environment, such as the transmitted power and the other characteristics of the wireless signal and channel used for communication with adjacent nodes in the WSN.

This results in complex designs for the hardware systems used in CR WSN systems, which increase the overall cost of the wireless sensors that need to be deployed. Since, in general, thousands of nodes are present in a typical WSN system, any incremental rise in the cost of the individual sensors in the system increases the overall cost of the system.

As we discussed earlier, the CR unit uses AI technology to adapt the communication parameters such as carrier frequency, transmitted power, and modulation technique based on the environmental conditions in order to choose the best channel for communication.

The transceiver unit needs to choose the best channel and release any channel which is required by a higher-priority node or for emergency communication. After the channel for communication is chosen, the transceiver unit is responsible for receiving and sending data.

The miscellaneous unit in a sensor is used to find the current location of the sensor using GPS or IPS services. Designing an intelligent sensor is a very challenging and expensive proposition, because the CR unit needs to fulfill basic principles such as observation, reconfiguration, and cognition. The choice of AI technique depends on the features of the CR sensors, such as responsiveness, design complexity, robustness, and reliability as well as the overall stability of the system.

Acknowledgments

I would like to acknowledge the support that Chandramouleeswaran Sankaran received from his family members and his friends, which have been invaluable in this work.

References

[1] Ibnkahla M 2013 *Wireless Sensor Networks: A Cognitive Perspective* (Boca Raton, FL: CRC Press) pp 21–66

[2] Bicen A O, Gungor V C and Akan O B 2012 Delay-sensitive and multimedia communication in cognitive radio sensor networks *Ad Hoc Networks* **10** 816–30

[3] Atakan B and Akan O B 2012 Biological foraging-inspired communication in intermittently connected mobile cognitive radio ad hoc networks *IEEE Trans. Veh. Technol.* **61** 2651–8

[4] Youssef W and Younis M 2008 A cognitive scheme for gateway protection in wireless sensor network *Applied Intelligence J.* **29** 216–27

[5] Joshi G P, Nam S Y and Kim S W 2022 Cognitive Radio Wireless Sensor Networks: Applications, Challenges and Research Trends *Sensors* **13** 2–13

[6] Kumar N and Makkar A 2020 *Machine Learning in Cognitive IoT* (Boca Raton, FL: CRC Press) pp 24–64

[7] Clark D D, Partrige C, Ramming J C and Wroclawski J T 2003 A knowledge plane for the internet *SIGCOMM '03: Proc. 2003 Conf. on Applications, Technologies, Architectures, and Protocols for Computer Communications* (New York: ACM) pp 3–10

[8] Kumar S, Raja R, Tiwari S and Rani S (ed) 2022 *Cognitive Behavior and Human Computer Interaction Based on Machine Learning Algorithm* (New York: Wiley) pp 2–25

[9] Thomas R W 2007 *Cognitive networks* Virginia Polytechnic and State University http://hdl.handle.net/10919/28319

[10] Al-Turjman F 2017 *Cognitive Sensors and IoT Architecture, Deployment, and Data Delivery* (Boca Raton, FL: CRC Press) pp 32–67

[11] Akan O B, Karli O B and Ergul O 2009 Cognitive radio sensor networks *IEEE Network* pp 34–40

[12] Shenai K and Mukhopadhyay S 2008 Cognitive sensor networks *IEEE 26th Int. Conf. on Microelectronics (MIEL)* (Piscataway, NJ: IEEE) pp 315–20

[13] Cavalcanti D, Das S, Wang J and Challapali K 2008 Cognitive radio based wireless sensor networks *Proc. 17th Int. Conf. on Computer Communication and Networks, ICCCN, 2008* (Piscataway, NJ: IEEE) pp 1–6

[14] Hwang K and Chen M 2017 *Big-Data Analytics for Cloud, IoT and Cognitive Computing* (New York: Wiley) pp 34–54 https://www.wiley.com/en-gb/Big+Data+Analytics+for+Cloud%2C+IoT+and+Cognitive+Computing-p-9781119247296

[15] 2003 *IEEE Recommended Practice for Information technology– Local and metropolitan area networks– Specific requirements– Part 15.2: Coexistence of Wireless Personal Area Networks with Other Wireless Devices Operating in Unlicensed Frequency Bands* IEEE Std 802.15.2-2003 IEEE

[16] Krüger D, Heynicke R and Scholl G 2012 Wireless sensor/actuator-network with improved coexistence performance for 2.45 GHz ISM-band operation *Int. Multi-Conf. on Systems, Signals & Devices* (Piscataway, NJ: IEEE) pp 1–5

[17] Olifer N and Olife V 2005 *Computer Networks: Principles, Technologies and Protocols for Network Design* (New York: Wiley) pp 4–32 https://www.wiley.com/en-gb/Computer+Networks%3A+Principles%2C+Technologies+and+Protocols+for+Network+Design-p-9780470869826

[18] Clark D D, Partridge C, Christopher Ramming J and Wroclawski J T 2003 A knowledge plane for the internet *SIGCOMM '03: Proc. 2003 Conf. on Applications, Technologies, Architectures, and Protocols for Computer Communications* (New York: ACM) pp 3–10

[19] Vijay G, Ben Ali Bdira E and Ibnkahla M 2011 Cognition in wireless sensor networks: a perspective *IEEE Sens. J.* **11** 582–92
[20] Boonma P and Suzuki J 2008 Exploring self-star properties in cognitive sensor networking *Proc. of IEEE/SCS Int. Symp. on Performance Evaluation of Computer and Telecommunication Systems (SPECTS), Edinburgh,* pp 36–43 https://ieeexplore.ieee.org/document/4667541
[21] Vijay G, Bdira E and Ibnkahla M 2011 Cognition in wireless sensor networks: a perspective invited paper *IEEE Sensors J.* **11** 2–65
[22] Iyengar S G, Varshney P K and Damarla T 2011 A parametric copula-based framework for hypothesis testing using heterogeneous data *IEEE Trans. Signal Process.* **59** 34–67
[23] Zhang M, Zhao H, Zheng R, Wu Q and Wei W 2012 Cognitive Internet of Things: concepts and application example *JCSI Int. J. Comput. Sci. Issues* **9** https://www.ijcsi.org/articles/Cognitive-internet-of-things-concepts-and-application-example.php
[24] Mitola J III and Maguire G Q 1999 Cognitive radio: making software radios more personal *IEEE Personal Communication* **6** 13–18
[25] Machado R and Tekinay S 2008 A survey of game-theoretic approaches in wireless sensor networks *Int. J. Comput. Telecommunications Networking* **52** 3047–61
[26] Gelenbe E, Liu P, Szymanski B K and Morrell C 2009 Cognitive and self-selective routing for sensor networks *Computat. Manag. Sci.* **8** 237–258
[27] Fischler M A and Firschein O 1987 *Intelligence: The Eye, the Brain, and the Computer* (Reading, MA: Addison-Wesley) pp 3–45
[28] Pfeifer R and Scheier C 1999 *Understanding Intelligence* (Cambridge, MA: MIT Press) pp 3–54 https://mitpress.mit.edu/9780262661256/understanding-intelligence/
[29] Akan O B, Karli O and Ergul O 2009 Cognitive radio sensor networks *IEEE Network* **23** 34–40
[30] Morrow R K 2004 *Wireless Network Coexistence* (New York: McGraw-Hill) pp 34–56
[31] Qiu R and Wicks M 2014 *Cognitive Networked Sensing and Big Data* (Berlin: Springer) pp 43–65
[32] Park J-H, Salim M M, Jo J H, Sicato J C S, Rathore S and Park J H 2019 CIoT-Net: A Scalable Cognitive IoT-based Smart City Network Architecture *Human-centric Comput. Inf. Sci.* **9** 29
[33] Yang D, Xu Y and Gidlund M 2011 Wireless coexistence between IEEE 802.11 and IEEE 802.15.4-based networks: a survey *Int. J. Distr. Sens. Network* **7** 912152
[34] Golmie N 2006 *Coexistence in Wireless Networks—Challenges and System-Level Solutions in the Unlicensed Bands* (New York: Cambridge University Press) pp 34–78 https://www.cambridge.org/gb/academic/subjects/engineering/wireless-communications/coexistence-wireless-networks-challenges-and-system-level-solutions-unlicensed-bands?format=HB
[35] Bdira E and Ibnkahla M 2009 *Performance Modeling of Cognitive Wireless Sensor Networks Applied to Environmental Protection GLOBECOM 2009 - 2009 IEEE Global Telecommunications Conf. 2009* (Piscataway, NJ: IEEE) pp 1–6

IOP Publishing

Cognitive Sensors, Volume 1
Intelligent sensing, sensor data analysis and applications
G R Sinha and Varun Bajaj

Chapter 6

Cognitive wireless sensor networks

P Saranya, N Jeevitha and V Senthil Kumar

Cognitive wireless sensor networks (CWSNs) can transform and adapt to the communication parameters that have the greatest divergence from those of predictable wireless sensor networks. With the assistance of underlay or overlay methods, it is possible to use CSWN technology and spectrum sensing to employ unoccupied frequencies for broadcasting. Spectrum sensing allows the performance to be evaluated by a disseminated or centralized system which involves separate sensor nodes, which in turn leads to efficient energy conservation for the entire network. This technique is also compact and has low complexity. Cognitive technology affords access to emerging spectrum in addition to defined spectrum bands, thereby providing enhanced broadcasting features. The improvements provided by cognitive wireless sensor networks enhance the communication flow range, reduce the number of the sensor nodes required to cover a given neighborhood, and improve power utilization. The crisis of spectrum scarcity can be conquered by developing contemporary WSNs into cognitive WSNs. The Sensing & Combining Decision approach Single Carrier (SCD-SC) spectrum-sensing technique permits performance to be upgraded by considering detection probability.

6.1 Introduction to wireless sensor networks

Wireless sensor networks (WSNs) are predicted to be the solution that will enable contextual intelligence. A WSN is a conscious grid of tiny sensor nodes that communicate with one another through radio signals and are distributed in large numbers to detect, monitor, and analyze the physical environment. Motes are Wi-Fi sensor nodes. The sensor device node is a multimodal, low-momentum wireless gadget. Motes have a wide range of industrial uses. A network of sensor nodes receives data from its surroundings to fulfill certain application goals. Transceivers can be used by motes to communicate with one another. The number of motes in a wireless sensor network might be in the hundreds or even thousands. WSNs serve as links between the physical and digital worlds and permit the observation of hitherto

doi:10.1088/978-0-7503-5326-7ch6

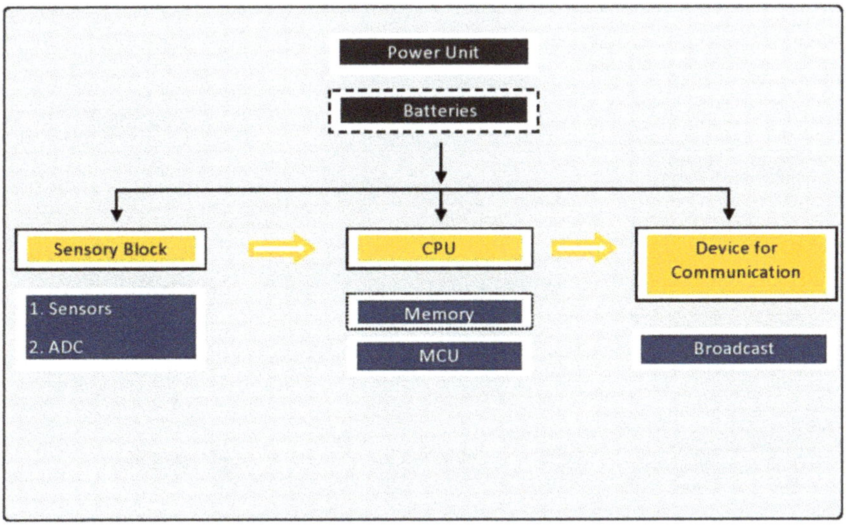

Figure 6.1. Wireless sensor node—basic units.

unobservable phenomena at a fine resolution over wide spatial and temporal scales [1]. They have several potential uses in industry, science, transport, civic infrastructure, and defense. Rapid advancements in processor power, storage capabilities, and broadcast technology have permitted the development of distributed networks built from compact, low-cost communication nodes. These nodes can sense and communicate, and they can be installed at a far smaller cost than that of a standard cable sensor system. Such systems are known as WSNs.

As shown in the figure 6.1, the various blocks involved are

1. Sensors: WSNs employ sensors to obtain environmental information and collect data. Electrical signals are derived from sensor signals.
2. Radio nodes: these gather data from the sensors and transmit it to the WLAN gateway. They comprise a controller, a communication module, secondary storage, and a power supply.
3. Wireless local area network (WLAN) access points: these accept data transmitted wirelessly by radio nodes, typically over the Internet.
4. Evaluation software: the information captured by the WLAN gateway is analyzed by software called evaluation software, which then presents the results to the users for additional data processing, evaluation, storage, and mining.

6.1.1 Defining wireless sensor networks

With the rise of novel remote innovations, for example, Zigbee and IEEE 802.15.4, business device interconnectedness has progressed, which is essential for guaranteeing versatility and low costs. The low-rate wireless personal area network (LR-WPAN) standard sets out the media and physical layers for radio transmission connections between WSN gadgets working in the industry, science, and medicine (ISM) bands at 2.4 GHz and 868/915 MHz [3]. As shown in figure 6.2, X-bee

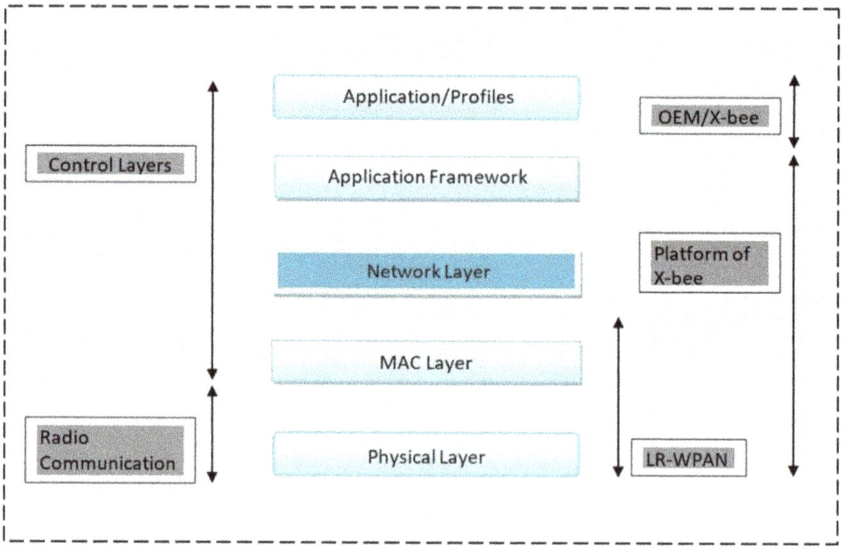

Figure 6.2. The LR-WPAN/X-bee protocol stack.

broadens the LR-WPAN standard by determining the organization and application user layers, notwithstanding a security specialist protocol (SSP). The X-bee Partnership, a coordinated effort of different firms, fostered the X-bee standard to improve intelligent, low-data-rate, and short-distance gadgets with minimal energy usage for WSN products [2]. Accordingly, the conventional WSN stack contains the X-bee stack, which incorporates LR-WPAN-based media and physical layers as well as remote LAN organization and application layers.

Figure 6.3 shows an example of multihop interaction in WSNs. Sensor device nodes transmit information to transmitting connections, which function as terminal nodes for the data. Compared to sensor networks, which are frequently driven by non-renewable energy sources, such intermediate nodes are more likely to use precise control points (essential series). The raw information is then relayed to a ground station, which might be in the deployment zone of the sensor network. The input is wirelessly sent from here and supplied to a final user who may analyze the observed data remotely [1, 5]. WSNs' qualities have made them exceedingly popular across a broad array of applications, including healthcare, the armed services, environmental control, and defense systems, and they are even regarded as the new pervasive computing paradigm.

6.1.2 Utilizing networks of wirelessly connected sensors

The benefits of WSNs include the following:

Web-based configuration may be used except in the case of transportable infrastructure.

They are ideal for inaccessible locations such as mountainous regions, seas, rural regions, and thickly forested areas.

They are adaptable if an extra workstation is required on an ad hoc basis.

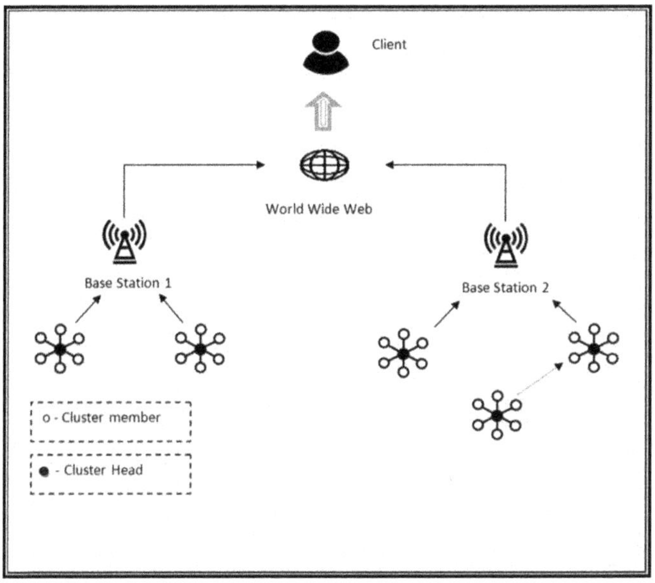

Figure 6.3. WSN multihop communication.

The deployment costs are reasonable.
They eliminate a lot of cabling.
They may accommodate new gadgets at any moment.
They can be accessed with the help of centralized monitoring.

Wireless sensor networks may monitor a wide range of environmental factors using sensors such as seismic, magnetic, temperature, optical, IR, radar, and acoustic sensors at lower sampling frequencies. Continuous monitoring, event recognition, event detection, and localized actuator control are all performed by sensor nodes.

The main applications of WSNs include:

1. The Internet of Things (IoT)
2. Surveillance and monitoring for the purposes of security and danger identification
3. The monitoring of temperature, humidity, and air pressure in the environment
4. The monitoring of environmental noise levels
5. Medical uses such as patient tracking
6. Agriculture
7. The detection of landslides.

6.1.3 Constraints of networks in wireless sensor systems

The following issues are the main areas of focus in studies of sensor network systems:

1. The difficulty of boosting the lifespan of power sensor device nodes in sensor network systems is addressed by network lifetime maximization, which determines optimal values for features such as data transfer power, frequency, and link scheduling.

2. Energy-saving routing. When information is sent by means of a wireless channel, a huge amount of power is needed for the transmission. Today, efforts are being made to determine the most effective routing plan for sensor network programs, with the aim of decreasing the quantity of electricity used while at the same time retaining the same amount of transmitted statistics and the same Internet connection.
3. Trustworthy incident detection and transmission: effective and influential detection and transmission are critical goals in event-based sensor networks. Some scholars focus on event-to-event reliable transmission techniques, while others focus on event-to-sink reliable transportation.
4. Optimizing numerous, contradictory goals: sensor device nodes may have multiple goals, such as primary cost and networking endurance. New strategies are always being researched to find ways to optimize network performance while striving to meet competing goals.
5. Increasing the adaptability of WSN application-specific architecture: Movable interfaces powered by software or middleware are employed to quickly launch new applications in sensor networks that are generally application-specific.
6. Energy consumption: one of the most significant difficulties experienced by WSNs is their high power utilization. Sensor nodes use batteries as their energy source. The sensor network is installed in hazardous areas, making it impossible to remove or recharge batteries [7]. Sensor node operations such as communication, detection, and information processing account for most of the energy usage. Communication consumes a significant amount of energy. Power consumption may be reduced at every layer by the use of appropriate routing protocols.
7. Sensor localization is a basic and critical problem in network operation. Sensor nodes are randomly placed and are oblivious of their location. The difficulty in determining the actual location of a sensor after it has been deployed is referred to as localization. A global positioning system (GPS), beacon nodes, and closeness localization can help with this issue.
8. Coverage: the wireless sensor network's cluster heads employ a coverage algorithm to detect data and route it to the sink. The sensor networks should be chosen such that they encompass the whole network. Efficient methods have been developed, including the least and largest exposed pathway techniques, in addition to the overall design protocol.
9. Clocks: synchronization of clocks is a vital feature in WSNs. The major goal of this synchronization is to establish a time standard among sensor network nodes equipped with local clocks. Several applications, including tracking and monitoring, require the synchronization of these clocks.
10. Computation: estimation is the sum of the data that passes through each node. The main problem with computers is that their resource usage must be reduced. If the life of the base station is shortened, data processing can be done by each node before the data is transmitted to the access point. If each node has limited resources, the complete computation should be executed at the sink.

11. Production costs: a WSN has many sensor nodes. As a result, if the cost of an individual node is exceptionally high, the price of the entire network will be rather high as well. Furthermore, the budget of each member node must be kept low. As a result, calculating the expense of each sensor device node in a wireless grid of devices is a tough task.
12. Hardware design: any sensor network's hardware, including the energy monitoring, microcontroller, and transceiver, must be energy-saving. Its design can be made in a way that uses less energy.
13. Quality of service: quality of service (QoS) simply means that data must be provided on time, because certain real-time sensor-based applications rely heavily on time. As a result, if the data is not sent to the receiver on time, it becomes worthless. There are several types of QoS issue in WSNs, such as the topology of the network, which can change quickly, and the availability of the information used for routing, which can be inaccurate.

6.2 An introduction to cognitive radio networks

Current networking technology is often limited in terms of a network's capacity to adjust to a variety of changing situations, which can result in suboptimal performance. Another severe issue stems from the present inflexible spectrum allotment, which results in resource waste and constrained business models. With few exceptions, network components' scope, status, and response capabilities are currently limited [5, 6]. This means that they are unable to make sound judgments. The adaptation strategies on offer today are often reactive, operating only when a problem has developed. This decreases the network's ability to supply smart and effective solutions, such as eco-friendly technology and favorable economic models.

Cognitive radio networks (CRNs) improve spectrum utilization by making use of idle or underused capacity. If the interference encountered by licensed users is modest, unauthorized people may retain access to the frequency band. This means that to use the bandwidth more productively (in the form of more cells/Hz/s/bits, for instance), not only are newer communication and networking technologies necessary, but also more complex ways to maximize spectrum utilization.

This poses several operational problems that should be resolved prior to the use of this approach. CRNs are entirely predicated on the use of cognitive radio devices (CRDs), which are responsible for configuring various criteria on the fly (e.g. probability sampling methods, spectrographs, system operations) based on the external environment, thereby exploiting underutilized parts of the spectrum, and avoiding the distinction between software defined radio (SDR) systems and access networking radio systems [13]. In contrast to broadcast radio technology, an SDR radio system conducts most of the radio and midlevel bandwidth operations digitally. As a result, SDR provides increased radio communication options.

'Cognitive radio' (CR), on the other hand, is the name given to systems that help SDRs to determine the type of operation and the precise specifications to use under specific network conditions. A cognitive radio networking system (CRN) is a network operated by a cognitive process system (CP). This refers to a system that continuously perceives and gathers contextual data about the communication system, analyzes it, and then determines the most appropriate course of action for the given situation [6]. The decision additionally considers the specified account information and the relevant set of recommendations applicable to the set of circumstances. These activities may pertain to numerous optimization tactics used at various levels of the TCP/IP routing protocol, including the media access control (MAC) protocol. These judgments concern the routing that occurs at the IP layer, which is presumed to be an ad hoc multidimensional version. The end-to-end (E2E) pathway is designed to consider several aspects, including location, bandwidth, energy, and duration. Because of this, the route is referred to as 'multi-dimensional,' which is a reference to the fact that multiple aspects are considered. The implementation of CP may take either a unified or dispersed approach, and each has its own advantages and disadvantages to offer.

A foundational concept called cognitive radio advanced technologies is being considered for incorporation into 5G, often known as the fifth generation of mobile networks. Standardization groups and major telecoms companies have not yet adopted the moniker '5G' as their official designation for the technology. It is most often viewed as a standard that addresses the challenges created by the widespread deployment of 4G technology. Considering that operators are already testing 3G Long Term Evolution (LTE) and that tests for 4G LTE are also underway: advanced (LTE-A) operators are expected to begin their network rollout in 2020 and considering that it typically takes around ten years to develop a new generation of wireless networks, 5G operators are expected to begin their rollouts in the future [18]. At this point in time, it seems as though 5G will place more of an emphasis on architectural and networking capabilities than on being a speedier carrier than 4G. It is anticipated that 5G will have several significant features, including pervasive wireless computers and telecommunication, cognitive radio advanced technology, Internet Protocol version 6 (IPv6), wearable electronics with artificial intelligence (AI) skill sets, and a unified worldwide standard. It is anticipated that Bluetooth and Wi-Fi protocols will be accessible in the form of radio modules by the year 2020, which will lead to the development of new advanced technologies as well as new future experiences.

Even though they use different algorithms and have different workloads, it is very important to keep in mind that the cognitive radio device systems and cognitive radio networks share the same CP. The observe, orientate, decide, and act loop is an emerging paradigm that is applied for CP. The specific steps are the primary emphasis of this loop:

- keep an eye on the radio environment;
- figure out how to adjust for previous errors;
- coordinate the process of adjusting; and
- engage in conversations inside the radio atmosphere.

Some examples of cognitive techniques include evolutionary algorithms, machine learning, game theory, SDR, and cross-layer policy design. The components of a system that take part in the transmission of a data flow are collectively referred to as end-to-end intentions. Examples of end-to-end components include routers, switches, interfaces, and waveforms. The second essential premise that CRNs adhere to is known as cross-layer design [25]. This is a method for communicating data across distinct levels of the TCP/IP stack. Improving execution is the aim of this endeavor. This technique is excellent for CRNs, since many applications require several parameters at low levels simultaneously.

6.2.1 Purposes of cognitive radio

There are some key differences between a CRD and a CRN, one of which is the fact that a CR offers secondary user (SU) access to temporarily available bandwidth, also known as a spectrum opportunity (SOP) or white space (WS) if the primary user (PU) is not using it at the time. Nevertheless, if the PU reuses an SOP, the SUs are required to swap to another SOP to keep in touch without interruption. Normally, CRDs must be able to function in very complex radio environments, which are mostly referred to as 'multi' environments, because the environments are usually very complex:

- they are broadband in nature
- they support multiple channels
- they support multiple modes
- they adhere to multiple standards
- they provide several services.

As a result of being expected to offer functionality for these instances, CRDs and CRNs have become quite complicated. Spectrum utilization, bandwidth allocation, spatial multiplexing, and mobility management are the four core processes required for management. A CRD should be able to:

- Identify accessible SOPs using certain criteria
- Dynamically choose a path to a specific destination CRD
- As a result, adjust distinct factors such as power, frequency/channel, and coding
- Modify the route depending on the network circumstances, i.e. multidimensional enhancement and routing techniques
- Avoid disrupting current PU conversations.
- Adopt safety precautions.

In other words, the CRD should be able to dynamically use all the different kinds of resource that are available for communication while also avoiding conflicts between the initial user and another communicating tertiary user (TU). Additionally, the interests of the user and the network operator are important factors that drive these activities. The customer is especially interested in inexpensive telecommunication services that provide higher throughput, more freedom in picking service providers and access technologies, better dependability and security, as well as quality of denial service (QoS) and quality of distributed denial service (QoD) [9, 17]. In contrast, the telecoms operator is concerned with managing lower levels of complexity, higher levels of flexibility, secrecy, and reusability and robust business models. As a direct consequence of this, certain tasks can become quite challenging.

6.3 Wi-Fi sensor network integration with cognitive radio

The term 'cognitive radio' refers to an advanced wireless communication system that gives the impression that it is aware of its surroundings (i.e. its environment). This type of system employs insight obtained by learning from its surroundings and changing its internal behavior using systematic modifications that depend on the obtained radio-frequency (RF) stimuli. These changes are effected by making the necessary adjustments in certain operational parameters in real time, with two crucial priorities. This framework was developed based on the methods used to manage the spectrum. This is essentially a programming model used to tie top-level criteria to the deep network using a cognitive system. Each of the coalition formation's (CF's) three levels has a unique role.

- The end-to-end gains (E2EG) layer tracks the transformation of user-requested E2E gains into cognitive-layer rules and suggestions.
- The cognition procedure (CP) has three distinct phases. These relate to information about the network sensors, the specification language (which transfers high-level user demands into a suitable cognition-layer language), and the cognition layer (which regulates cognitive processes).
- Software applied networking (SAN) has two components: (i) Based On Digital technology for Application Software, which offers an appropriate interface to the cognition layer, and (ii) changeable information about smart sensors, which delivers input from the server to the cognitive stack. CF might be centralized, distributed, or hybrid.

In the development and construction of CRNs, a few significant concerns must be addressed. Dynamic spectrum access (DSA), networking diversity, and the aim of providing communication with a defined QoS and QoD are the most notable of these challenges [24]. A CRN/CRD can continuously detect and acquire a wide range of data about the wireless medium, such as the behavior of traffic, the situations in which interference occurs, and the local restrictions that are in place. This information may be broken down into three distinct classes: articles, biographies, and (regional) regulations. The input from the defined profile, together with the policies that are currently in effect, are used by CRNs/CRDs to provide a variety of different alternatives for the deployment of data services with the required QoS and QoD [9]. This is also due to the employment of appropriate optimization algorithms to provide high-quality service and experience. At the culmination of the process, these evaluations are incorporated into a variety of learning procedures that enhance the performance of the decision-making system.

The following is a list of the typical cognitive strategies that may be utilized in CFs to manage CRNs:

- Evolutionary methods based on dynamic programming
- Processing of signals using Bayesian methods
- Theorizing through playing games

- Machines that learn with the help of feedback
- Frequency modulation with a dynamic range
- SDR systems

Another important characteristic of CRNs is their cross-layer architecture. There are currently two different techniques that may be used for a cross-layer design. The terms 'implicit' and 'explicit' are used to describe the two distinct sorts of cross-layer architecture. The architecture of the reference layers is not changed during the implicit design process; however, the stacked topology may be altered during the explicit design process. The goal is to increase one's capacity to monitor and alter the properties of low-layer environments. To achieve this goal, it is possible to use specialized optimization techniques, the majority of which are associated with the lower layers of the network stack [4]. These techniques include data link-layer tuning, network management, and multinetwork optimization.

CRN designs are now classified into two broad groups at the network level. These are founded on the idea of TV white spaces (TVWS), also known as vertical spectrum sharing. Its primary goal is to make use of the VHF and UHF television frequencies that will be available after the transition from analog to digital TV is complete. The USA and the United Kingdom have recommended this as a solution to the problem. TVWS is currently in the process of being recognized by the IEEE 802.22 working group. The basic problems are not associated with spectrum detection; rather, they lie in the inability to understand what is sensed [13, 15]. This occurs because the available spectrum changes depending on location. This suggests that temporal and spatial information must first be obtained and evaluated in a variety of settings before one can build space–time maps, which are often referred to as geospatial databases. Using these maps as a guide, a secondary user might potentially utilize the licensed bandwidth for communication purposes, as long as licensed broadcast television services are not adversely affected. However, interruption reduction is a process that involves a lot of complexity.

The horizontal sharing of radio frequencies over several channels is sometimes referred to as CRN, even if the technology is not available. This design is used more often in Europe, and the IEEE P.1900 working group is now working to standardize it. The major objective of simulating this technology is to test the boundaries of what is now possible with 4G technology in terms of DSA, software, and prototypical hardware to enable the provision of CR services. Given the circumstances of CR users, this not only requires solutions for technical issues, but also the establishment of acceptable income models for telecoms operators. As a result, additional answers are required to the many questions about the telecoms operators' control over the infrastructure's resources. CRN necessitates the resolution of several technical problems, which may be found in a variety of contexts and pose differing degrees of difficulty.

6.4 The structure of a cognitive wireless sensor network

Cognitive radio comprises two essential modules: a cognitive unit that makes decisions based on data and a modifiable SDR unit with flexible software. Radio resource architecture often includes an energy detection subsystem to detect competing services or clients. Subsystems do not usually define a solitary piece of equipment, but may incorporate network-wide components. Cognitive radio systems are sometimes called functional networks.

The cognition component is split into two halves, as shown below. The 'intelligent engines' solve or maximize a performance target depending on the radio's internal state and operating environment. The 'policy engine' guarantees that the intellectual engine's solution complies with external rules and regulations. Secondary users who do not have spectrum allocation rights can now temporarily use the primary users' unused licensed frequency bands due to the advancement of CR technology [10]. As a result, as exemplified in figure 6.4, the mechanisms of a CR web design include both a secondary network and a primary group. A primary network comprises a group of legitimate users and one or more crucial assistance locations. Under the jurisdiction of the principal base stations, certified frequency bands may be used by the primary users. Their transmissions must not be interfered with by secondary networks. In general, CR functionalities are not offered to the primary users and the core base stations [21]. Consequently, if a tertiary system occupies a certified frequency channel that contains a key system, the tertiary network must rapidly spot the presence of a legitimate user and redirect the tertiary spread to other available spectrum to avoid conflicting with the primary network's availability, in addition to detecting spectrum white space and utilizing the optimal spectrum band.

A tertiary system is a network consisting of distinct users, either with or without a tertiary access point. Secondary users may only use the wireless channel when a primary user does not utilize it. The variable access range of the TUs is often

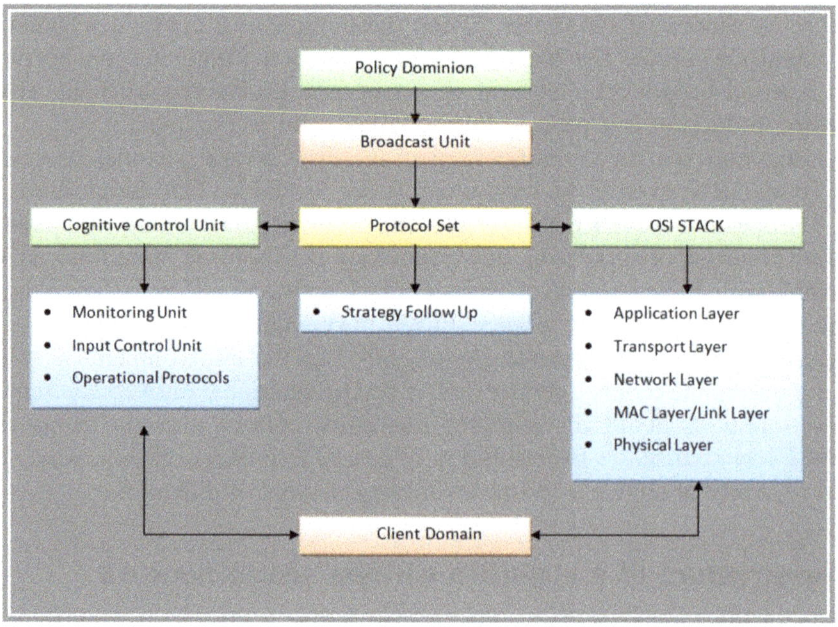

Figure 6.4. Block diagram of the operating modes used by CR.

monitored by a tertiary access point, a permanent architectural element that functions as a secondary master unit. Secondary users and tertiary access points are both equipped with CR [16]. If many tertiary networks use the same carrier frequency, their wavelength consumption may be controlled by a frequency band broker; this leads to a centralized network organization. The frequency band broker collects information about how each tertiary network works and shares network resources to make frequency disclosure fair and cost-effective (figure 6.5).

Figure 6.6 displays the core structure of a cognitive transceiver. Its three parts are an RF range, an analog-to-digital (A/D) converter, and core computing.

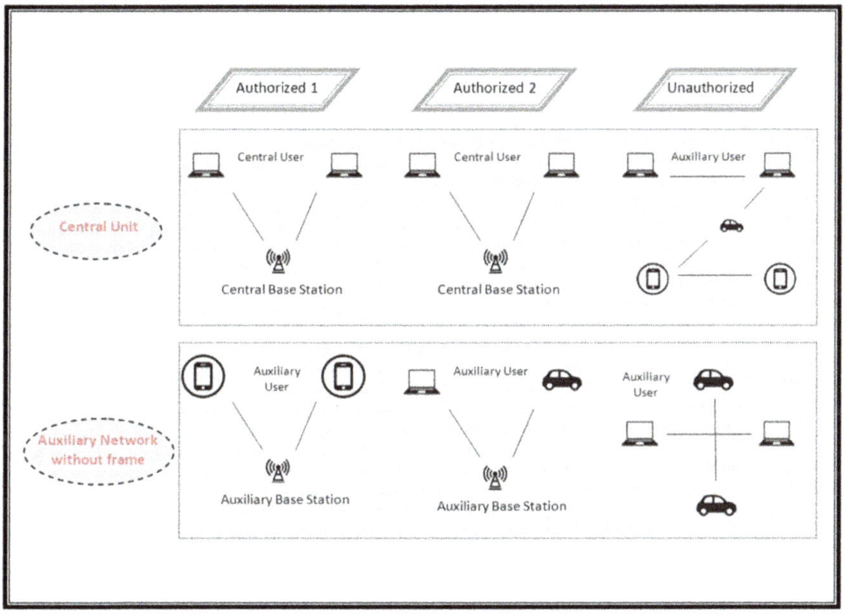

Figure 6.5. Network architecture of CR.

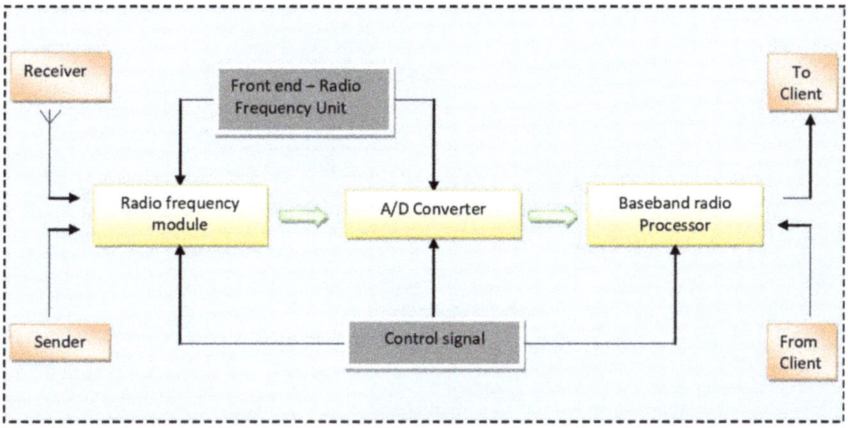

Figure 6.6. The basic structure of the cognitive transceiver.

The adaptability of some of these components is dependent on the wireless communication system. In CR systems, just the RF front edge with spectrum sensing is adaptive and distinct from a conventional receiver (RX). Adaptive baseband processing is also required for fully CR. This is achieved most efficiently by performing the baseband processing in software on a digital signal processor (DSP).

As a result, many of its supporters refer to fully CR as software radio. Figure 6.7 depicts an example of an RF RX front-end structure. To enable CR functionality, this front end must be adaptable. It is made up of the following components: an RX filter, a low-noise amplifier (LNA), a remote amplifier, an RX downward translator, an RX low-pass filter, an automatic gain control (AGC) amplifier, and RX A/D converters. The following explains the functions of the different components:

The RX RF filter: this makes a random choice from the established bandwidth. The adjustable frequency band matches the throughput of the received signal. To avoid saturation of the low-noise amplifier, out-of-band signals are disregarded.

The low-noise amplifier boosts the signal received by the RX filtration system, decreasing the impact of noise added by subsequent RX loop components on the

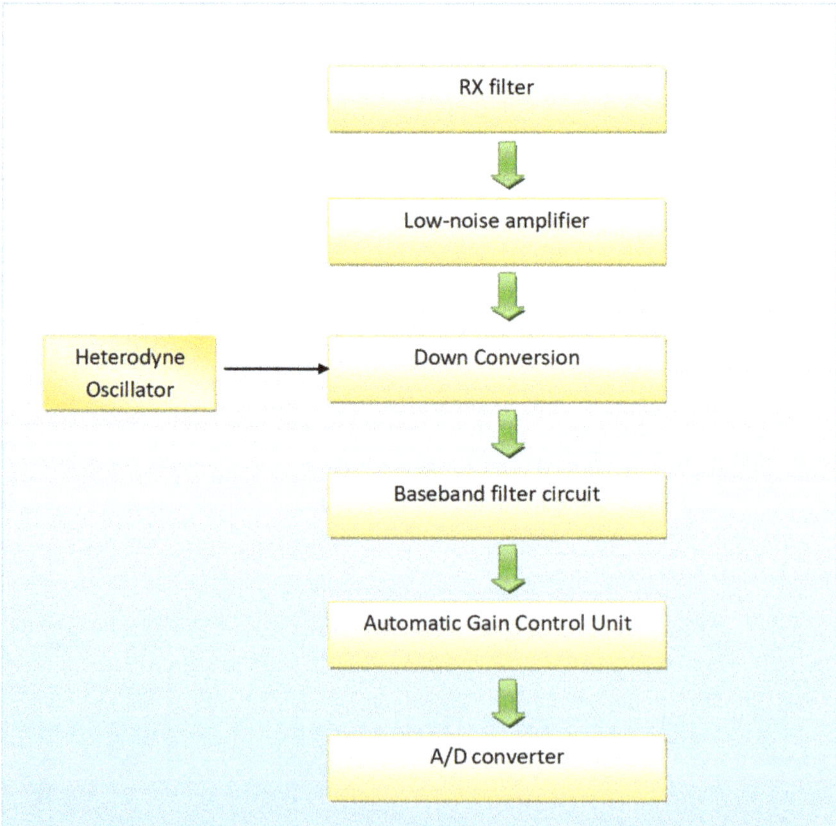

Figure 6.7. Typical CR RF RX front-end structure.

transmission ratio. The succeeding phases of the downconversion result in more amplification.

The local oscillator (LO): this provides harmonic signals that correspond to the information transmitted by the RX filtration system, enabling the CR to perform downconversion and acquire the required signal [26]. The frequency of the local oscillator (LO) may be upgraded using a voltage-controlled oscillator (VCO) to guarantee that the LO generates a signal that, after downconversion, produces the desired reference voltage.

The RX downconverter converts the message data to a baseband frequency. The signal is provided as a sophisticated analog waveform in the base band.

The RX low-pass filter: This allows the user to choose their favorite radio frequencies (in contrast to the receiver bandpass filter, which selects the frequency mode in which the service provider operates). It reduces disturbances from adjacent channels and cacophony. This filtration ought to have as little effect as possible on the target signal.

The AGC ensures that the signal is at the right level for the ADC to digitize it.

The RX A/D converter system converts the analog data to quantitative values that represent duration and amplitude. The ADC's action is mostly influenced by the changing aspects of the subsequent wave processing. If the testing theorem's criteria are met, the sample rate has little significance. Oversampling increases the ADC's overheads but makes signals simpler to process later.

The following are the primary needs of this CR structure:

Broadband operation: the transmitters, the LNA, the resonant frequency and AGC components of the RF receiver front end must all be able to function at all potential CR operating frequencies.

Tunability: cognitive wireless systems need a reconfigurable RF front end that can choose a channel for operation. This may be achieved with the help of an adjustable local oscillator.

Intrusion denial: because stream allocation occurs only after the downstream transformer (mixing) in CR, RX saturation by strong out-of-band signals is a serious challenge. A transmission that is currently in the aggregate coverage region of the CR and is on a wavelength that the receiver wants to reject at that specific time may overwhelm RF components like the LNA [3]. The use of a tunable RF notched filter in front of the LNA helps to decrease noise. The physical value of these filters is substantial, as the surface modification is limited.

Cognitive communications improve spectrum efficiency and provide more bandwidth because CRs can perceive, discover, and stay updated on the RF levels in their surroundings. They can also alter their operational traits based on the conditions. Furthermore, inexpensive scale sharing can alleviate the costs of consolidated band management.

However, one might wonder why the key clients would let inferior clients utilize their range, i.e. one for which they have paid. Possible explanations include:

Profit: network operators may be able to charge TUs for the actual duration of use by a certain incidence group, even for a fleeting period, using CR technology.

This is useful for certain purposes, but it may also be infeasible due to the possibility that the costs of monitoring and charging for this service might exceed the proceeds from spectrum sales.

Emergency services: at times of emergency, services that would typically be deemed 'primary users' must relinquish wavelengths to rescue services, such as those responding to environmental hazards.

• Supervisory obligations: the frequency controller has the authority to mandate that cognitive devices may operate inside a certain frequency range if they do not cause interference for the primary users. Such an approach has the potential to be used for portions of the spectrum, such as television, for which the main users never pay [27]. In the United States and globally, television stations did not have to pay for the spectrum they used; rather, they were given it for free since it was generally accepted that these channels provided a public service. Because of this, the frequency regulator has the ability to quickly demand that television stations collaborate with other firms 'for the public good.'

6.5 Spectrum-sensing device approaches in cognitive wireless sensor networks

The following are the four basic operations of CWSNs:

The spectrum-sensing device is the CR element that detects, perceives, acquires, and is conscious of a group of wireless environmental parameters, including bandwidth, power availability, radio channel characteristics, and network architecture. Spectrum-sensing techniques include energy detection, waveform detection, cyclization detection, cooperative detection, interference detection, and prediction detection.

Spectral management is the act of selecting the optimum SOPs from the available spectral possibilities to construct the most viable E2E route for an SU transmission with a specified QoS/QoE. To do this, one must initially model and describe the standard operating procedures [21, 23]. This also requires a comprehensive grasp of PU behavior. Finally, appropriate choice procedures are necessary.

Spectral sharing pertains to the resolution of numerous conflicts that might develop in CRNs, notably when many SUs must share the licensed bandwidth at the same time. In this scenario, SU access must be managed in accordance with a predetermined standard. This is mostly a data MAC problem. Spectrum accessibility and resource allotment are the two kinds of spectral access control system. Various control strategies, such as cooperation or non-cooperation mechanisms, overlaid or underlaid access, and mixed mechanisms, may be used to overcome these challenges.

Spectrum mobility is the process of relocating a communication channel in response to a change in operating circumstances. This may include situations in which a TU is required to change the radio band because a key user has requested a communication channel [28]. Spectrum sensing requires extremely complex regulatory mechanisms that function at several network protocol layers. The objective is to maintain the desired QoS despite the change in operational circumstances. These four fundamental techniques are responsible for communication in CRNs.

The spectrum-sensing techniques are classified as shown in figure 6.8.

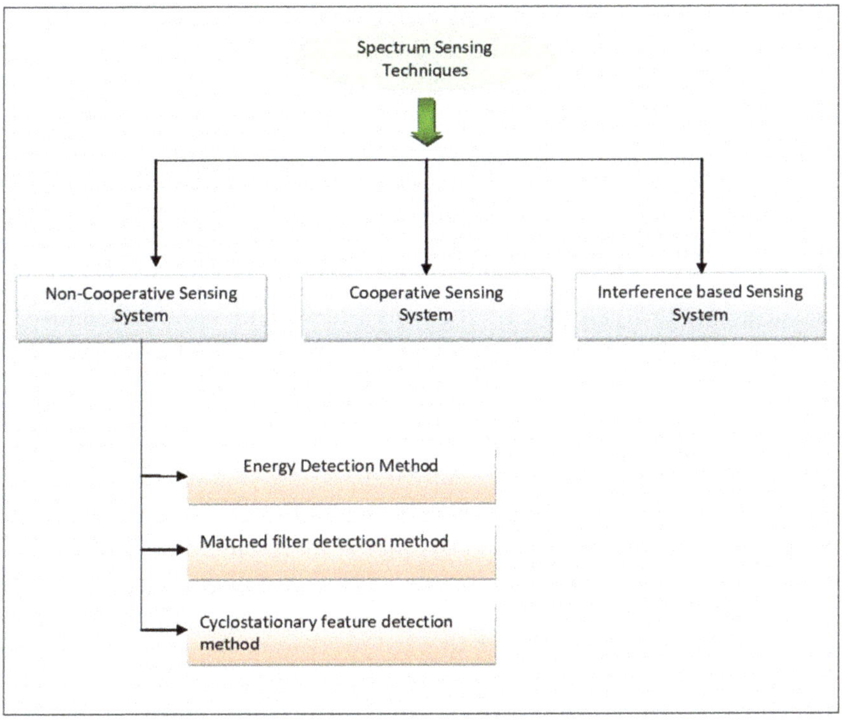

Figure 6.8. Classification of spectral detection strategies.

6.5.1 Non-cooperative system sensing

6.5.1.1 Detection of energy

The energy detector is a non-coherent system, since no previous knowledge of the signal is needed. It is also uncomplicated. As seen in figure 6.8, in order to choose a channel, the input is processed by a bandpass filter before being incorporated over a predetermined amount of time. The result is then evaluated using different methods of measuring energy.

The momentum detector has the drawback of disregarding motion structure and operating poorly in environments with minimal signal-to-noise ratios (SNRs). Considering the effectiveness of power spectral density in terms of first-type failures (fault alarms) and second-type failures (skipped detection), the cutoff used to determine H1 and H0 must be modified, as there is a trade-off relationship between first-type and second-type failures if the energy detector's efficiency remains unchanged. It is not possible to reduce skipped detections and fault alarms simultaneously.

Two methods can be used to enhance the effectiveness of a momentum detector.

1) Increasing the SNR of the established indication.
2) Boosting the size or adaptability of the signal breathing space accepted.

In a real-world setting, it is difficult to increase the received SNR due to noise unpredictability, masking, and multipath fades, the effects of which are neither foreseeable nor compensable. Consequently, we focus on increasing the freedom level of the signal space received. In contrast, when an SU obtains a greater number of inspection samples, they are integrated by a cumulative assessment and the final choice is more trustworthy. Figure 6.9 illustrates that we have N degrees of freedom in the time domain if the TU contributes N tests of established validity during a sensing interval. Nonetheless, as N increases, the frequency segment that the SU may essentially use for data broadcast diminishes, restricting the SU's ability to exploit the degree of flexibility within the temporal domain more efficiently [19]. The extended level of flexibility pertains to the use of several antenna arrays for energy detection.

6.5.1.2 Detection using matched filters

Once the input to be detected consists only of flavors of white Gaussian interference, a matched filter is the optimal choice. Since a matched filter is an altered momentum and time-delayed version of the message signal, an example or framework of the signal to be recognized is required to create a matched filter. In radio communications, a matched filter is often used to transmit a known input and retrieve its reflections. The matched filter is then utilized to identify the similarity between the sent and intercepted signals, enabling deductions to be made about the reflection's source. The objective of using a matched filter as a detection device in the context of a CRD is to identify the activity of the key user in the signal, while less focus is placed on deciphering and obtaining the message content [2, 8]. A matched filter is a version of a to-be-detected signal sample that has altered momentum and is time delayed. If $r(t)$ is a random signal, it is first changed to $r(-t)$ to time change it, and then it is relocated to match the original signal's time interval by making it equal to

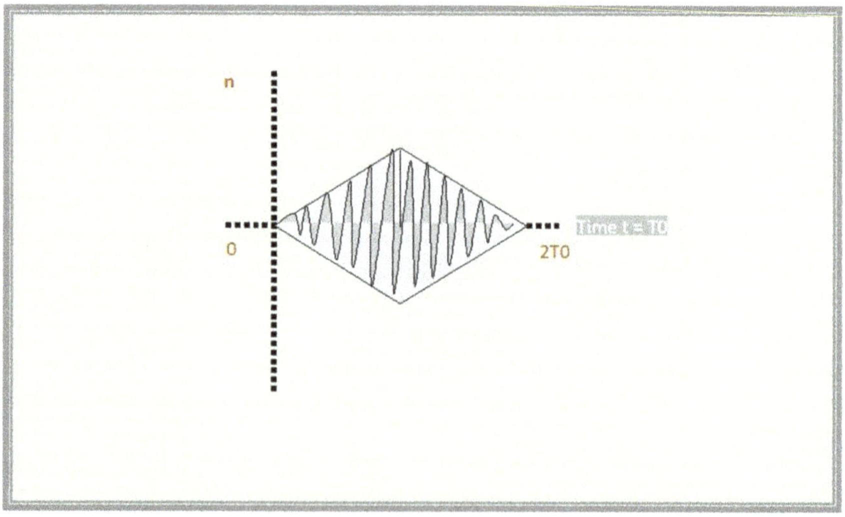

Figure 6.9. Matched filter output.

$T - t$, where T is the transmitter period. Thus, the signal impulse retort $h(t)$ for $r(t)$ of the matched stream is $r(T - t)$. The result is processed at $t = T_0$ or $t = T$, where T is the extent of the original signal.

The purpose of sampling the output of the tallied stream at $t = T$ is to get the greatest SNR at T. The following is an example of a matched filter output. After sampling is complete, a selection device may be used to choose the bits to be sent, depending on a predefined limit. The detection of bits is irrelevant for the identification of the primary user. In this instance, just the presence of the primary user must be recognized, which may be done by establishing a threshold that corresponds to the lowest observed value at T when the transmitted signal is detected. If the output goes below this threshold, the decision device decides that the primary user is absent. The matched filter selection instrument pulse-coded modulation (PCM) wave y sample has a value of one if y is greater than $r(t)$. The sample has a value of zero if y at time $t = T_b$. Threshold white Gaussian noise (t). The diagram illustrates an example of a matched filter output when the primary user was identified, or when the matched filter was matched to the transmitted signal [22]. The input signal has a duration of T, whereas the yield of the matched filter has a duration of $2T$. At $t = T$, the peak value reflects the output with the greatest SNR. Thus, sampling at time $t = T$ produces the maximum value of the output signal. The reasons for sampling at T are described in detail below.

Since precise detection takes just one sample at $t = T$, matched filters provide a quick detection time. The mathematical model is a part of the matched filter that is crucial in order to achieve the highest SNR. Peak (SNR) = $(2{*}E_o)/(N_o)$.

This demonstrates that the detection of a maximum SNR by a matched filter depends purely on the energies E_o and noise. The input signal values, structure, and waveform are irrelevant to matched filter detection, allowing the designer to choose the optimal waveform. This choice may be made for any purpose, such as producing a waveform with a smaller bandwidth footprint. On the other hand, a matched filter fails when the principal signals originate from many sources. In such cases, it is difficult to obtain prior knowledge about the incoming signal [7, 14]. Even if the many major systems work with the cognitive system to exchange the information required for the operation of a matched filter, the recipient system must be equipped to cope with the decoding complexities, which makes the matched filter an expensive detection method.

6.5.1.3 Cyclostationary feature detection

This method achieves its objectives by utilizing the innate periodicity of modulated signals. The modulated signal is often combined with sinusoidal carriers, resulting in a periodicity that is intrinsic to the signal. If a cognitive system recognizes this periodicity, it may identify the primary user's presence. Periodicity is seen in oscillatory carriers, pulsing trains, hopped patterns, and repetitive prefixes of main signals. These periodic signals thus include periodical characteristics and spectral correlations that are absent from interference signals. Thus, cyclic properties can be used to differentiate between unwanted noise and modulated signals, and cyclical detection should perform better even in situations with minimal SNRs [16, 23].

Unlike previous strategies, this method does not need perfect synchronicity, nor does the cognitive user need to remain quiet throughout the spectrum sharing process.

This method is less prevalent than others, such as energy detection, due to disadvantages such as high computational complexity and longer sensing durations. For practical uses of cyclostationary feature detection, the following characteristics have to be noted:

- Double-sided sine-wave carriers
- Data rate (symbol period)
- Modulation type.

6.5.2 System-wide cooperative sensing

The performance of all non-cooperative detection approaches may be enhanced by merging the outcomes of separate techniques for objects into clusters, leading to better decisions. Consequently, the development of individual techniques and that of cooperative sensing go hand in hand. Fuzzy logic-based detection (FLD) is a plausible implementation of such a cooperative sensing method. The balanced outputs of the matched filter, energy detection, and cyclostationary pattern detector are sent into a fuzzy logic system, which classifies the option as either robust or weak, resulting in a considerable improvement in outcomes. Consequently, in addressing the real application of spectrum sensing, it is essential to consider the feasibility of cooperative sensing, as it may enhance the output without significantly increasing the computational costs or time taken [8, 13]. Selfish wireless sensing has the option of employing either cooperative or local spectrum sensing, which is a drawback of cooperative sensing. A self-centered wireless sensor picks the most profitable option. The objective is to arrive at a desirable decision result that improves spectrum use while complying with self-attention boosting and PU constraints. The advantage is the conveyance of data, while the disadvantages are the latency and power expenditure. The decision-making process used by radio receiver devices is described as potentially dangerous competition, and the ideal solution communicates via a balanced set of choices such that no sensing device is inclined to diverge from it unilaterally.

6.5.3 The interference-based sensor technique

This method of detection will not be simulated, but it must be described to provide the learner with a better knowledge of alternatives to spectrum access. Until recently, we determined whether the key client is active or not based on a set of criteria [29]. This idea is invalidated by the SNR wall when the SNR value is low. The investigation of diversity now includes the identification of event bases. In this instance, the hypothesis is the discovery of energetic edges. If we can identify events in which the energy levels change suddenly, we may deduce information about the PU's arrival or exit from the spectrum [5]. Thus, even in very low SNR conditions, it is possible to detect sudden fluctuations in energy levels, enhancing detection performance. This method is known as distortion-driven sensing or incident-driven sensing because it depends on the constant monitoring of spectrum interference.

6.6 Implementing cognitive wireless sensor networks

There are hardware limitations for CR wireless sensors in terms of computing capacity, memory space, and power utilization. In contrast to conventional wireless sensors, these sensors must detect channels, analyze, determine, and respond. Wireless CR sensors should be able to alter their settings or transmitters in response to environmental interactions.

Sensing units include both A/D converters and sensor devices. The analog wave from the sensor is transformed into a numerical format and delivered to the central controller. The CR wireless node sensors should make it possible to consider deploying cutting-edge AI algorithms. This ability is included in the CR approach. CR units must automatically modify communication properties such as carrier signal, signal intensity, and modulation. The unit must choose the optimal channel, share the frequency with many other users, and control frequency band portability, i.e. release the channel if the PU desires to utilize it. A transmitter and receiver unit manages data reception and transmission. Due to the rapid development of power storage mechanisms in wireless sensing nodes, power conversion or recharging devices are now customizable and sensor-specific devices [15]. Examples of multifunctional units include destination modules, power-gathering units, and mobility units.

Developing intelligent technology for CR WSNs is a challenging endeavor. Numerous AI-based solutions have been created to satisfy the CR model of surveillance, reconfiguration, and cognition, which is the primary premise of CR. Some examples include synthetic neural networks (artificial neural networks), meta-heuristic optimization methods, Markov chain models (hidden Markov models (MMs)), regulation systems, annotation systems (ontology base stations (BSs)), and instance systems (ISs). Aspects that impact the selection of AI techniques include reaction time, complexity, privacy, durability, and consistency [6, 30]. However, it is unknown how often sophisticated hardware for WS CR networks is sufficient, and no limits have been established.

Changes in topology: The network lifespan of WSNs is directly influenced by the network topology. Depending upon the requirements, a CR wireless sensing device may be statically or dynamically positioned. Component failure is prevalent in all types of WSN due to malfunctioning circuitry and resource depletion. CR WSN modules may be identical to those used in regular WSN topologies; however, they may undergo more frequent changes than ad hoc CR networks. Four kinds of CR WSN topology exist: (i) ad hoc CR WS networks, (ii) clustered CR WS networks, (iii) heterogeneous and hierarchical CR WS Networks, and (iv) mobile CR WS networks.

The smallest amount of electrical output needed to send a communication to a remote location is proportional to n. With low-lying stations and narrow stations, the factor n is typically nearer to four, as is usual with sensor network wireless connectivity. Therefore, routes with more bounces but smaller bounce distances can be more energy economical.

In the construction of static sensor networks, it is not always possible to identify such a route. In order for a CR WSN to achieve adaptability, decrease momentum utilization, and enhance web execution, identifying and adjusting a topology mechanism is essential [18]. Techniques for adaptive self-configured topologies outperform stable regional topologies, even if they are difficult to build and configure.

Self-assembly, self-configuration, and self-healing should be features of CR WS networks. In other words, any time any nodes or connections fail, an alternative route must be generated that avoids the faulty node or link. Software or hardware failures, as well as natural calamities such as fires, floods, explosions, volcanic activity, or tsunamis, among others, may lead to malfunctions in CR WSNs. A CR WS network should always be ready for such eventualities. There are numerous defect categories, including nodal faults, network faults, and sink faults. One of the challenges of CR WSNs is their lack of fault tolerance. Protocols designed for CR WSNs should provide data consistency so that the WSNs continue to operate normally.

The expense of manufacture: in general, the use of CR wireless sensors has been widespread. Consequently, the expense ought to be modest. In contrast to ordinary WSNs, which need smaller amounts of storage and computational capacity, CR WSNs expect a sufficient amount of storage and processing power [31]. To reduce computer hardware expenses, it is necessary to develop algorithms that require less computational memory and processing power. Developing such programs is a challenging task. In addition, CR wireless sensors must integrate complex radios, software navigation systems (e.g. GPS navigation), power storage devices, and other components, which all increase the costs of manufacture.

6.7 Spectral optimization and new technology spaces

The electromagnetic spectrum is nature's greatest gift. The amount of accessible radio spectrum cannot be increased, but rather could be utilized more effectively. With the exception of the ISM radio bands, the use of radio bands needs authorization from the national government. Due to the high cost of bandwidth authorization, several research institutions and hardware firms have concentrating on building appliances for spectrum ensembles. Consequently, the ISM frequency ranges are crowded, which hinders the advancement of innovative technology. In contrast, several of the certified ensembles have been either underused or compartmentalized [8, 10]. WSs, or unused spectrum, may be accessed by cognitive spectrum-sensing sensors without disturbing the permit holders. The free or low-cost usage of these bands by unlicensed spectrum users enables the development of further technologies for these bands.

Utilization of numerous channels: the majority of traditional WSNs communicate through a particular channel. When an event is detected, sensor nodes in WSNs produce bursts of packet traffic. With densely deployed WSNs, several WSN nodes inside the event zone attempt to use the same channel simultaneously. This raises the risk of conflicts and reduces the global consistency of communication due to packet loss, resulting in excess power requirements and network congestion. To solve this potential problem, CR WSNs use several channels opportunistically.

Energy conservation: in WSNs, packet relaying wastes a large amount of energy due to packet loss. To adapt to the channel conditions, wireless sensor devices using cognitive frequency bands might be capable of changing their driving constraints. Consequently, the energy usage caused by congestion and signal retransmission may be decreased.

Universal operational capability: every nation has a unique set of spectrum control rules. A band that is accessible in one country may not be accessible in another. Conventional mobile nodes that have a fixed data rate may not work correctly when distributed wireless sensors with a variable operating frequency are introduced [32]. However, if nodes are outfitted with cognitive broadcast spectrum-sensing expertise, they may resolve a variety of conflicts by altering their broadcast transmission range. Consequently, CR wireless sensors are potentially applicable anywhere on the Earth.

Utilization of application-oriented spectrum band: recent developments have increased the variety of applications for wireless sensors. The data streams in WSNs are often temporally and spatially connected. WSNs generate packet bursts when an incident occurs; otherwise, they remain quiet. These spatial and temporal connections provide a design challenge for WSN communication systems. Due to the complex communication protocols used by CR WSNs, a wireless node of sensors with the same objective may utilize the frequency bands of different carriers in geographically overlapping space [21]. This is possible due to spectrum sharing amongst SUs, which mitigates electromagnetic interference issues.

Renting or leasing provides financial benefits to incumbents: wherever and whenever certain permitted spectrum bands are not in use, registered users may contract their use to SUs for a low price. This is possible while preserving the incumbents' access to the spectrum bands as required. This is an attractive option for those who cannot get a primary license for a particular frequency due to contractual or financial constraints. This dynamic benefits both newcomers and SUs.

Preventing threats: unlike cognition RF WSNs, most commercially available wireless sensors are limited to certain radio frequencies. As a result of their diverse spectrum use, SUs in CR WSNs can avoid a variety of attacks.

6.8 Applications and issues in cognitive wireless sensor networks

CR WSNs can be utilized in various applications, as shown in figure 6.10. Indeed, a CR WSN may generally be installed instead of a WSN. Management systems, equipment surveillance and predictive maintenance, precision farming, medicine and health, logistical support, detection and tracking, telemetry, intelligent roadside units, security systems, motion control, the general upkeep of complex systems, and the tracking of environments, both indoors and outdoors, are some examples of potential applications for CR WSNs [6, 11]. This section gives examples of some of the essential areas in which CR WSNs can be deployed.

6.8.1 Public safety and military applications

Traditional WSNs are utilized for a variety of applications for the armed forces and public safety, including (i) chemical, biological, radiological, and nuclear (CBRN) attack detection and investigation; (ii) command and control; (iii) gathering combat

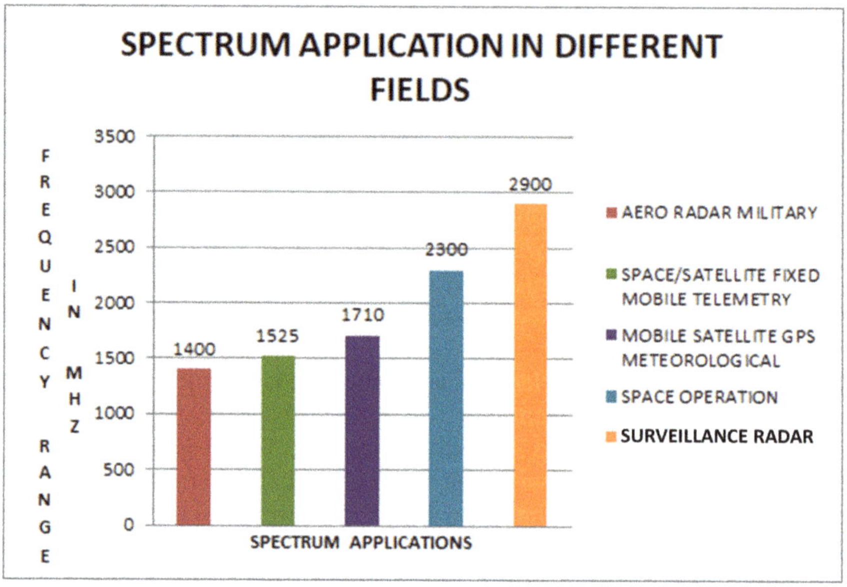

Figure 6.10. Analysis of the spectrum used by applications in different fields.

rescue operation information; (iv) battlefield observation; (v) support with intelligence in cryptography; (vi) directing unmanned vehicle to protect network,etc. An opponent may use jamming techniques to interfere with broadcast communications on the battleground or in disputed territories [33]. Since CR WSNs may hand off frequency bands on a large scale, CR WSNs can utilize other radio frequencies in such cases, thereby avoiding the effects of the disruptive broadcast. Moreover, many products intended for military use need to provide high throughput, secure network connectivity, and low communication latency. CR WSNs may be the optimal solution for such applications.

6.8.2 Healthcare

The use of wearable body sensors for healthcare purposes, such as telemedicine, is on the rise. Numerous wireless sensor nodes are implanted in patients to capture vital information for remote patient monitoring by medical professionals. In 2011, a draft reference for body area networking (BAN) technology was adopted by the wireless personal area network (WPAN) Working Group 6. Several remote regions of emerging nations, such as India and Nepal, have already implemented wireless healthcare systems. Wi-Fi BAN for medical practices is appropriate in regions where the number of healthcare professionals is declining. Therapeutic data is significant, time-sensitive, and miscalculation prone. Consequently, the inadequacy of ordinary WSNs restricts the potential of telemedicine [17]. If the operating spectrum band for 'eHealth with BAN' is crowded, the QoS may not attain an acceptable level. The introduction of CR body-worn wireless sensors may solve problems related to spectrum, interference, and global maneuverability, hence enhancing dependability.

6.8.3 Bandwidth-intensive applications

Multimedia applications, including the streaming or real-time streaming of video, sound, and images, are very challenging to implement over resource-constrained WSNs due to their enormous bandwidth requirements. Additional WSN applications, including hospital IoT networks, vehicle WSNs, monitoring, and surveillance, demonstrate substantial spatial and temporal variations in signal density that are proportional to network size. These applications have high bandwidth consumption, unacceptable latency, and a bursty nature. Because the SUs in CR WSNs may utilize several pathways when possible and necessary, CR WSNs are appropriate for such bandwidth-intensive applications.

6.8.4 Transport and automobile networks

IEEE 1609.4 mandates multiplex operations for wireless connectivity in automobiles (WCA). The WCA system utilizes 75 MHz of bandwidth within the 5.9 GHz frequency band, with one host controller and six distribution channels. To transmit data at the 5.9 GHz frequency, all users of vehicles will have to compete for the available channels. However, spectrum shortage concerns persist [12]. This spectrum scarcity problem and WCA's CR requirements have been studied.

The development of vehicular wireless sensing systems requires a novel network architecture for proactive monitoring data collection in metropolitan settings. CR WSNs can be beneficial in this area. As a result of further research in this field, several suggestions for improving highway safety using CR WSNs have been created.

6.8.5 Virtual surveillance applications

Virtual domestic surveillance applications, such as vehicle tracking, species diversity mapping, grassland surveillance, weather forecasting, monitoring climatic conditions affecting agriculture and livestock, water management, submerged WSNs, inventory tracking, rescue missions, and bridge or tunnel tracking, require the shortest possible network connectivity and data transmission delay [20]. Certain virtual surveillance applications are particularly delay-sensitive and need high levels of reliability. Multihop WSNs may experience a delay due to connection failure if the channel status is degraded. In contrast, in the case of CR WSNs, WS networks move to a different channel if they detect a dormant transmitter with superior conditions [34]. Channel clustering coupled with the concurrent use of several routes can be used to increase the channel bandwidth in CR WSNs.

6.9 Summary

The electromagnetic spectrum is one of nature's most valuable and limited commodities. Because of its unique and crucial function in wireless communications, it is a strictly controlled resource. With the development of remote services, the degree of interest in using RF bands is constantly creating a shortage of spectrum range assets. Spectrum sensing has recently received a lot of attention as a vital technique for developing CR. Spectrum-sensing performance has been improved by the use of both multi-antenna and cooperative sensing systems. Spectrum sensing can permit the concurrent use of spectrum by authorized clients and unregistered IoT gadgets, leading to ideal and dynamic range use by CR-based IoT connections if they are designed properly. A sensing throughput trade-off strategy can be designed to boost SU throughput by allocating sensing and transmission time more efficiently. The SU's range of spectrum use is boosted by optimizing the appropriate detection time and SU choice.

Acknowledgments

This endeavor could not have been accomplished without the involvement and help of a huge number of people, some or all of whose names may not be included here. Their help is very much respected and warmly recognized, but the group would like to thank and express its debt of gratitude to everyone who helped us.

Finally, we want to express our gratitude to everyone in our families for all the love, support, and encouragement they have shown us.

We are thankful to the editors, Professor GR Sinha and Dr Varun Bajaj, for providing useful feedback and critical suggestions during the development of the manuscript.

Above all else, the authors wish to express my gratitude to the great Lord God, who is the source of all wisdom and knowledge, for the immense love he has shown me.

References

[1] Urkowitz H 1967 Energy detection of unknown deterministic signals *Proc. IEEE* **55** 523–31

[2] Cabric D, Misra S M and Brodersen R W 2004 Implementation issues in spectrum sensing for cognitive radio *Conference Record of the Thirty-Eighth Asilomar Conference on Signals, Systems and Computers* **1** 772–76

[3] Cordeiro C, Ghosh M, Cavalcanti D and Challapali K 2007 Spectrum sensing for dynamic spectrum access of TV bands *2007 2nd Int. Conf. on Cognitive Radio Oriented Wireless Networks and Communications* (Piscataway, NJ: IEEE) pp 225–33

[4] Sutton P D, Lotze J, Nolan K E and Doyle L E 2007 Cyclostationary Signature Detection in Multipath Rayleigh Fading Environments *2007 2nd Int. Conf. on Cognitive Radio Oriented Wireless Networks and Communications* (Piscataway, NJ: IEEE) pp 408–13

[5] Xing Y, Mathur C N, Haleem M A, Chandramouli R and Subbalakshmi K P 2007 Dynamic spectrum access with QoS and interference temperature constraints *IEEE Trans. Mob. Comput.* **6** 423–33

[6] Bater J, Tan H-P, Brown K N and Doyle L 2007 Modelling interference temperature constraints for spectrum access in cognitive radio networks *2007 IEEE Int. Conf. on Communications* (Piscataway, NJ: IEEE) pp 6493–98

[7] Ganesan G and Li Y 2007 Cooperative spectrum sensing in cognitive radio, part II: multiuser networks *IEEE Trans. Wireless Commun.* **6** 2214–222

[8] Yucek T and Arslan H 2009 A survey of spectrum sensing algorithms for cognitive radio applications *IEEE Communications Surveys & Tutorials* **11** 116–30

[9] Su H and Zhang X 2008 CREAM-MAC: an efficient cognitive radio-enabled multi-channel MAC protocol for wireless networks *2008 Int. Symp. on a World of Wireless, Mobile and Multimedia Networks* (Piscataway, NJ: IEEE) pp 1–8

[10] Xu G, Lu Y, He J and Hu N 2008 Primary users detect for multiple-antenna cognitive radio based on blind source separation *2008 Int. Symp. on Intelligent Information Technology Application Workshops* (Piscataway, NJ: IEEE) pp 777–80

[11] Chen R, Park J M and Reed J H 2008 Defense against primary user emulation attacks in cognitive radio networks *IEEE J. Sel. Areas Commun.* **26** 25–37

[12] Sharma M, Sahoo A and Nayak K D 2008 Channel modeling based on interference temperature in underlay cognitive wireless networks *2008 IEEE Int. Symp. on Wireless Communication Systems* (Piscataway, NJ: IEEE) pp 224–28

[13] Heinzelman W B, Chandrakasan A P and Balakrishnan H 2002 An application-specific protocol architecture for wireless microsensor networks *IEEE Trans. Wireless Commun.* **1** 660–70

[14] Zahmati A S, Abolhassani B, Shirazi A A B and Bakhtiari A S 2007 An energy-efficient protocol with static clustering for wireless sensor networks *Int. J. Electronics, Circuits and Systems* **0.0** 135–38

[15] Moghaddam N M, Zahmati A S and Abolhassani B 2007 Lifetime enhancement in WSNs using balanced sensor allocation to cluster heads *2007 IEEE Int. Conf. on Signal Processing and Communications* (Piscataway, NJ: IEEE) pp 101–4

[16] Sai Shankar N, Cordeiro C and Challapali K 2005 Spectrum agile radios: utilization and sensing architectures *1st IEEE Int. Symp. on New Frontiers in Dynamic Spectrum Access Networks, 2005. DySPAN 2005* (Piscataway, NJ: IEEE) pp 160–69

[17] Cavalcanti D, Das S, Wang J and Challapali K 2008 Cognitive radio based wireless sensor networks *2008 Proc. 17th Int. Conf. on Computer Communications and Networks* (Piscataway, NJ: IEEE) pp 1–6

[18] Gao S, Qian L and Vaman D R 2008 Distributed energy efficient spectrum access in wireless cognitive radio sensor networks *2008 IEEE Wireless Communications and Networking Conf.* (Piscataway, NJ: IEEE) pp 1442–47

[19] Gao S, Qian L, Vaman D R and Qu Q 2007 Energy efficient adaptive modulation in wireless cognitive radio sensor networks *2007 IEEE Int. Conf. on Communications* (Piscataway, NJ: IEEE) pp 3980–86

[20] Byun S S, Balasingham I and Liang X 2008 Dynamic spectrum allocation in wireless cognitive sensor networks: improving fairness and energy efficiency *2008 IEEE 68th Vehicular Technology Conf.* (Piscataway, NJ: IEEE) pp 1–5

[21] Mitola J III 2000 An integrated agent architecture for software defined radio *PhD Dissertation* Royal Inst. Technology, Stockholm, Sweden http://www.diva-portal.org/smash/get/diva2:8730/FULLTEXT01.pdf

[22] Haykin S 2005 Cognitive radio: brain-empowered wireless communications *IEEE J. Sel. Areas Commun.* **23** 201–20

[23] IEEE 802.22 draft standard, IEEE P802.22TM/D0.4 Draft Standard for Wireless Regional Area Networks, https://www.ieee802.org/11/ November 2007

[24] IEEE Computer Society LAN MAN Standards Committee 2000 IEEE Standard for Information Technology - Telecommunications and information exchange between systems - Local and Metropolitan networks - Specific requirements - Part 11: Wireless LAN Medium

Access Control (MAC) and Physical Layer (PHY) specifications: Higher Speed Physical Layer (PHY) Extension in the 2.4 GHz band *IEEE Std 802.11b-1999* pp 1–96
[25] Willig A 2008 Recent and emerging topics in wireless industrial communications: a selection *IEEE Trans. Ind. Inf.* **4** 102–24
[26] Rajaravivarma V, Yang Y and Yang T 2003 An overview of wireless sensor network and applications *Proc. 35th Southeastern Symp. on System Theory, 2003* (Piscataway, NJ: IEEE) pp 432–36
[27] Raghavendra C S, Sivalingam K M and Znati T (ed) 2006 *Wireless Sensor Networks* (Berlin: Springer)
[28] Rawat P, Singh K D, Chaouchi H and Bonnin J M 2014 Wireless sensor networks: a survey on recent developments and potential synergies *J. Supercomputing* **68** 1–48
[29] DeBardelaben J A 2003 Multimedia sensor networks for ISR applications *The 37th Asilomar Conf. on Signals, Systems & Computers, 2003* vol 2 (Piscataway, NJ: IEEE) pp 2009–12
[30] Diamond S M and Ceruti M G 2007 Application of wireless sensor network to military information integration *2007 5th IEEE Int. Conf. on Industrial Informatics* vol 1 (Piscataway, NJ: IEEE) pp 317–22
[31] Vijay G, Bdira E B A and Ibnkahla M 2010 Cognition in wireless sensor networks: a perspective *IEEE Sens. J.* **11** 582–92
[32] Han J A, Jeon W S and Jeong D G 2011 Energy-efficient channel management scheme for cognitive radio sensor networks *IEEE Trans. Veh. Technol.* **60** 1905–910
[33] ITU I 2009 *Definitions of software defined radio (SDR) and cognitive radio system (CRS)* SM.2152-0 (2009) ITU https://www.itu.int/pub/R-REP-SM.2152-2009
[34] Zhou G, Stankovic J A and Son S H 2006 Crowded spectrum in wireless sensor networks *IEEE EmNets* https://www.cs.wm.edu/~gzhou/files/CS_EmNets06.pdf

IOP Publishing

Cognitive Sensors, Volume 1
Intelligent sensing, sensor data analysis and applications
G R Sinha and Varun Bajaj

Chapter 7

Applications and challenges of IoT-based smart healthcare systems that use cognitive sensors: an overview

Devanand Bhonsle, Yogiraj Bhale, Anu G Pillai, Shruti Tiwari, Vishal Moyal and Chih-Peng Fan

The Internet of Things (IoT) is a heterogeneous and variable network, due to which, it is less intelligent; as a consequence it does not provide the expected performance for various applications. To overcome this limitation of the IoT, a new concept has been introduced called the cognitive IoT (CIoT). This is made possible by the use of cognitive sensors. It analyzes the collected knowledge, which makes the system intelligent and allows it to take adaptive actions; hence, its performance increases. In today's world, life has become very busy and difficult, which affects the human body adversely; therefore, it is necessary to monitor people's health conditions. The CIoT can be used to create a healthcare system which communicates with the smart devices, sensors, and various stakeholders in the healthcare environment. When the CIoT is used for healthcare, it can take intelligent decisions based on the physical state of the patient. This is very helpful for the patient, as it provides timely information about the patient at low cost. The Internet of Medical Things (IoMT) can be used to monitor the health conditions of patients. It makes the healthcare system smart, i.e. all the required data is available in the network and can be accessed by authorized persons. During a pandemic such as COVID-19, this could be helpful for detection, remote health monitoring, contact tracking, surveillance, etc. The CIoT provides many services in various application areas; however this chapter is restricted to a discussion of the healthcare system. Various technologies are used to send (retrieve) data to (from) the network. Although the CIoT provide many facilities for patients, it is prone to various problems, viz. accuracy, privacy, and security. Various privacy and security measures may be applied to avoid these issues. Hence we can say that the CIoT may be a promising technology with which to establish better healthcare systems in the near future.

doi:10.1088/978-0-7503-5326-7ch7

7.1 Introduction

Today, the healthcare system [1, 2] is actively evolving in developed countries; as a result, life expectancy has been increased by about thirty years. As a result, the number of old people has increased. As life expectancy increases, chronic health issues are overloading healthcare systems all over the world [3]. The biggest obstacle in the healthcare system is the shortage of resources, which deteriorates the overall system and makes it inaccessible for the low-income groups in society. To overcome this gap and increase the efficacy of healthcare systems, the IoMT has been introduced [4–10]. It combines the functionality and protection provided by conventional medical equipment with the IoT, which consists of advanced and portable technology. It can handle diseases which may persist for a long period. One of the best features of the IoMT [11, 12] is that it may be used for the real-time monitoring of patients in their daily lives; hence, it makes home care possible. It provides an interconnection between many medical instruments and commercial firms or organizations that provide health experts. The IoMT [13–17] is growing day by day using innovative technologies, but it has numerous challenges [18]. The IoMT is sometimes referred to as the Healthcare IoT [1, 19], which is a collection of various software-based medical devices linked via a network. Healthcare professionals use various software packages to optimize diagnosis results, reduce time delays, and control illness. From an economic point of view, it enhances customer service and reduces price. In this chapter, we discuss various subsets of the IoMT. Figure 7.1 shows an example of cognitive sensors which may be wearable, implementable, or attached to

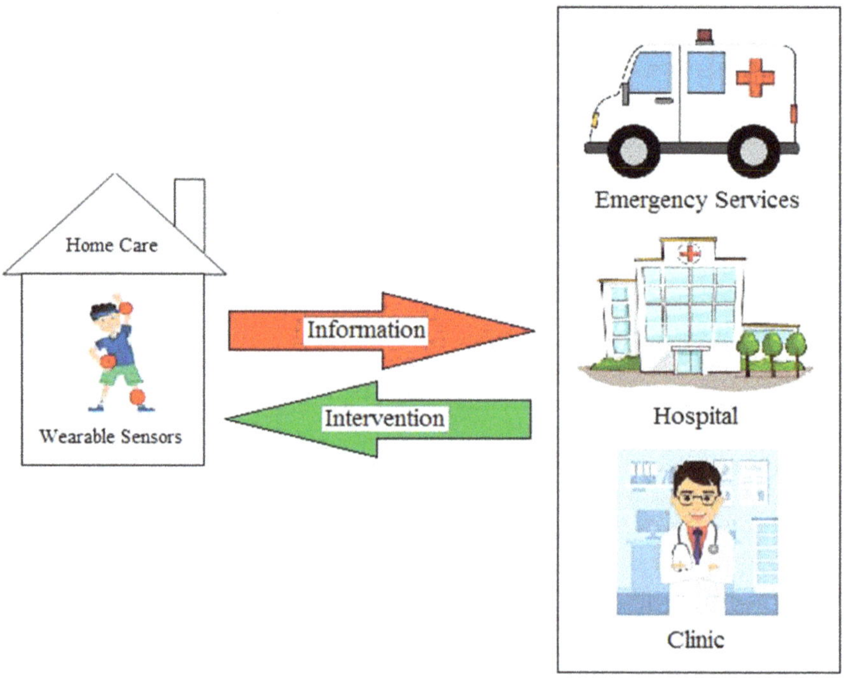

Figure 7.1. Cognitive sensors used in IoT-based smart healthcare systems.

or installed in the human body, wheelchair, or bed of a person so that his/her various parameters may be collected for health monitoring.

Figure 7.2 illustrates various aspects of the research trends in the field of smart healthcare systems. Examples of some important services provided by such systems are remote health monitoring [20], community healthcare, automated prediction and diagnosis, and so on. Smart healthcare systems include many applications [21], viz. wheelchair monitoring, smartphone healthcare solutions, medication management, fitness, rehabilitation, single conditions, etc. Single conditions include the monitoring of various human body parameters such as body temperature, changes of electrical potential occurring during the heartbeat, blood pressure, glucose level [22], oxygen level, respiration rate, etc. Various technologies are used in IoT-based

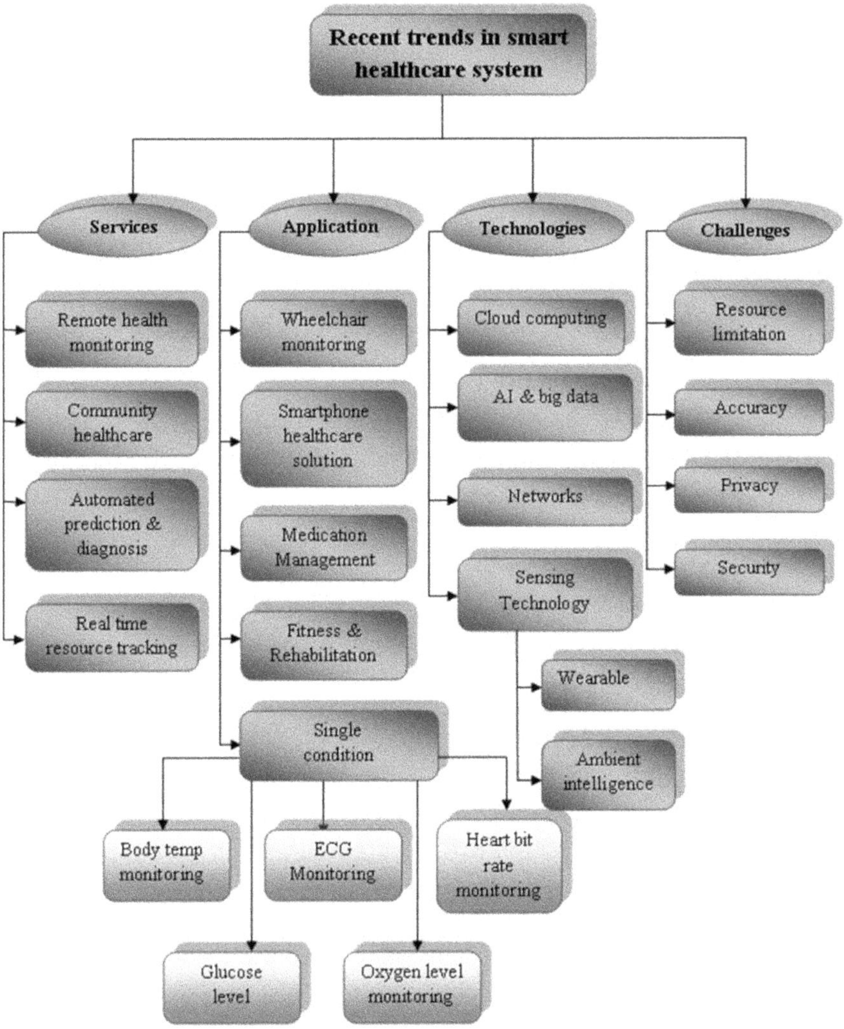

Figure 7.2. Recent trends in smart healthcare systems.

medical healthcare systems [23]. Examples of these technologies are cloud computing [24, 25], artificial intelligence (AI), deep learning [25, 26], networks (such as 4G, 5G, etc.), sensing technologies, etc. Sensing technologies can be divided into either wearable or ambient intelligence. However, many smart healthcare systems [27] face many challenges [28], which create obstacles for the smooth operation of these applications. Some of these challenges are resource limitations, accuracy, privacy, security, etc [29]. This chapter discusses all the points mentioned above; it is organized as follows: section 7.2 discusses various types of sensors which may be used for various purposes. Section 7.3 explains a smart healthcare framework that uses cognitive sensors. Section 7.4 explains various services provided by IoT-based smart healthcare systems. Section 7.5 discusses applications of the system. Section 7.6 discusses various technologies used in healthcare systems. Section 7.7 describes the challenges and risks of these systems. Section 7.8 explains the security measures used to provide privacy and security in IoT-based healthcare systems. Section 7.9 offers the conclusions.

7.2 Types of sensor

In general, sensors are devices that sense a particular entity and produce a corresponding output [35]. In the healthcare field, it is desirable to know the patients' conditions in order to provide the appropriate treatment; therefore, many devices are used to monitor patients' body conditions. We can divide the sensors into different categories based on their applications, the physical quantities to be measured, etc. The following sections describe the classification of sensors into different categories.

7.2.1 Sensors classified according to type

- Sensors used to measure body temperature
- Sensor used to measure the percentage of glucose in the blood
- Sensors used to measure blood oxygen
- Electrocardiographs
- Image sensors
- Motion sensors
- Sensors used to measure blood pressure

7.2.2 Sensors classified according to application

- Sensors for diagnostics purposes
- Sensors for monitoring purposes
- Sensors for medical therapeutics
- Sensors for imaging
- Wellness and fitness sensors

7.2.3 Sensors classified according to sensor placement

- Wearable sensors
- Strip sensors
- Ingestible sensors

- Implantable sensors
- Invasive sensors, non-invasive sensors

We can also divide biomedical instruments into other categories. These sensors are different from other sensors [35]. Any sensor can be used for biomedical applications with some modifications. However, biomedical sensors must have some special features. Some of these are listed below:

- As biomedical sensors use immobilized biological active materials as catalysts together with expensive reagents, it should be possible to use them repeatedly to detect the same biological parameters.
- Biomedical sensors should have strong specificity, which means that they only sense a specific substance and are not affected by the color or concentration of the material from which the measurement is made.
- The processing must be very fast, i.e. they should analyze the results rapidly.
- Their accuracy must be very high, i.e. the relative error must be very small.
- The analysis system must be simple for any person to use.
- They should be very cost-effective.

7.3 Smart healthcare using cognitive sensors

This section describes a framework for a smart healthcare system which uses cognitive sensors [30]. These sensors are useful to detect and monitor various functions and parameters of the human body, which is required in order to know about the health conditions of patients, or sometimes this information may be used by the user for fitness purposes. For example, a smart wristwatch can calculate the distance covered while its owner exercises, his/her heartbeat rate, and other information. Figure 7.3 shows the framework of a smart healthcare system which has the following stages:

- Cognitive sensors
- Low-power, short-distance interface
- Hosting layer

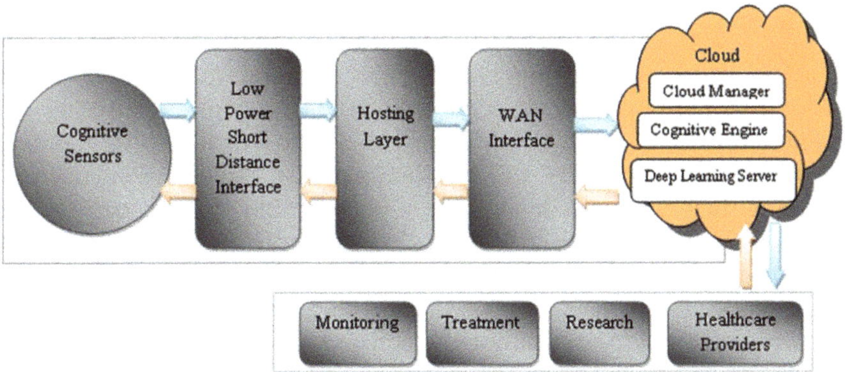

Figure 7.3. Framework of a smart healthcare system that uses cognitive sensors and the IoT.

- Wide area network (WAN) interface
- The cloud

The following is an explanation of the system architecture of this framework, which is able to make the healthcare system smart using the IoT. In this system, multimodal signals are acquired by cognitive sensors which provide health-related data. Smart IoT sensors may be wearable sensors or fixed sensors which can measure the different human body signals used for medical purposes [35]. The most common signals are body temperature, facial expression, body movements, heartbeat rate, respiration rate, brain signals, oxygen level, glucose level, blood pressure, etc. These sensors are embedded and installed in the patients' surroundings. These sensors can communicate with other devices using the IoT. These data are sent to the low-power, short-distance interface unit, which further transmits these data to another layer called the hosting layer. This hosting layer may consist of smart devices, for example, laptops, computers, or smartphones. This layer is called the hosting layer because it can store the acquired signal and forward it to the WAN. The WAN layer employs an advanced communication network, viz. 4G, 5G, or Wi-Fi. These networks are used to transmit the data to the cloud [25]. The cloud contains three different units viz. the cloud manager, the cognitive engine and the deep learning server. The cloud manager is responsible for data flow and security [5, 7]. It authenticates the patients' data and sends it to the cognitive engine, which processes the data and produces information about the patient's condition; it then makes an intelligent decision and sends it to the deep learning server. This unit sends back the detection result to the cognitive engine for a final decision about the health condition of the patient, which is further sent to the relevant medical person who analyzes the results and takes the necessary action for the betterment of the patient. [31].

7.4 Services

IoT-based systems provide many healthcare services; a few of them are discussed below.

- **Remote health monitoring:** using the remote health monitoring technique, any patient can be monitored from any place, viz. home, remote areas, etc. These systems use wearable sensors so that real-time conditions can be monitored and the appropriate action taken. This technique may be very helpful for chronic conditions and especially when the patient is suffering from an infectious disease such as COVID-19 [32–34]. IoT devices offer the facility for healthcare professionals to monitor patients' health. Patients can also monitor their own health. A variety of wearable IoT devices are available which provide many benefits to healthcare professionals and their patients. IoT devices automatically collect data related to body temperature, heart rate, blood pressure, and many more parameters. These metrics are forwarded to software applications which provide data sets. This data can be viewed by healthcare professionals and patients. This reduces the cost of treatment, saves time, and benefits the patient, as he/she may be at home or another place to rest and need not be physically available in the hospital or clinic. Many algorithms have been developed which recommend treatments or generate alerts accordingly [11]. For example, if an IoT sensor detects unusual low blood pressure or a low heart rate, it sends an alert to the healthcare professional, who can intervene.

- **Community healthcare:** This refers to the health services delivered by laypersons outside clinics and hospitals. Community health workers and volunteers play an important role, and they are the main practitioners who provide primary healthcare services. This service has to cover a large geographical area. IoT-based community healthcare systems make this easy by handling large volumes of data about many patients. These services may be classified into four broad categories:
    - **Preventive health services:** these include the treatment of hypertension and diabetes, cancer screening, and chemoprophylaxis for tuberculosis.
    - **Promotive health services:** these include family planning, vaccinations, health education, nutritional supplements, etc.
    - **Curative health services:** these include the treatment of malaria, pneumonia, lice infestations, etc.
    - **Rehabilitative health services:** these include social work, prosthetics, physical therapy, and other mental health services.
- **Automated prediction and diagnosis:** IoT devices are used to collect patients' data, which may be processed using software. These processed data may be further used to predict and/or diagnose disease. Many algorithms have been developed for this purpose.
- **Real-time resource tracking:** many vaccines, tablets, drugs, biological products, bandages, monitoring devices, instruments, apparatus, appliances, etc. are required for treatment purposes [36]. In the event of natural disasters, train or road accidents, fire hazards, and other emergency scenarios, it is essential to track these resources so that healthcare services can be provided as soon as possible. This is possible using IoT systems which are connected to the global network.

7.5 Applications of the IoT in healthcare

There are many fatal diseases which cause high death rates worldwide. These death rates can be reduced by providing appropriate care and effective treatments. For this purpose, various AI models have been used for the detection and treatment of disease [37, 38]. With the introduction of the IoT into the medical field, healthcare is becoming cheaper than the conventional techniques which made life easier by improving patient health. The most desirable aim of the healthcare field is the real-time diagnosis of disease. Sometimes, diagnoses performed using conventional healthcare techniques take a long time; however, the introduction of IoT has made diagnosis possible much quicker and sometimes in real time. The use of the IoT makes monitoring effective, as a sudden drop in the health of the patient sends an immediate alarm to healthcare professionals, who can take the necessary actions.

It also informs family members about the patient's medication. User experience is the most crucial factor which is helping the IoT to develop at present, as the IoT is simple to use, affordable, and easy to use with less training. It also helps doctors to maintain patient records, saving time and energy and reducing the time required for patient records to be retrieved and updated. Some of the most important applications which may be used in the medical field to monitor the patient health are listed below.

- **Wheelchair monitoring:** a chair with wheels used to carry a patient from one place to another is called a wheelchair. It is helpful for disabled persons who are not able to walk. Manually operated wheelchairs are moved by either another person or by the patient. Using the latest technologies, wheelchairs are equipped with cognitive sensors and various applications. These wheelchairs may also collect data about the health conditions of patients, which may help medical personnel or doctors to provide the required treatment. Some wheelchairs can be operated by patients whose whole bodies are paralyzed. This advancement has become possible due to the brain–computer interface (BCI).
- **Smartphones:** the main task of a cell phone (mobile) is voice calling and messaging services, but today, many extra features have been added to mobile phones; hence, they are referred to as smart phones. With the advancement in networks, video calling facilities are also available. Many smartphones support various applications which may be used for health monitoring and fitness; for example, they can record periods of walking or running. The cognitive sensors used in smartphones provide environmental data such as temperature, humidity, etc. In some cases, they may be used to monitor sleep quality or the restlessness of patients. Their built-in cameras may be used for telemedicine, i.e. a patient can take the doctor's advice and prescription, even if the patient is not physically present in the hospital or clinic.

Since smartphones carry so much confidential information, it is desirable for them to be secure so that they cannot be accessed by unauthorized persons, thereby maintaining privacy. Many features are available for security and privacy purposes, viz. fingerprint detection, face-lock systems, personal identification numbers (PINs), password protection, etc [11, 14]. Using these features, whole smartphones and/or individual applications can be locked. Figure 7.4 illustrates some of the features available in smartphones that have cognitive sensors.

- **Healthcare solutions:** a healthcare solution is used to provide end-to-end consultancy for patient health. Such a solution may provide an integrated service by knowing the root cause of health-related issues. Many healthcare solution companies are available worldwide. They provide the best consultations provided by experienced doctors in their specialized fields. They maintain the records and data for individual patients, which may be accessible by family members or the patient. These data are confidential; hence, these companies maintain privacy and provide security [20]. Some of the top healthcare companies in India are Apollo Hospitals Enterprise

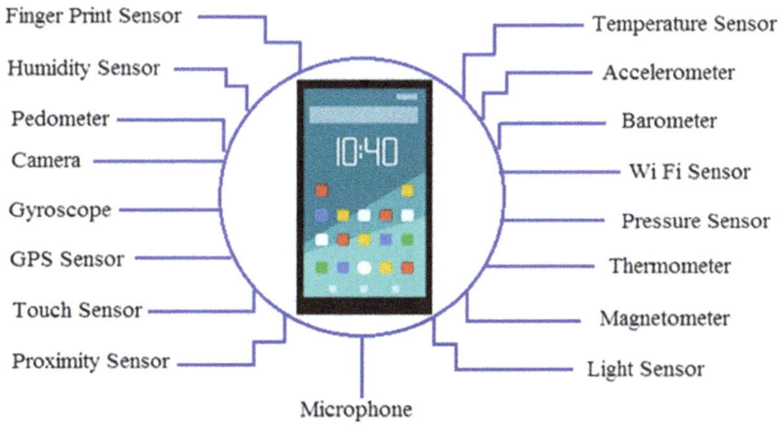

Figure 7.4. Different kinds of sensor available in a typical smartphone.

Limited, Fortis Healthcare Limited, Zydus Lifesciences, Dr Reddy's Laboratories, Wockhardt Limited, Kauvery Hospitals, Metropolis Healthcare Limited, Poly Medicure Limited, Consure Medical, Ajanta Pharma Limited, Aurobindo Pharma Limited, etc.

- **Medication Management:** medication management is a strategy for engaging with patients and their caregivers to create an accurate and complete medication list. The main goal of medication management is to get the desired outcome for the patient. In this scenario, it is desirable to use a medicine dispenser, as it reduces the burden on health workers and provides a contactless service, so that virus transmission due to physical contact can be avoided. Medicine dispensers use infrared sensors. They remind patients about their medicines using a buzzer and indicate when the medicine box is empty. Such systems is designed to use the IoT for communications between human and machine. This service can contribute to improving patient use of all their medications, improving the numbers of patients meeting their healthcare goals, reducing the side effects of drugs, etc. The IoT may provide the best approach for medication management [39–41]. A smart medicine box is a box which is one of the best examples of medication management using the IoT. Elderly people often forget to take the right medicine at the right time. A smart medicine box gives an indication via an alarm when it is time to take a medicine, and the medicine is dispensed by the box at the right time. This system may be remotely tracked by medical personnel.
- **Single condition:** this category includes those applications in which only one parameter of the human body is measured using a specific sensor device. The parameter chosen may be body temperature, heartbeat rate, blood pressure, glucose level [22], oxygen level, etc. Those parameters which are often monitored to track the health of a person are listed below.
  - **Body temperature monitoring:** body temperature is one of the most common physiological parameters; it is measured by temperature

sensors. The body temperature means the temperature of the skin, which may give the indication of health conditions. Changes in temperature may result from the symptoms of medical illness or stress [42–44], which might lead to abnormal conditions such as stroke, heart attacks, etc. Measuring the of temperature of the human body is necessary to determine the physiological condition of a human being. It may also be useful for other analyses, for example, harvesting energy from body heat, activity classification, etc.

- **Heartbeat rate monitoring:** another common physiological parameter in human body is the heart rate of a patient who is being medically monitored. The heartbeat rate is a precisely regulated variable. It plays an important role in health as well as in the identification of disease. The heartbeat rate can be measured using various methods such as photoplethysmography (PPG), sound, changes in the brightness of a person's face, and so on.
- **ECG monitoring:** wearable ECG sensors are used for the short-term assessment of cardiovascular diseases when the patient has chronic heart problems. They provide very useful information about the rate and regularity of heartbeats, which are used in the diagnosis of cardiac diseases.
- **Glucose level monitoring:** Glucose monitoring is an essential requirement for diabetic patients. Traditional methods are difficult and inconvenient, because they are manually operated and have to be used at an exact time. The results may fluctuate if precautions are not taken. To avoid this problem, IoT-based devices can be used to provide automatic and continuous monitoring of the glucose in a patient's body. These devices do not keep a record of the glucose level in the body but rather sound an alert signal if the glucose level goes outside the threshold levels [22].

Many other monitoring systems which use cognitive sensors and indicate the health condition of the patient are available, for example, oxygen level monitoring, respiration rate monitoring, etc.

7.6 Technologies used in IoT-based healthcare systems

- **Cloud computing:** cloud technology is used to store huge data sets; such data, which is available via networks, must only be accessed by authorized persons. This data could provide knowledge about various analyses of a patient's body, the patterns of various diseases, and other related information. All this valuable information is retained in the cloud. In recent years, much research has been done in the field of smart healthcare systems, which has concluded that cloud technology is a boon to IoMT systems as they require large data sets which can only be stored using cloud-based storage.

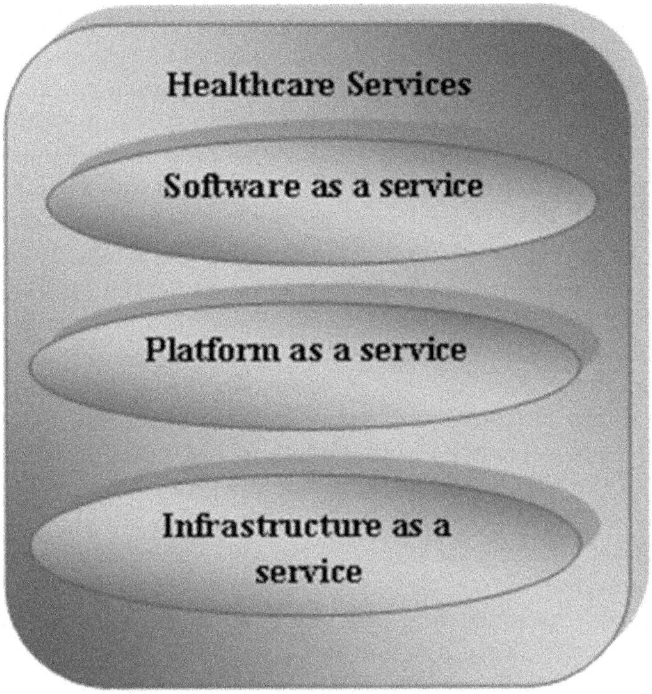

Figure 7.5. Major services of cloud computing.

Cloud storage technology provides three major services to smart healthcare IoMT systems, viz. software as a service (SaaS), platform as a service (PaaS), and infrastructure as a service (IaaS) (see figure 7.5).

- SaaS provides software to healthcare providers online, which enables them to use the software and perform relevant tasks.
- PaaS provides various tools for different services such as networking, virtualization, database management systems, etc.
- IaaS provides physical infrastructure, i.e. hardware such as storage, servers, etc.

The abovementioned three services, namely, Saas, PaaS, and IaaS are provided by the cloud module. The software applications are delivered by SaaS, which may only be accessed by authorized persons or the organization that provides the healthcare system [45]. Virtualization and database management tools are delivered by PaaS and infrastructure, i.e. hardware is provided by IaaS. The main objective of the healthcare system is to create and store health records. Such stored data can be used by health practitioners, medical students, and medical professionals for study purposes. Using this information, improvements in the health of patients can be analyzed and studied through visualization. The patient may generate his/her profile using some personal information. The patient is authorized to control who can access his/her profile and the uses that can be made of the medical data. From the above discussion, it is obvious that big data management systems and the cloud play pivotal roles in the field of IoMT-based smart healthcare systems.

- **AI and big data:** AI is one of the best tools for making healthcare systems smart by supporting preventive medicine and the discovery of new drugs. AI-based healthcare systems have the ability to learn, think, and make decisions or take actions. Patient health-related data and other information can be used by AI systems, which help doctors and medical personnel to provide accurate diagnosis of diseases and treatment plans to cure them. Using AI, healthcare systems become more predictive and accurate by analyzing big data related to patient health.
- **Networks:** the success of IoT systems depends upon various factors, but without a reliable connection between devices, sensors and IoT platforms cannot even be considered in the design of IoT-based healthcare systems. In the case of the IoT, the network is wireless. Four types of IoT are discussed below, as shown in figure 7.6.
    - **Cellular:** this network type is based on the networks used by smart-phones. IoT devices can use cellular networks for communication. This type of wireless network provides reliable and secure communication for cell phones as well as for the data used by IoT systems. However, cellular connectivity is weak and sometimes unavailable in some places, for example, in elevator shafts, utility closets, basements, etc. This network type is costly compared to the other networks.
    - **Local and personal area networks (LANs/PANs):** LANs and PANs are cost-effective networks but are sometimes unreliable. These networks use Wi-Fi and Bluetooth technologies for IoT connectivity.
    - **Low-power wide area networks (LPWANs):** IoT devices send small packets of information when they use LPWANs. This type of network has a greater range than those of LANs/PANs.
    - **Mesh network:** In this type of network, all the sensor nodes cooperate to exchange data and try to reach the gateway.

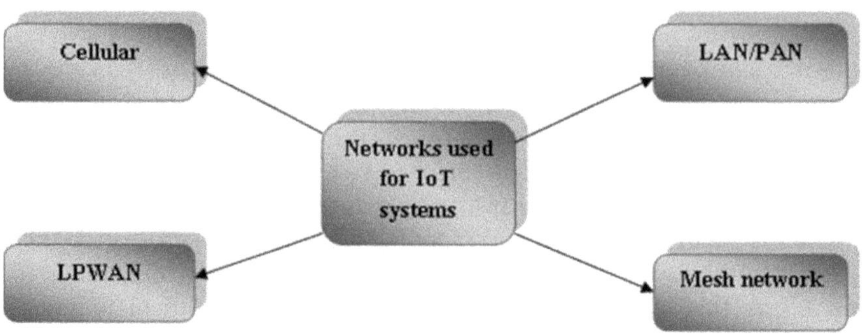

Figure 7.6. Types of network used in IoT systems.

- **Sensing technologies:** the IoMT is a group of various sensors and devices connected to one another which form a smart network. This smart system consists of different types of body sensor, surgical instruments, and, most importantly, the network which help to establish the connection between the patient and medical personnel. A brief discussion of various devices, viz. wearable devices, implantable devices, ambient devices, and stationary devices is given below.
- **Wearable devices:** These are basic components of the healthcare system which measure various body parameters, such as the oxygen level, the heartbeat rate, the blood pressure, the blood glucose level, the respiration rate, and other body-related signals [45]. These parameters are measured by various sensors and sent to an AI- and machine learning (ML)-based system, which processes these signals, combines them, and reaches useful conclusions about the health of the patient. These devices are either positioned over the clothes or in direct contact with the body. Wearable devices provide continuous and accurate data for real-time applications. Respiratory sensors are used to measure the respiration pattern of the patient, location sensors monitor the position of patient, monitoring systems check the heartbeat with the help of an electrocardiogram (ECG), thermometer sensors measure body temperature, blood pressure sensors check the pressure of blood in the body, glucose sensors measure the blood glucose level, acceleration sensors record patient recovery, etc [22]. In addition to these sensors, which are used to get information about the patient, other sensors are also used. Biometric sensors are used to provide authenticated access. Biometric information includes fingerprints, palm prints, retinal scanning, etc. Trackers are used for fitness activities.
- **Ambient intelligence**: this is used to detect the ambient environment of the patient, such as access to the toilet, quality of sleep, and behavioral changes. If any suspicious activity is detected, a warning is to be sent to the medical personnel or the caregiver. These devices make the environment smarter and healthier for patients. Some examples of ambient sensors are temperature sensors used to sense the room temperature, motion sensors that sense the motion of the patient, vibration sensors that sense the body vibrations of patient, daylight sensors that sense ambient light, etc.
- **Implantable devices:** these devices are implanted inside the body of the patient; e.g. embedded cardiac monitors gather information and transmit it via a radio channel to a neighboring network. A cardioverter–defibrillator is placed beneath the skin to keep track of the patient's heart rhythm.
- **Stationary devices:** imaging devices fall into this category as these devices are non-portable. Examples of these devices include computed tomography scanner, ultrasound scanners, magnetic resonance imaging (MRI) scanners, and x-ray machines. These imaging systems are used to obtain medical images of different body parts. These images are used by the radiologist or doctors to examine the functioning of body parts or any abnormality in internal body parts.

7.7 Challenges in IoT-based healthcare systems

Since IoT components transmit and receive data wirelessly, there is a high degree of risk to privacy and security in healthcare systems built using such components [50]. An intruder may either steal useful information or corrupt the network, which may hamper the whole healthcare system. Hence, it is essential to provide high security in the wireless sensor network (WSN). IoMT systems are connected to the Internet; therefore, while it may be easy to access the system, this also increases the risk of attacks by unauthorized persons. In terms of the authentication used, it is necessary it provide full protection to the system, thereby ensuring the security and confidentiality of the smart healthcare system [39, 41]. Various concerns relate to IoMT systems, viz. accuracy concerns, privacy concerns, security concerns, and trust concerns.

- **Resource limitations:** also known as capacity limitations; this problem is associated with the limits of the required medical healthcare resources. This condition may become worse during a pandemic, for example, during the peak of COVID-19 cases [46–49], there was a shortage of almost all medical materials, including drugs, vaccinations, beds, oxygen cylinders, etc. Even if these resources were available, in some cases they did not reach the places where they were urgently required. This problem can be overcome and the healthcare system can become a smart system by using the IoT. Since all the medical healthcare information is available in the network, any authorized person can access the information and ask for the medical materials required.
- **Accuracy concerns:** these are the most important concerns, as malfunctions may lead to partial or permanent injuries to the patients, or adverse effects on their health. Accuracy problems may take the form of a false disease diagnosis, a wrong medical prescription, etc.
- **Privacy concerns:** any intruder may attack the system and gather or disclose information about patients which may be related to their identity or any other sensitive or confidential information. In some cases, the patient or his/her family members do not want to disclose the medical conditions or history, as this may be harmful for them socially, economically, or emotionally. Privacy issues may be a massive threat to the patient, because an intruder is able to identify the patient and his/her health conditions. Hence, it is desirable that privacy should be one of the primary concerns in IoMT systems. To address privacy concerns, it is necessary to maintain the secrecy of the sensitive information, which must not be disclosed to unauthorized persons.
- **Security concerns:** since IoMT devices use open wireless networks for communication, they are prone to attacks; therefore, IoMT networks must be secured in order to avoid any attack. Any attacker may intercept the

incoming or outgoing information [51]. If the security measures are insufficient, any intruder can enter the network system and access personal and confidential information about patients, which may further be converted into threats, ransom demands, or blackmail. This activity may destroy user trust in IoMT systems. Hence, security is a big concern in all IoMT network systems.

7.8 Security measures

This section discusses various challenges and issues in the field of IoMT. Privacy and security measures present the biggest challenges. Another challenge is the lack of the knowledge, awareness, and training required to make the IoMT popular in the present era. Many studies have investigated possible security solutions [50]. These solutions may be categorized into two major classes, namely, cryptographic and non-cryptographic solutions. As soon as the security level increases, the system performance deteriorates; thus there is a tradeoff between them. In the following, we discuss a security solution with five different layers which are used for the detection and prevention of attacks [51]. They also reduce the damage caused by attacks and preserve the privacy of the patient. IoMT devices constantly track the health of the patients in real time, while supporting physical mobility and flexibility. The available types of monitoring include the heartbeat rate, glucose level, fitness level, respiration rate, etc. These features provide the flexibility that the patient may stay at home and not in the hospital. However, IoMT devices are prone to cyber attacks because they do not have security features. This is the major issue and the biggest challenge in the field of securing smart healthcare systems that use IoMT devices. Figure 7.7 shows the various security measures used to provide high security for IoMT systems. They are broadly classified into two categories, viz. technical security measures and non-technical security measures.

In the following, we discuss both categories and explain all the measures contained within them.

7.8.1 Technical security measures

Technical security measures are used to ensure the end-to-end security of IoMT systems, which can be achieved using the various techniques discussed below.

- **Identification and verification:** in order to restrict unauthorized access to IoMT systems, it is necessary to provide a robust technique for identification, which verifies whether a person trying to access the system is authorized or not. Sometimes, unauthorized persons, known as intruders, may try to access the system, either to steal information or for other illegal reasons. One of the best technical solutions is to use a biometric system. Such a system needs various types of database for authorized personnel. These may be further divided into two groups, viz. physical biometrics and behavioral biometrics.

Figure 7.7. Security measures.

- **Physical biometrics:** this technique is based on the physical traits of individuals, which include fingerprint detection, facial recognition, retinal scanning, iris scanning, etc.
    - **Fingerprint detection:** this is the most common biometric identification in many applications. The fingerprint of an adult does not change as an individual ages.
    - **Facial recognition:** this is another preferred way to recognize authorized persons. Every individual has unique facial features which are different from those of other individuals.
    - **Retinal scanning:** this technique focuses on examining the area of blood vessels behind the human eyes. This is one of the best biometrics; it provides accurate verification and it is considered to be the most reliable test.
    - **Iris scanning:** this technique uses a high-contrast image of the human iris taken in the visible or near-infrared band. This technique gives precise and accurate measurements; therefore, it is considered to be one of the best biometrics for recognition and verification.

- **Behavioral biometrics:** behavioral biometrics include gait identification, speech, signature, keystroke analysis, cognitive biometrics, etc. The precision of this category can be determined using two parameters, viz. the false acceptance rate (FAR) and the false rejection rate (FRR).
- **Honeypots:** These devices are used to detect attackers, their targets and method or tool used to attack. They can be attached to the system for the security purpose.

7.8.2 Non-technical security measures

These include training given to staff members and information technology (IT) professionals who are involved with the healthcare system. The purpose of the training is to secure patients' personal information. The training can be accomplished in three ways, viz. raising awareness, technical training, and raising the educational level.

- **Awareness:** this is mandatory for all staff, especially the IT professionals involved with the healthcare system. They should be able to identify the occurrence of attacks in the IoMT network.
- **Training:** along with the awareness training, technical training must be given to all staff, because it helps them to deal with security issues and mitigate the adverse effects of attacks. Technical training may be given after the teaching session so that practical cases of any kind of attack may be handled. This is one of the best ways to provide security in IoMT-based smart healthcare systems. Technical training may include various tasks, viz. identification, confirmation, classification, reaction, containment, investigation, and enhancement. These tasks are discussed below:
    - **Identification:** it is important for IT professionals to identify suspicious behaviors among normal routines.
    - **Confirmation:** this is the ability to confirm the occurrence of an attack on the system.
    - **Classification:** this is the ability to identify or recognize the type of attack that has occurred in the IoMT system.
    - **Reaction:** this is the ability of the response team to react to the attack as soon as possible, so that any attack can be stopped or its effects minimized.
    - **Containment:** this task involves containing attacks and overcoming them.
    - **Investigation:** in this process, an investigation is performed using forensic evidence. The purpose of such an investigation is to identify the reason for an attack, its impact, and the damage it has done.
    - **Enhancement:** this includes the process of learning from the previous incidents so that similar attacks can be handled wisely in the future.
- **Raising the education level:** this is the most important process, which is a must for all the IT professionals involved with IoMT healthcare systems. IT professionals must have knowledge about various attacks, their effects, and preventative measures.

7.9 Experimental prototype of a smart healthcare system using the IoT

From the above discussion, it is clear that various types of application can be developed for smart healthcare systems using IoT-based cognitive sensors. To understand the application of the IoT in healthcare systems, a device has been developed, called an IoT-based medicine dispenser, which is an arrangement of five pill compartments used to hold medicines. Each compartment is arranged in such manner that it can help the patient to separate medicines for the morning, afternoon, and night. The use of IoT makes it contactless. The following is a list of possible use cases for the dispenser:

Case I. It helps patients to manage their medications and thereby reduces the risk of chronic disease.

Case II. It is automatic device which avoids the worries associated with mixed medicines.

Case III. It helps visually impaired people to take the right medicine at the right time.

Case IV. Alzheimer patients may take multiple does and forget which pill has already been taken.

For the above cases, a prototype has been developed which can be used to guide patients to take their medications, e.g. to guide them regarding the times at which to take the medicines, to take the right pill at the right time, and to avoid unwanted multiple doses. Figure 7.8 shows the block diagram of the pill dispenser system.

Figure 7.9 shows the prototype automatic pill dispenser. It works on the principle that when the whole circuit is 'ON,' power is provided to the system and the control element comes into play. It is connected to a Wi-Fi module which sends data to a phone. As soon as the circuit is completed, the servomotor adjusts itself to its initial position. The system generates health alerts for higher temperatures and sends notifications to the patient's carer via the cloud using the Blynk app. In the worst-case scenario, the patient can send an emergency message to their carer via the

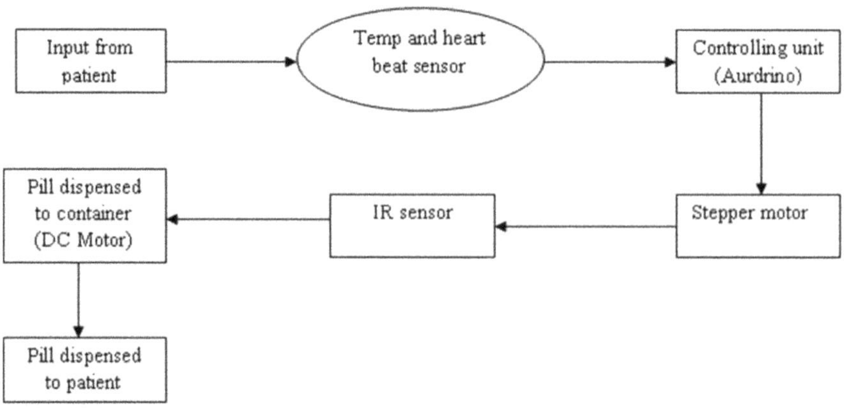

Figure 7.8. Block diagram of the pill dispenser system.

Figure 7.9. Prototype automatic pill dispenser.

keypad to avoid any mishaps. A worker will then place the medicine in the compartment and insert the details of the pills with the help of the keypad and the timer is set. When the allotted time comes, the system notifies the patient with the help of a buzzer. When the patient nears the tray to pick up their medicine, the IR sensor checks for the proximity of an object and the LM35 temperature sensor which is attached to the tray notes down the temperature and displays it on the liquid-crystal display (LCD). The allotted sequence of medicines is taken at the correct times and the worker is able to monitor the schedule.

7.10 Conclusions

This chapter depicts the use of cognitive sensors and the IoT for healthcare systems, which represents the beginning of a new era of smart and efficient systems in the field of healthcare. These systems use various types of modern sensor which produce huge amounts of data. In the healthcare system, data are various parameters which help medical personnel to diagnose and treat diseases. The most common data types are body temperature, heartbeat rate, respiration rate, oxygen level, glucose level, etc. These data may be collected by conventional methods, but such methods are manually operated. Cognitive sensors are used to collect these data automatically; such sensors may be either installed in clothes or implanted in the patient's body. These sensors produce a large amount of data; hence, cloud technologies or big data systems are used. These data are processed and integrated using AI- and ML-based modeling systems. These modeling systems use various algorithms which can suggest precautions that need to be taken or treatments to cure the disease. Various networks are used to access the data. These networks are secured so that the personal and confidential data of patients can not be accessed by any intruder. Hence, privacy and security are prime concerns of IoT-based smart healthcare systems. These systems produce highly precise and accurate information. This information can be analyzed by experts and treatment can be given to patients accordingly. However, there is still room to modify IoMT systems to make them more robust for healthcare purposes. From the above discussions, it can be seen that by applying various measures, systems can be made highly secure and privacy of the data can be maintained. Various research has been completed and is still continuing to popularize IoMT systems in society, so that such systems may be provided to the maximum number of patients. The prime concerns of future research may be to make such systems cost-effective, reliable, and highly secure, so that the IoMT can create a revolution in the healthcare field.

References

[1] Yang F *et al* 2021 Internet-of-things-enabled data fusion method for sleep healthcare applications *IEEE Internet of Things J.* **8** 15892–905

[2] Kumar M and Chand S 2020 A secure and efficient cloud-centric internet-of-medical-things-enabled smart healthcare system with public verifiability *IEEE Internet of Things J.* **7** 10650–9

[3] Kakhi K, Alizadehsani R, Khosravi A, Nahavandi A and Islam S M S 2021 *Challenges of Internet of Medical Things for Electronic Healthcare 2021 IEEE 4th Int. Conf. and Workshop Óbuda on Electrical and Power Engineering (CANDO-EPE)* pp 19–24

[4] Fuke R P and Mahajan R P 2022 *Pragmatic Analysis of IoMT Network Modelling Techniques from a Statistical Perspective 2022 Int. Conf. on Electronics and Renewable Systems (ICEARS)* pp 531–42

[5] Ghubaish S T, Zolanvari M, Unal D, Ali A and Jain R 2021 Recent advances in the Internet of Medical Things (IoMT) systems security *IEEE Internet of Things J.* **8** 8707–18

[6] Almogren M I, DinI U, Almajed H and Guizani N 2021 FTM-IoMT fuzzy-based trust management for preventing sybil attacks in Internet of Medical Things *IEEE Internet of Things J.* **8** 4485–97

[7] Rizk D, Rizk R and Hsu S 2019 *Applied Layered-Security Model to IoMT 2019 IEEE International Conference on Intelligence and Security Informatics (ISI)* p 227

[8] Yuldashev Z, Sergeev A and Nastueva N 2021 *IoMT Technology as the Basis of Wearable Online Monitors for Space Distributed Monitoring Systems for Pregnant Women 2021 Wave Electronics and its Application in Information and Telecommunication Systems (WECONF)* pp 1–4

[9] Vishnu S, Ramson S R J and Jegan R 2020 *Internet of Medical Things (IoMT) - An overview 2020 5th Int. Conf. on Devices, Circuits and Systems (ICDCS)* pp 101–04

[10] Yanambaka V, Mohanty S, Kougianos E, Puthal D and Rachakonda L 2019 *PMsec: PUF-Based Energy-Efficient Authentication of Devices in the Internet of Medical Things (IoMT) 2019 IEEE Int. Symp. on Smart Electronic Systems (iSES) (Formerly iNiS)* pp 320–1

[11] Ahmed J, Nguyen T N, Ali B, Javed A and Mirza J 2022 On the physical layer security of federated learning based IoMT networks *IEEE J. Biomedical Health Informatics* https://doi.org/10.1109/JBHI.2022.3173947
[12] Dutta P E, Neog H and Medhi N 2021 Health monitoring in Internet of Medical Things (IoMT) using machine learning (ML) approaches *IEEE Globecom Workshops (GC Wkshps)* pp 1–6
[13] Singh P, Devi K J, Thakkar H K and Kotecha K 2022 Region-based hybrid medical image watermarking scheme for robust and secured transmission in IoMT *IEEE Access* **10** 8974–93
[14] Sun Y, LoF P W and Lo B 2019 Security and privacy for the Internet of Medical Things enabled healthcare systems: a survey *IEEE Access* **7** 183339–55
[15] Limaye A and Adegbija T 2018 HERMIT: a benchmark suite for the Internet of Medical Things, *IEEE Internet of Things J.* **5** 4212–222
[16] Meng W, Cai Y, Yang L T and Chiu W Y 2021 Hybrid emotion-aware monitoring system based on brainwaves for Internet of Medical Things *IEEE Internet of Things J.* **8** 16014–6022
[17] Zhang T *et al* 2020 A joint deep learning and Internet of Medical Things driven framework for elderly patients *IEEE Access* **8** 75822–5832
[18] Wazid M, Singh J, Das A K, Shetty S, Khan M K and Rodrigues J J P C 2022 ASCP-IoMT: AI-enabled lightweight secure communication protocol for Internet of Medical Things *IEEE Access* **10** 57990–8004
[19] Boutros-Saikali N, Saikali K and Naoum R A 2018 *An IoMT platform to simplify the development of healthcare monitoring applications 2018 Third Int. Conf. on Electrical and Biomedical Engineering, Clean Energy and Green Computing (EBECEGC)* pp 6–11
[20] Kaushal C, Md Islam K, Singla A and Amin M A 2022 An IoMT-based Smart Remote Monitoring System for Healthcare *IoT-Enabled Smart Healthcare Systems, Services and Applications* ed S Rani *et al* (New York: Wiley) pp 177–98
[21] Sun L, Jiang X, Ren H and Guo Y 2020 Edge-cloud computing and artificial intelligence in Internet of Medical Things: architecture, technology and application *IEEE Access* **8** 101079–01092
[22] Joshi A M, Jain P and Mohanty S P 2020 *iGLU: Non-Invasive Device for Continuous Glucose Measurement with IoMT Framework 2020 IEEE Computer Society Annual Symp. on VLSI (ISVLSI)* (Piscataway, NJ: IEEE) pp 598–99
[23] Wei K, Zhang L, GuoY and Jiang X 2020 Health monitoring based on Internet of Medical Things: architecture, enabling technologies, and applications *IEEE Access* **8** 27468–7478
[24] Khan M A and Algarni F 2020 A healthcare monitoring system for the diagnosis of heart disease in the IoMT cloud environment using MSSO-ANFIS *IEEE Access* **8** 122259–69
[25] Siddiqui S Y *et al* 2021 IoMT cloud-based intelligent prediction of breast cancer stages empowered with deep learningin *IEEE Access* **9** 146478–46491
[26] Ahmed I, Ahmad A and Jeon G 2021 An IoT-based deep learning framework for early assessment of Covid-19 *IEEE Internet of Things J.* **8** 15855–62
[27] Casino F, Patsakis C, Batista E, Postolache O, Martínez-Ballesté A and Solanas A 2018 *Smart Healthcare in the IoT Era: A Context-Aware Recommendation Example 2018 International Symposium in Sensing and Instrumentation in IoT Era (ISSI)* pp 1–4
[28] Wazid M, Das A K, Rodrigues J J P C, Shetty S and Park Y 2019 IoMT malware detection approaches: analysis and research challengesin *IEEE Access* **7** 182459–82476

[29] Hatzivasilis G, Soultatos O, Loannidis S, Verikoukis C, Demetriou G and Tsatsoulis C 2019 Review of security and privacy for the Internet of Medical Things (IoMT) *2019 15th Int. Conf. on Distributed Computing in Sensor Systems (DCOSS)* pp 457–64
[30] Joshi S and Joshi S 2019 *A Sensor based Secured Health Monitoring and Alert Technique using IoMT 2019 2nd International Conference on Intelligent Communication and Computational Techniques (ICCT)* (Piscataway, NJ: IEEE) pp 152–6
[31] Scrugli M A, Loi D, Raffo L and Meloni P 2022 An adaptive cognitive sensor node for ECG monitoring in the Internet of Medical Things *IEEE Access* **10** 1688–705
[32] Amin S U, Hossain M S, Muhammad G, Alhussein M and Rahman M A 2019 Cognitive smart healthcare for pathology detection and monitoring *IEEE Access* **7** 10745–0753
[33] Liu J, Miao F, Yin L, Pang Z and Li Y 2021 A noncontact ballistocardiography-based IoMT system for cardiopulmonary health monitoring of discharged COVID-19 patientsin *IEEE Internet of Things J.* **8** 15807–5817
[34] Zhang T, Liu M, Yuan T and Al-Nabhan N 2021 Emotion-aware and intelligent Internet of Medical Things toward emotion recognition during COVID-19 pandemic *IEEE Internet of Things J.* **8** 16002–6013
[35] Rahman M A and Hossain M S 2021 An internet-of-medical-things-enabled edge computing framework for tackling COVID-19 *IEEE Internet of Things J.* **8** 15847–5854
[36] Trigo J D *et al* 2020 Patient tracking in a multi-building, tunnel-connected hospital complex *IEEE Sensors J.* **20** 14453–4464
[37] Alshehri F and Muhammad G 2021 A comprehensive survey of the Internet of Things (IoT) and AI-Based smart healthcarein *IEEE Access* **9** 3660–678
[38] Sharma H K, Kumar A, Pant S and Ram M 2022 *Artificial Intelligence, Blockchain and IoT for Smart Healthcare* (Gistrup: River Publishers)
[39] Jing Q, Vasilakos A V and Wan J *et al* 2014 Security of the Internet of Things perspectives and challenges *Wirel. Netw.* **20** 2481–501
[40] Al-Fuqaha A, Guizani M and Mohammadi M 2015 Internet of Things a survey on enabling technologies protocols and applications *IEEE Commun Surv Tutor* **17** 2347–76
[41] Alsubaei F, Abuhussein A and Shiva S 2017 *Security and Privacy in the Internet of Medical Things: Taxonomy and Risk Assessment 2017 IEEE 42nd Conf. on Local Computer Networks Workshops (LCN Workshops)* (Piscataway, NJ: IEEE) pp 112–20
[42] Rachakonda L, Mohanty S P and Kougianos E 2020 *Stress-Lysis: An IoMT-Enabled Device for Automatic Stress Level Detection from Physical Activities 2020 IEEE International Symposium on Smart Electronic Systems (iSES) (Formerly iNiS)* (Piscataway, NJ: IEEE) pp 204–05
[43] Rachakonda L, Bapatla A K, Mohanty S P and Kougianos E 2021 SaYoPillow: blockchain-integrated privacy-assured IoMT framework for stress management considering sleeping habits *IEEE Trans. Consum. Electron.* **67** 20–9
[44] Rachakonda L, Mohanty S P, Kougianos E and Sundaravadivel P 2019 Stress-lysis: a DNN-integrated edge device for stress level detection in the IoMT *IEEE Trans. Consum. Electron.* **65** 474–83
[45] Jaleel A, Mahmood T, Hassan M A, Bano G and Khurshid S K 2020 Towards medical data interoperability through collaboration of healthcare devices *IEEE Access* **8** 132302–32319
[46] Parah X S A *et al* 2021 Efficient security and authentication for edge-based internet of medical things *IEEE Internet of Things J.* **8** 15652–62

[47] Rizvi T, Bhonsle D, Rahangdale R and Bagga J 2022 A survey on behavioral change during the COVID-19 outbreak in India *Computational Intelligence and Applications for Pandemics and Healthcare* ed S S Kshatri *et al* (Hershey, PA: IGI Global) pp 184–203
[48] Rizvi T, Bhonsle D and Uzma R 2022 Analysis and comparison of psychological constraints among various countries during COVID-19 *Computational Intelligence and Applications for Pandemics and Healthcare* ed S S Kshatri *et al* (Hershey, PA: IGI Global) pp 248–68
[49] Tai Y, Gao B, Li Q, Yu Z, Zhu C and Chang V 2021 Trustworthy and intelligent COVID-19 diagnostic IoMT through XR and deep-learning-based clinic data access *IEEE Internet of Things J.* **8** 15965–76
[50] Ghubaish X A, Salman T, Zolanvari M, Unal D, Al-Ali A and Jain R 2021 Recent advances in the Internet-of-Medical-Things (IoMT) systems security *IEEE Internet of Things J.* **8** 8707–18
[51] Ud Din I, Almogren A, Guizani M and Zuair M 2019 A decade of Internet of Things: analysis in the light of healthcare applications *IEEE Access* **7** 89967–79

Chapter 8

Redundancy issues in wireless sensor networks

Veena I Puranikmath, Sridhar Iyer and Rahul Pandya

Wireless sensor networks (WSNs) face multiple issues related to energy consumption and lifetime. Redundancy is a key factor that affects the efficiency of a WSN. Due to redundancy, data is collected at the cluster head (CH), creating multiple copies, which results in increased consumption of energy by the WSN. The elimination of redundant data occurs only at the sensor nodes or at the CH. In this chapter, we detail various types of redundancy and present a survey of the most recent articles which have aimed to minimize the effect of redundancy on WSN performance. For the articles considered in the survey, we provide a summary, the method or algorithm implemented, its limitations, and its future scope. We also provide directions for future research on the topic of redundancy within WSNs. Our survey reveals that timely solutions are required to (i) eliminate redundant data without affecting the networks' time period, (ii) implement a mobile CH that uses redundancy elimination method to save energy, (iii) allocate specified time durations for appropriate communication between the sensor nodes and the CH in order to save power, and (iii) formulate new similarity check algorithms to help the CH to avoid storing the same data that exists at the nodes.

8.1 Introduction

Over the past decade, WSNs have been deployed for different applications including the military, medicine, etc [1]. The main aim of WSNs is to gain control over the large areas in which they are constructed from small sensor nodes, which help collect information from a defined region [1]. Each sensor node collects and transmits collected data and aggregates it using appropriate protocols. In WSNs, the data is gathered via various sensor nodes deployed in a specified area, and then sent to the base station (BS). The sensed data of these nodes is aggregated at the CH and then transmitted to a sink node (see figure 8.1). It can be observed from figure 8.1 that the data is collected from sensor nodes within a defined cluster, and then, before the data is transmitted to the BS, it is gathered at the CH. A node is formed by combining

doi:10.1088/978-0-7503-5326-7ch8

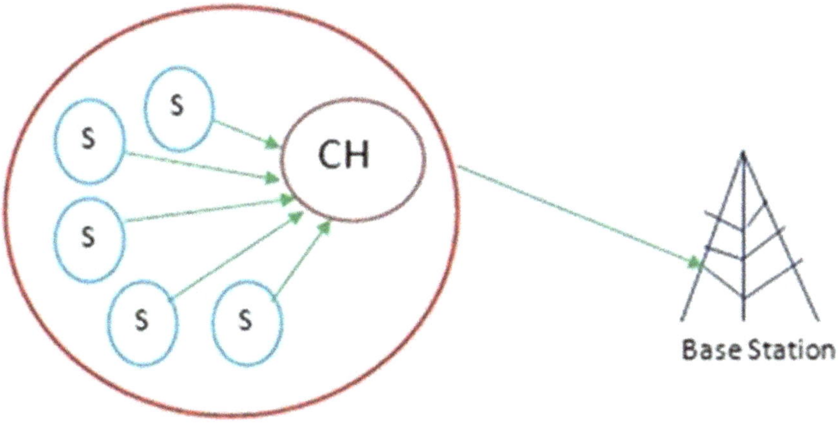

Figure 8.1. Architecture of a WSN.

various subunits, which all contribute to the operation of the sensor node. Various units are used for sensing, recording, monitoring, and analyzing the information which is gathered from the physical conditions. Sensor nodes are not only responsible for sensing the data, but they also need to perform operations such as processing the sensed data, communicating with remote servers and storage units, etc. Other than these characteristics, the sensor nodes are also responsible for the collection of information and for correlating and fusing the data gathered by different sensors with the available data and the analysis performed by the network. The selection of the CH depends on the node's energy, and this process varies with the type of aggregation. The lifetime of a network is reduced if its energy consumption increases. Also, transmitting information from the sink node to the base node reduces the energy of the node; in turn, the network energy reduces [2]. Hence, to develop an efficient WSN, multiple parameters including energy, reliability, scalability, quality of service, etc. need to be considered.

Sensor units are divided into four key units, viz. the sensing unit, the processing unit, the communication unit, and the power unit. All these units are responsible for the proper operation of the sensor nodes. The first unit is the sensing unit, which includes a sensor and an analog-to-digital converter (ADC). Analog data collected from environment is converted into digital form with the help of an ADC. The second unit is the processing unit, which performs data processing and manipulation of the information; the processing unit is usually a microcontroller or a microprocessor. The communication unit is responsible for smooth communication between the transmitter and the receiver. It consists of a radio system and an antenna; radio transceivers are used for the transmission and reception of data, and the antenna aids in the transmission and reception of the communication signals.

The data collected from the different environment by the sensor nodes is transmitted to the sensor network, which includes the BS and the router. The BS acts as a bridge between the sensor nodes and the users. Data gathered via different nodes is collected at the central station or the BS, which serves as a channel to other networks reached via the Internet. The data collected at the BS undergoes some processing, and this processed information is sent to the end user via the Internet. These BSs are at central locations; in other words, we may also call them servers. Another special component of the WSNs is the router. WSN routers are designed to be capable of computing, calculating, and distributing the routing table within the network. Multiple routing techniques for finding the most effective routing path between a source node and a BS have been proposed and are being researched. The main aspects to be considered when designing a routing protocol are power, the availability of resources, the channel's time variation, the probability of packet loss, and the related latency [2] (figure 8.2).

To monitor a sensor's behavior within the node and to communicate with the BS, networking is required. The role of a network architecture is to monitor the area where multiple sensor nodes have been deployed for a particular application purpose. Communication has to take place between the sensor nodes as well as between the BSs; for this communication, the sensor network is used in the wireless mode, and hence, the name given to it is wireless sensor network—WSN. The sensor node operates under the control of the BS at the same time as the other network sensor nodes.

The two main types of WSN architecture are single-hop (SH) and multihop (MH). In the SH network architecture, all the sensor nodes are directly connected to the BS, and it also permits transmission over a longer distance; this results in high energy consumption by the WSN. This affects the performance of data aggregation and data prediction. Hence, to overcome the drawback of this architecture, the MH network architecture is adopted, in which the presence of intermediate nodes reduces the traffic between the sensor nodes and the BS. It is usually implemented in flat and hierarchical network architectures.

The key issues that occur in WSNs include power consumption during communication and data redundancy. Redundancy occurs in WSNs when the same data

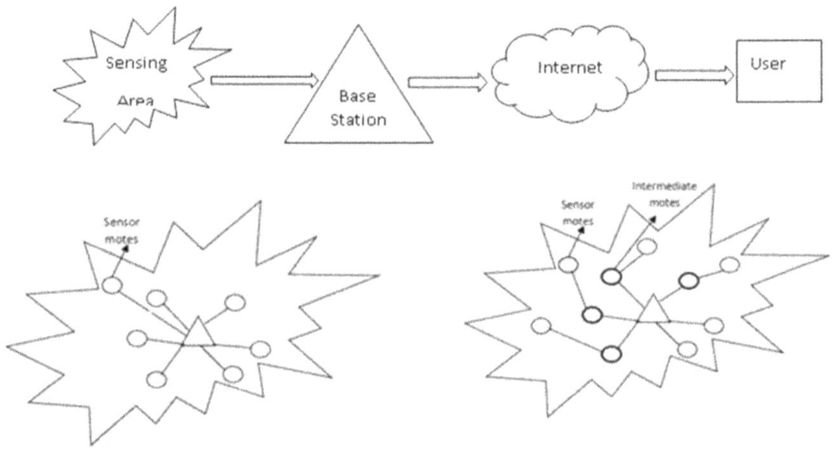

Figure 8.2. Network architecture of a WSN.

exists at multiple nodes, leading to the inconsistencies in the data [3, 4]. To eliminate data duplication at the BS, data originating from the sensor nodes needs to be monitored and filtered before it is transmitted to the BS. Non-redundant data ensures the reliability of the data and reduces the energy consumed by the sensor nodes, in turn increasing the lifetime of the network [5].

Redundancy is a Latin verb stemming from '*redundare*' which means to overflow, and it may be used to describe a positive effect such as an overflow of money or a negative effect such as ballast. In turn, this ambiguity generates an interest in the domain of redundancy, since it has a high intrinsic degree of uncertainty. In WSNs, this domain is represented by the WSN applications. Furthermore, the redundancy level required is a key factor which influences most of the techniques used to design WSNs. Specifically, it is not beneficial to have low redundancy rate, as this makes WSNs prone to extreme errors; on the other hand, increasing the redundancy results in the WSN becoming more fault tolerant.

Hence, redundancy in WSNs can be an ally or an enemy, which challenges researchers to enhance the favorable aspects and minimize the unfavorable aspects. Over the past ten years, a key research area has been that of exploiting redundancy so as to enhance the accuracy of the data, the reliability of the sensing, the lifetime of the system, and the security of the WSN, regardless of whether this causes an increase in the network cost, system complexity, and evaluation times. For instance, if we consider the sensor to be the individual unit of communication and sensing, even though the removal of redundancy enables immense energy savings, combining sensors is expected to provide further energy savings. A probable combination could include a few redundant nodes within the sensing layers which are chosen initially, following which, redundancy within the communication layers is removed.

Figure 8.3. shows a general redundancy diagram, from which it can be observed that the data is collected at the node from different sensor nodes deployed at various places. The data collected at the node is then verified for information duplication and redundant data is eliminated using various methods. The base station receives the non-redundant data, and in this manner, network energy is saved, simultaneously increasing the network's lifetime. Data consistency, accuracy, and reliability in WSNs are ensured by using one of the existing types of redundancy, as illustrated in figure 8.4.

Redundancy is the condition of replicated information, which provides similar results. It is hard to find a general definition for redundancy in wireless sensor networks, as it needs to take account of issues such as sensor coverage, communication, data collection measurement and many more. It must also be mentioned that redundancy has not been defined in a formal manner in all respects; specifically, in the WSN domain. Redundancy can be also used to improve the measurement consistency within a WSN by either (i) collecting reports over long periods of time

Figure 8.3. System model for redundancy elimination.

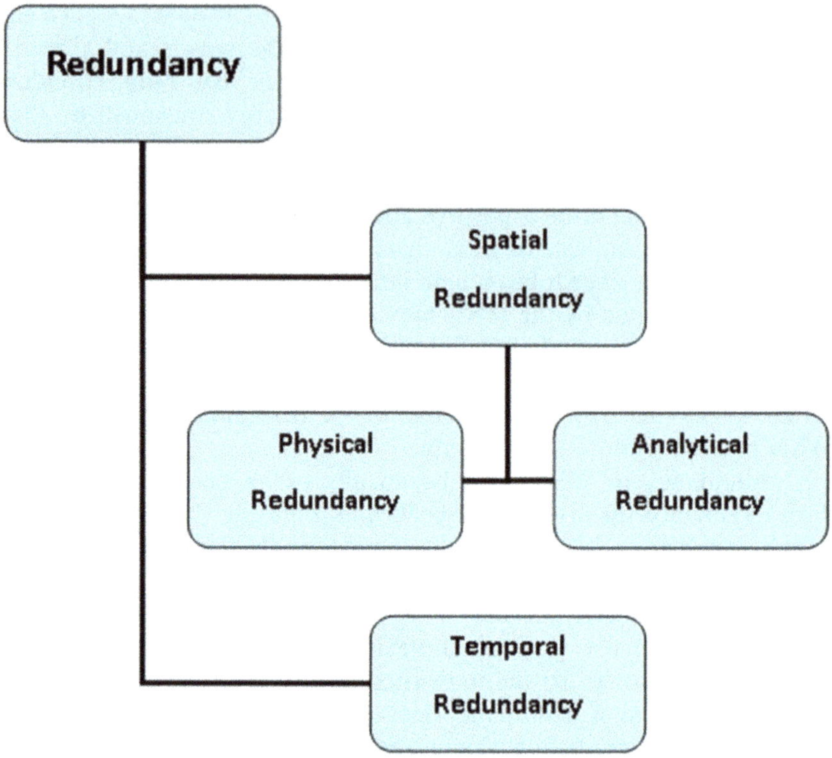

Figure 8.4. Types of redundancy.

via different neighboring sensor nodes, or (ii) obtaining multiple reports over a period via the same sensor node, or (iii) the inclusion of data checking bits within the messages transmitted through the WSN.

Figure 8.4 shows the techniques or types of redundancy method used in WSNs. Using the methods listed above improves the redundancy, either by

1. Collecting the data from more than a single sensor node or using spatial redundancy, which relates to sensing the information and is rarely implemented for communication.
2. Collecting multiple data points from similar sensor nodes or implementing temporal redundancy, which relate to both sensing and communication.
3. Including message checking bits within the network data or implementing information redundancy, which relate to (a) the replication of the data structure which is exchanged within the network, and (b) the amount of measurement storage which exists at the BS.

8.1.1 Spatial redundancy

Spatial redundancy provides the possibility of accessing the data for a specific place via various sources. It depends on the geographical allocation of sensor nodes and includes the duplication of information or data within the coverage area of the network. It is a proven fact that wireless sensor networks have more spatial redundancy, which means that the data for a particular location may be provided by multiple sensors. Spatial redundancy is used to create fault tolerance, to enhance the reliability of the collected data, and to secure the data. It increases the probability of occurrence of redundant resources like sensor nodes, connections used for communication, or models which are mathematically oriented, and it allows varied data to be monitored using the replicated data. Two possible types of spatial redundancy exist, depending on the validation steps of the measurements which occur during experiments, which are provisioned via a set redundancy, i.e. redundancy of the physical or the analytical type.

Physical redundancy: this is a WSN feature which measures a variable in a specified location using multiple sensors. Physical redundancy is also called direct redundancy or hardware redundancy, and it is a common method for ensuring a system's reliability. It comprises independent nodes deployed over a wide range to cover a particular area. In this redundancy method, the aggregation process uses supplementary sensors and selects data which appears similar for the majority of nodes. This redundancy method is used for the deployment method or to cover the area when some sensors are faulty. It secures the network against hardware/software node failure during attacks. The disadvantage of physical redundancy is that if source nodes are deployed near each other, redundant data generates voluminous traffic over the channel, which is wasteful of bandwidth and causes high energy consumption in the network.

Analytical redundancy: this is a WSN attribute which has the ability to estimate an individual variable at a particular place using mathematical models; it uses the real-time sensing data which is provisioned via the closest sensor nodes. This provides a way to compare the real sensor data value with an estimated sensor data value, helps to improve operation, and detects any sensors which are malicious in nature. Analytical redundancy is an approach which depends on a mathematically driven model; such a model helps to predict the sensor data value by considering the previous and the current data values of the closest sensor nodes.

8.1.2 Temporal redundancy

Temporal redundancy (or time redundancy) performs an individual action multiple times using a skewed time pattern; this is followed by a check of the results with the aim of enhancing the reliability of the system. This approach can be sensing related, communication related, or may include a relation which combines both aspects.

Temporal sensing redundancy: this type of redundancy is used to obtain multiple items of information from the same sensor node within a specified time interval. This type of redundancy is mostly implemented for video surveillance, and supports different codecs via special methods related to data compression.

Temporal communication redundancy: this type of redundancy is used for transmitting similar data sets multiple times over a skewed timeline. The temporal redundancy mainly occurs in the network communication by retry and failover methods. Methods such as retries and timeouts are utilized by various communication protocols.

8.1.3 Information redundancy

Information redundancy uses data which is redundant in the form of excessive bits; this allows lost data to be reconstructed. In this type of redundancy, excess data is added to detect and correct any faults which may appear. The bits, which are used for parity checking, are included for error detection, and this forms one of the subparts of this type of redundancy. This redundancy type is also used to recover an original message without retransmitting the data.

To solve the problem of redundancy, several methods have been implemented by various studies described in the research literature. In this chapter, we survey different methods and algorithms for minimizing redundancy within WSNs described in the literature. For the articles considered in the survey, we provide a summary, the method or algorithm implemented, its limitations, and its future scope. We also list the open areas of research related to the redundancy issue in WSNs, which require timely solutions.

Our survey reveals that timely solutions are required to (i) eliminate redundant data without affecting the networks' time, (ii) implement a mobile CH that includes a redundancy elimination method to save energy, (iii) allocate specified time durations for communication between sensor nodes and the cluster head to save power, and (iv) formulate new similarity check algorithms to help the CH to avoid duplicating the data that exists at the nodes.

The rest of this chapter is arranged as follows: in section 8.2, we present a detailed literature review of studies focused on the redundancy issues in WSNs. Section 8.3 presents a proposed algorithm which is intended to save network energy and increase the network lifetime. In section 8.4, we present the related quality factors. Section 8.5 details the various research avenues. Finally, section 8.6 offers the conclusions.

8.2 Related work

To minimize the duplication of data in WSNs, the zoom-in zoom-out (ZIZO) mechanism has been implemented, in which the network is divided into two levels [1]. In the first level, index bit encoding (IBE) is used to aggregate similar kinds of data, and data is subsequently removed at the CH level using the sampling rate adjustment process. To transmit the processed data and minimize the information capacity consumed, the iterative identification method was used in the study reported in [2]. In this method, a circular topology is used to generate a link between the CH and the nodes. To minimize the distance between the nodes and the power consumption, two adjoining relay nodes are connected.

In addition to ensuring the elimination of duplicate data, it is equally important to maintain data privacy; this was addressed in [3] via the collection tree protocol (CTP). In the CTP, the routing tree and path for data selection are selected, and if any data is missed, it is restored during the transmission. The gray wolf optimization

(GWO) and distributed particle swarm optimization (D-PSO) optimization methods were reported in [4], which assign the CH to time-division multiple access (TDMA) slots for the sensor nodes so that no data duplication occurs at the CH.

For data aggregation from sensor nodes, the spatial correlation-based data redundancy elimination (SCDRE) protocol for data aggregation protocol was described in [5], which works at the source and aggregator levels. At the source level, this protocol collects data and checks for similarity, whereas, at the aggregator level, the protocol eliminates similar data using a correlation coefficient and aggregates independent data for further processing. To ensure the accuracy of the aggregated data, exponential moving average and threshold-based mechanisms were used at the node level in [6], which improved accuracy and forwarded data to the base station; the Euclidean distance function was applied at the CH level, where it measured the distance between the node and CH.

The enhanced low energy adaptive clustering hierarchy (E-LEACH) and an energy hole mitigation scheme based on WEdge MERging (WEMER) protocol were implemented in the study described in [7] to (i) create a cluster in the network and (ii) a CH in the cluster, and (iii) to select the CH. The local tree reconstruction method aggregated and removed the redundant data in [8]. This method is divided into two steps; in the first step, the algorithm divides the nodes into several layers, following which, energy levels are set for the layers in the second step.

To secure the data, the asymmetric key encryption method was used in [9]; this method generates and manages keys using a media access control (MAC) generation algorithm, simultaneously maintaining end-to-end encryption. The fuzzy-attribute-based joint integrated scheduling and tree formation (FAJIT) method was adopted in [10] to remove redundancy from the network. In this method, the parent node is selected if the nodes are equal in number. Further, the selected parent node is responsible for the deduplication of the collected data. In the hierarchical method used in [11], an intermediate node is set between the CH head and nodes. This intermediate node then removes duplicated data by comparing it with the previously collected data. Once the data is independent, it is then sent to the CH.

In [12], the fuzzy C-means algorithm was implemented for the clustering process. To aggregate the data and remove redundancy or errors from the message, this algorithm uses the effective support degree function (ESDF) method. Checking for repeated data in the network cluster is an issue in WSNs. To resolve this issue, the study described in [13] implemented the data redundancy-controlled energy-efficient multihop (DREEM) method via statistical tests, including a degree of confidence algorithm, so that redundant data was identified, and the CH was segregated to remove data before it was forwarded to the base station. To overcome the difficulty of selecting a CH, the studies described in [14, 15] implemented the spatiotemporal multicast dynamic CH role rotation method, which selects the CH based on the strength of the node. Further, the proposed method used a data mining approach to collect data from the sensor nodes, which was then verified. Subsequently, duplicate data was removed, and the relevant information was made consistent with the temporal collected data by means of the sensed information from leaf and intermediate sensor nodes.

In the convolutional neural network (CNN) method presented in [16], nodes retransmit the data before forwarding it to the CH, which secures the CH and the data from redundant and malicious nodes. The redundancy elimination for data aggregation (REDA) algorithm was implemented in [17] to remove the redundant data from each node via the differential data collected by the sensor nodes. Further, each data generates a pattern code for the retransmission of data; however, if the data already exists, it is eliminated [18–21]. Finally, if the network cluster area is recognized, the fuzzy matrix method is used to identify the similarity index.

The sleep scheduling algorithm described in [22] ensures that if a sensor node has already contributed the collected data to the CH, it remains in the sleep mode until its turn occurs again. The redundancy-aware topology control protocol (RTCP) was implemented in [23], in which the complete network is divided into groups to monitor the redundant nodes. The RTCP assigns a redundancy degree and a threshold connectivity degree to each group and maintains the redundant nodes in the sleep mode.

Finally, for enhanced readability, a summary of the reviewed articles is presented in table 8.1.

8.3 Proposed algorithm

In this section, we present the details of a proposed algorithm which aims to reduce the network energy and increase the network lifetime. The motivation for formulating such an algorithm is that an increase in the network lifetime can be ensured by reducing the energy consumed by the nodes or the network.

Once the nodes have formed a network, it is necessary to select a CH using an efficient CH selection algorithm based on the desired criteria. The selected CH acts as a bridge between the BS and the leaf nodes, which in turn saves leaf node energy. Once the selection of a CH is completed, the distance between the CH and the leaf nodes (D_{LH}) must be calculated. This enables the shortest path to the BS to be selected and reduces the energy consumption of the network. For the threshold distance denoted by D_{TH}, if $D_{LH} < D_{TH}$, then the specific path is chosen for communication; this path may use the single-hop method or a multihop method. Otherwise, if $D_{LH} > D_{TH}$, then the path is neglected.

In order to remove the redundant data (R_d), the existing data set available at the CH must be checked. Then, if $R_d = 0$, the data is further communicated; otherwise, if $R_d = 1$, then the data is discarded by the leaf node itself. Since network energy and network lifetime are directly proportional to each other, this proposed algorithm saves energy and increases the lifetime. The aforementioned algorithm is shown in the flowchart of figure 8.5. In addition, in a future study, this algorithm will be implemented on a WSN.

8.4 Quality factors

Redundancy can be exploited to improve the various data quality factors, as shown in figure 8.6. These include the following:

Table 8.1. Summary of the articles reviewed in the study.

Reference	Year	Summary	Method/algorithm	Limitation/future scope
[1]	2021	Index bit encoding (IBE) is used to aggregate similar data The sampling rate adjustment process is used to sample the data at the CH level	Zoom-in zoom-out (ZIZO) mechanism	Improve the compression ratio of IBE Minimize difference between original readings and decompressed readings set at the CH
[2]	2021	For the link between the CH and the sensor node, a circular topology is used. To transmit processed data and minimize information capacity used, an iterative identification method was proposed	Topology optimization and an iterative parameter identification method	For large-area and real-time applications
[3]	2021	Privacy of data is accounted for The algorithm selects the routing tree and path for data selection It restores the data if data is lost	Collection tree protocol (CTP)	Reducing communication overhead and adopting homomorphic encryption
[4]	2021	E-LEACH creates cluster and CH GWO and D-PSO optimization methods reduce time consumed in selecting CH CH assigns TDMA slots for sensor nodes	E-LEACH, energy-aware TDMA scheduling, and dynamic fuzzy-based best-channel selection	Large-scale environment for a particular application
[5]	2020	Works at two levels: the source-level uses the data similarity function, and the aggregator level uses the correlation coefficient to eliminate redundancy and aggregate data	Spatial correlation-based data redundancy elimination (SCDRE) for data aggregation protocol	Adapting these methods to multidimensional data with minimum energy consumption.
[6]	2020	EMA and threshold-based mechanisms are implemented at the node level to improve the data aggregation accuracy The Euclidean distance function is applied at the CH level to aggregate and forward data toward the BS	EMA, threshold-based mechanism, and Euclidean distance function	Check the security issues

[7]	2020	Various data aggregation algorithms are discussed	Data agglomeration techniques	Development of a decision system which is enabled via knowledge with the aim of removing malicious nodes from the network
[8]	2020	The complete network is divided into clusters, and the CH is chosen based on the distance and energy of the node	WEMER protocol	Security issues
[9]	2020	Step 1—algorithm divides the nodes into different layers Step 2—nodes are assigned different energies in the various layers	Local tree reconstruction	Security-related issues need to be considered
[10]	2020	Generates and manages key Uses end-to-end encryption Designed MAC generation algorithm	Asymmetric key encryption	Data transmission rate to be studied
[11]	2020	Applied if dynamic neighbors are equal in number Selection of the parent node is made	Fuzzy-attribute-based joint integrated scheduling and tree formation (FAJIT)	Other parameters of a network need to be considered
[12]	2019	The network hierarchy is the intermediate node between the CH and the sensor nodes The data collected by nodes is received by an intermediate node, where comparison and removal of duplicate data take place; data then moves to the CH	Hierarchical approach	Not implemented for the Internet of Things (IoT)
[13]	2019	Clustering process uses the fuzzy C-means approach. For further aggregation and to remove errors, an effective support degree function is used	Fuzzy C-means approach	Not implemented for multiple dimensions, detection mode, architectural structure, and correlation extraction

(Continued)

Table 8.1. (*Continued*)

Reference	Year	Summary	Method/algorithm	Limitation/future scope
[14]	2019	Network clustering check repeated data through statistical tests with a degree of confidence CH segregates and removes redundant data before forwarding it to the BS Spatiotemporal multicast dynamic CH role rotation selects the cluster head based on the strength of the node	Data redundancy-controlled energy-efficient multihop (DREEM)	Mobile for data collection
[15]	2019	Using the data mining approach, data collected by the sensor nodes is verified and duplicated data removed. This information is then made consistent with temporal collected data by means of sensed information from leaf and intermediate sensor nodes	Novel data mining (NDM)	Real-time sensor data can be put to the test with this method
[16]	2019	Sensors retransmit before forwarding data to the CH This secures the CH and the data from redundant and malicious nodes	CNN strategy	Implemented for heterogeneous networks
[17]	2018	Parameters used to select CH node should include distance to BS, node certainty, and node degree	Fuzzy logic and genetic algorithm. Fuzzy logic using if-then rules and probability between zero and one	Fuzzy logic and a genetic algorithm can be combined in a single network
[18]	2018	Redundant data from each node is removed using differential data collected by nodes Each data item is given a pattern code; using this approach, retransmission of the same data is avoided	Redundancy elimination for data aggregation (REDA) based on pattern generation approach; retransmission strategy originated from the Pareto principle and the scale-free property	Study of the REDA based on pattern code generation for large areas and different data collection
[19]	2018	The node with the least coverage of the region of interest is removed from the network	Coverage-preserving algorithm	The lifetime of the network is reduced

[20]	2018	Using tree node structure hierarchy, received packets are compared with the blocks of incoming packets to remove redundancy	Rabin Karp hashing algorithm	Execution time is high
[21]	2018	Specific data fusion techniques and their effects on various parameters	Data fusion techniques	Reliable and secure data
[22]	2018	Once the sensor area is recognized, a fuzzy matrix is used to recognize the similarity index; a correlation function based on fuzzy theory can be defined to divide sensor nodes The redundant nodes are in a sleep state for the next round	Sleep scheduling with similarity measures	Heterogeneous sensor networks are composed of different types of sensors
[23]	2018	Divides the network into groups to keep track of redundant nodes RTCP assigns redundancy degree and threshold connectivity degree to each group Keeps redundant nodes in sleep mode	Redundancy-aware topology control protocol (RTCP)	Not addressing the accuracy and energy consumed by nodes
[24]	2017	Checking the accuracy of predicted data and removing redundancy	FTDA Algorithm	To select the parameters of the time series prediction method
[25]	2017	Forming a grid and selecting the best path for communication	Grid Model	Reduce the network traffic
[26]	2017	Redundant data was removed by considering overlaps of the sensing areas	Overlapping rate of sensing area	Occurrence of multiple events at the same time

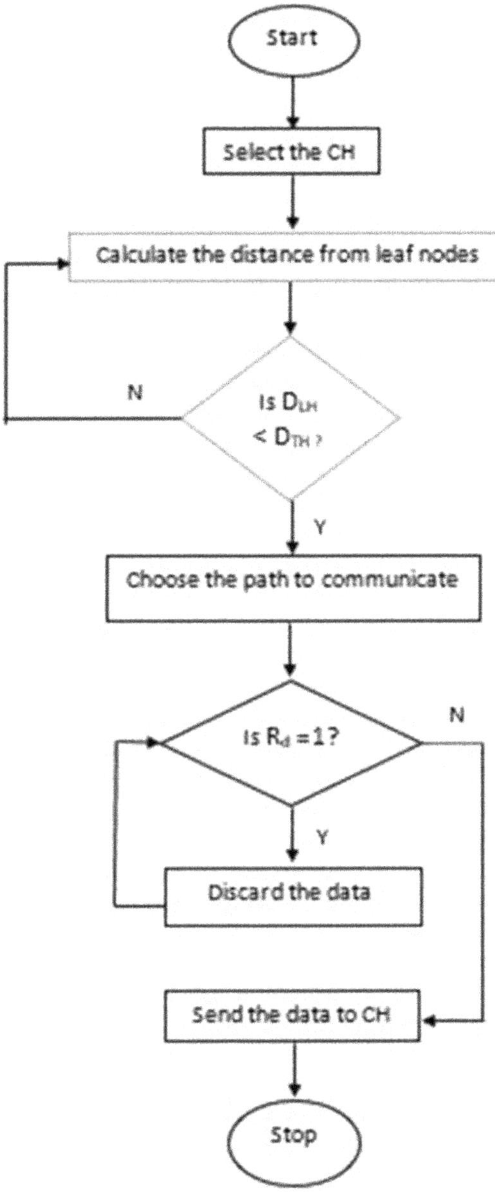

Figure 8.5. Flowchart of the proposed algorithm.

1. **Reliability:** redundancy ensures reliability in WSNs. Compared to the single-entry information received by the sensor motes, redundant data is more accurate. In some cases, sensors may provide improper data, events, or wrong measurements because of their environments. Hence, collecting information from a single sensor may not always produce reliable information. The reliability of the system can be increased by using redundant data from the sensor motes.

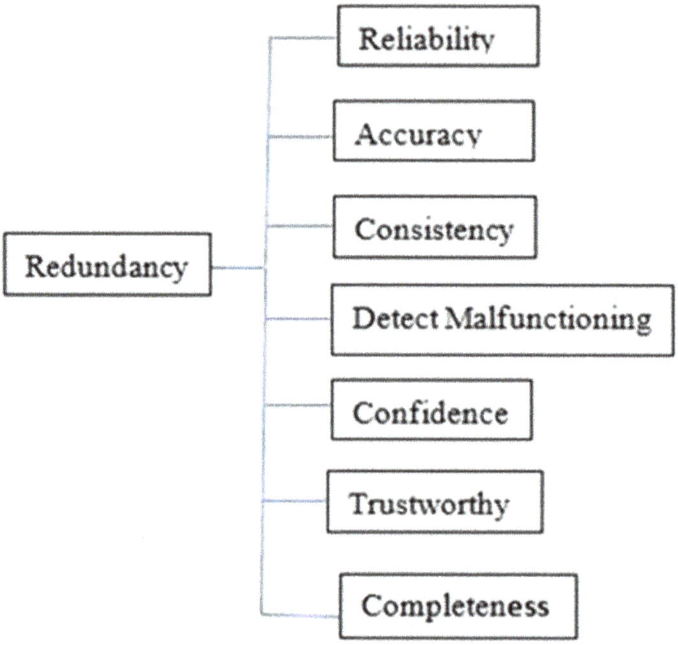

Figure 8.6. Quality factors.

2. **Accuracy:** reliable data is accurate data. To improve the accuracy of a sensor system or a network, a temporal redundancy method is used. As more than one independent source collects the same information over a period, this increases data confidentiality and helps to provide accurate information.
3. **Consistency:** the different information collected may lead to conflicting data. The conflicts can be resolved using spatial, temporal, or informational redundancy.
4. **The detection of malfunctioning or malicious nodes:** one of the redundancy methods, analytical redundancy, is used to discover faulty or corrupt nodes in the network. The mathematical model used in this method helps to predict the values obtained by the sensor nodes using the past and present values obtained by the neighboring sensor nodes. These predicted values are then compared with real sensor values obtained from nodes to find erroneous sensors.
5. **Confidence:** the confidence one has in information is a key feature which originates from a rigorous observation of the sensor data, for which quality does not matter as much. Confidence in information shows how much one can trust the data received from a source. The level of confidence one has in information is improved if the same information is received from two or more sensors.
6. **Trustworthiness or validity:** if the level of confidence in data is high, then the collected information is more trustworthy, and it remains valid for a long time period compared to other information obtained from sensor nodes.

7. **Completeness:** the data collected must form a data set and must be stored. If any data is missing, this may affect the system. Redundancy may compensate for a lack of completeness.

In regard to the key benefits of WSN redundancy, the following have demonstrated immense promise:

1. **Energy savings:** this is one of the most important issues in the WSNs, the nodes of which mostly rely on limited battery power. In general, in the active mode, any sensor node is able to operate with approximately 90–100 AAA battery pairs; however, in harsh environments, as is the case in real-time situations, it is not possible to recharge the batteries or to replace them. Hence, energy conservation is necessary to enhance WSN lifetimes [28].

   Further, a key part of the energy-saving technique is the exploitation of the spatial sensing redundancy which is inherent to the WSNs; this can be ensured through the definition of the node sub-sets which are active at different time instants, which permits sensors to save energy when they are inactive. If WSNs are deployed in dense environments, the most efficient method for saving energy is to place multiple redundant nodes into sleep mode, thereby ensuring that such nodes are able to use only part of the energy when they are in the active mode [29].

   Further, topology control is a way of conserving energy which is based on the spatial communication redundancy technique; it minimizes the number of nodes amount that take part in the forwarding and routing of packets created by other nodes, simultaneously minimizing the connectivity and coverage of the network. Such a technique is practical in WSNs because nodes are mostly densely deployed, thereby creating a very high degree of redundancy [30].

   Finally, any energy-conserving mechanism must reach its decisions using a distributed mechanism so as to avoid any unnecessary data overhead; such overheads are mainly incurred in implementations that use a centralized mechanism.
2. **Improving the reliability:** redundancy can also be used to provision data which is highly robust and fault tolerant. This is highly desirable when individual sensor nodes are malfunctioning or malicious. In harsh environments, sensors may provide incorrect readings if they are damaged and can also influence the correctness of the other properly functioning sensors. Therefore, in order to obtaining correct results from a WSN, it is mandatory for the WSN use cases to have access to all reliable data [31].

   Further, it has already been established that different nodes which observe the same place at similar times provide high quality; in such cases, readings from nearby sensors can be used to analyze the correctness of the data [32].

   Overall, considering an individual sensor, redundant data can be used to improve the reliability of the results, following which, these results can be aggregated using the methods detailed in [33]. This, in turn, ensures an increase in the reliability of the data.

3. **Upgrade in security:** redundancy can also be used to provide improvements in the security of the data. WSNs are typically deployed in harsh environments, due to which the sensors fail or provide incorrect data. In addition, existing mechanisms such as agreement protocols prove ineffective, since they require large resources and are not scalable.

WSNs are exposed to significant attacks owing to the multiple constraints imposed on their functioning. Attacks such as cryptography attacks, spoofing attacks, forwarding attacks, Sybil attacks, etc cannot be prevented, even by cryptographic methods [34]. Amongst these, the best-known attack is that of node capture, wherein a hacker obtains complete control of the sensor nodes via direct physical access [35]. Such an attack is different from other attacks, as it is not reliant on gaps in security protocols, broadcasts, etc. Instead, it is dependent on the deployment of the sensor nodes.

Overall, the main aim of security in WSNs is to develop methods which are specific and which exploit the redundancy feature of WSNs. Such approaches must rely on the fact that any sensor node may not operate according to its original design, even if it transmits correct data; i.e. it is still liable to send incorrect messages. Further, by using the spatial redundancy technique and comparing the readings with those obtained via redundant sensors, it may be possible to develop a knowledge-enabled decision system which will aim to remove malicious sensor nodes from WSNs.

8.5 Research avenues

Based on our survey, we have identified various avenues of research into redundancy issues in WSNs which require timely solutions.

These open research problems are as follows:

1. The elimination of redundant data without affecting the time of the network is a challenge, as the time required to eliminate redundant data is high, which in turn affects the network lifetime.
2. Static CHs lose energy while collecting data from the sensor nodes deployed at various places in a cluster. Implementing a mobile CH with a redundancy elimination method can help to save energy and is an open area of research.
3. Allocating a specified time duration for sensor nodes to communicate with the CH helps to save energy at the nodes and avoids repetitive data collection at the CH. Such an allocation requires further design.
4. The use of a similarity check algorithm at the CH while it collects data from sensor nodes would help the CH to avoid identical data in the nodes. Such an algorithm needs to be formulated.

8.6 Conclusions

This survey chapter provided an overview of the various redundancy-related challenges in WSNs. In this chapter, we discussed the types of redundancy and surveyed the most recent studies published in the literature which have aimed to minimize the impact of redundancy in WSNs. We also listed various research problems related to WSN redundancy issues, which require timely solutions. Our survey reveals that timely solutions are required to (i) eliminate redundant data without affecting the networks' period, (ii) implement a mobile CH that uses a redundancy elimination method to save energy, (iii) allocate specified time durations for communication between sensor nodes and the CH to save power, and (iii) formulate new similarity check algorithms to help the CH to avoid identical data which exists at the nodes.

Acknowledgments

The authors wish to thank Professor Rajarshi Khanai of the Department of CSE, KLE Technological University, MSSCET, Belagavi, and Professor Dinesha HA, of the Department of CSE, Nagarjuna College of Engineering, Bengaluru for proof-reading the chapter, which helped the authors to improve the language and the content.

References

[1] El Sayed A, Harb H, Ruiz M and Velasco L 2021 ZIZO: A Zoom-In Zoom-Out Mechanism for Minimizing Redundancy and Saving Energy in Wireless Sensor Networks *IEEE Sens. J.* **21** 3452–62

[2] Lu Z, Wang N and Yang C 2022 A novel iterative identification based on the optimized topology for common state monitoring in wireless sensor networks *Int. J. Syst. Sci.* **53** 25–39

[3] Li P, Xu Chao and Xu H 2021 Data privacy protection algorithm based on redundant slice technology in wireless sensor networks *Int. J. Info. Security and Privacy* **15** 190–212

[4] Bhusan S, Kumar M, Kumar P, Stephan T, Shankar A and Liu P 2021 FAJIT: a fuzzy-based data aggregation technique for energy efficiency in wireless sensor network *Complex & Intelligent Systems* **7** 997–1007

[5] Sinde R, Begum F, Njau K and Kaijage S 2020 Lifetime improved WSN using enhanced-LEACH and angle sector-based energy-aware TDMA scheduling *Cogent Engineering* **7** 1795049

[6] Tang X, Xie H, Chen W, Niu J and Wang S 2017 Data aggregation based on overlapping rate of sensing area in wireless sensor networks *Sensors* **17** 1527

[7] Curiac D-I and Volosencu C *et al* 2009 Redundancy and its applications in wireless sensor networks: a survey *WSEAS Transactions on Computers* **8** 706–14 http://www.wseas.us/e-library/transactions/computers/2009/29-149.pdf

[8] Rhesa M J and Revathi S 2020 An exploration of current data agglomeration technique in wireless sensor network *Int. J. Eng. Adv. Technol. (IJEAT)* **9** 732–46

[9] Bello A D and Lamba O S 2020 Energy efficient for data aggregation in wireless sensor networks *Int. J. Eng. Res. Technol. (IJERT)* **9** 110–31

[10] Zhang Z, Li J and Yang X 2020 Data aggregation in heterogeneous wireless sensor networks by using local tree reconstruction algorithm *Complexity* **2020** 3594263

[11] Qi X, Liu X, Yu J and Zhang Q 2020 A privacy data aggregation scheme for wireless sensor networks *Science Direct Procedia Computer Sci.* **174** 578–83

[12] Nalayini P and Prakash R A 2019 Hierarchical data redundancy elimination approach (HIDREA) for wireless sensor network with energy saving approach *IOSR J. Comput. Eng.*

(IOSR-JCE) **21** 27–32 https://www.iosrjournals.org/iosr-jce/papers/Vol21-issue5/Series-3/D2105032732.pdf

[13] Wan R, Xiong N, Hu Q, Wang H and Shang J 2019 Similarity-aware data aggregation using fuzzy c-means approach for wireless sensor networks *EURASIP J. Wireless Communication and Networking* **2019** 59

[14] Ahmed G, Zhao X and Fareed M M S *et al* 2019 Data Redundancy-Control Energy-Efficient Multi-Hop Framework for Wireless Sensor Networks *Wirel. Pers. Commun.* **108** 2559–83

[15] Kumar S and Chaurasiya V K 2019 A strategy for elimination of data redundancy in internet of things (IoT) based wireless sensor network (WSN) *IEEE Syst. J.* **13** 1650–57

[16] Yücel T and Altın-Kayhan A 2019 A copy-at-neighboring-node retransmission strategy for improved wireless sensor network lifetime and reliability *J. Oper. Res. Soc.* **70** 1193–202

[17] Saadaldeen R S M, Osman A A and Ahmed Y E E 2018 Clustering for energy efficient and redundancy optimization in WSN using fuzzy logic and genetic methodologies a review *2018 International Conference on Computer, Control, Electrical, and Electronics Engineering (ICCCEEE)* (Piscataway, NJ: IEEE) pp 1–5

[18] Khriji G, Vinas Raventos G, Kammoun I and Kanoun O 2018 Redundancy Elimination for Data Aggregation in Wireless Sensor Networks *2018 15th International Multi-Conference on Systems, Signals & Devices (SSD)* (Piscataway, NJ: IEEE) pp 28–33

[19] Biradar S and Mallikarjuna Shastry P M 2018 Redundancy elimination with coverage preserving algorithm in wireless sensor network *Int. J. Commun. Networks and Information Security (IJCNIS)* **10** 1193–202

[20] Priya D and Enoch S 2018 The effect of packet redundancy elimination technique in sensor networks *J. Comput. Sci.* **14** 740–46

[21] Verma N and Singh D 2018 Data redundancy implications in wireless sensor networks *Procedia Computer Science* vol 132 (Amsterdam: Elsevier) pp 1210–17

[22] Wan R, Xiong N and Loc N T 2018 An energy efficient sleep scheduling mechanism with similarity measure for wireless sensor networks *Human–Centric Comput. Info. Sci.* **8** 18

[23] Zebbane B, Chenait M, Benzaid C and Badache N 2018 RTCP: a redundancy aware topology control protocol for wireless sensor networks *Int. J. Info. Commun. Technol.* **12** 271–98

[24] Yang M 2017 Data aggregation algorithm for wireless sensor network based on time prediction *2017 IEEE 3rd Information Technology and Mechatronics Engineering Conference (ITOEC)* (Piscataway, NJ: IEEE) pp 863–67

[25] Mohansundaram R, Thenmozhi M, Sreekumar A and ArunAdithiyan A 2017 Reduction of data redundancy using data aggregation in wireless sensor networks *Int. J. Pure and Applied Mathematics* **117** 299–312 https://acadpubl.eu/jsi/2017-117-15/articles/15/25.pdf

[26] Maivizhi R and Yogesh P 2020 Spatial Correlation based Data Redundancy Elimination for Data Aggregation in Wireless Sensor Networks *2020 Int. Conf. on Innovative Trends in Information Technology (ICITIIT)* (Piscataway, NJ: IEEE) pp 1–5

[27] Jan S R U, Khan R and Jan M A 2021 An energy-efficient data aggregation approach for cluster-based wireless sensor networks *Ann. Telecommun.* **76** 321–9

[28] Wang L and Xiao Y 2005 Energy saving mechanisms in sensor networks *2nd International Conference on Broadband Networks, 2005* vol 1 (Piscataway, NJ: IEEE) pp 724–32

[29] Perrig A, Szewczyk R and Tygar J *et al* 2002 SPINS: Security Protocols for Sensor Networks *Wireless Networks* **8** 521–534

[30] Gao Y, Wu Kui and Li F 2003 Analysis on the redundancy of wireless sensor networks *WSNA '03: Proceedings of the 2nd ACM international conference on Wireless sensor networks and applications* (New York: ACM) pp 108–14
[31] Yan T, He T and Stankovic J A 2003 Differentiated surveillance for sensor networks *SenSys '03: Proc. 1st Int. Conf. on Embedded networked sensor systems* (New York, NY: ACM) pp 51–62
[32] Cloqueur T, Ramanathan P, Saluja K K and Wang K-C 2001 Value-Fusion versus Decision-Fusion forFault-tolerance in Collaborative Target Detection in Sensor Networks Fusion *2001 4th Int. Conf. on Information Fusion* (Paris: International Society of Information Fusion) https://isif.org/events/conference/fusion-2001 TuC24
[33] Karlof C and Wagner D 2003 Secure routing in wireless sensor networks: attacks and countermeasures *First IEEE Int. Workshop on Sensor Network Protocols and Applications, 2003* (Piscataway, NJ: IEEE) pp 113–27
[34] Becher A, Benenson Z and Dornseif M 2006 Tampering with Motes: Real-World Physical Attacks on Wireless Sensor Networks *Int. Conf. on Security in Pervasive Computing SPC 2006* (Berlin: Springer) pp 104–18
[35] Curiac D-I, Volosencu C, Doboli A, Dranga O and Bednarz T 2007 Discovery of malicious nodes in wireless sensor networks using neural predictors *WSEAS Trans. Comput. Res.* **1** 38–44

Chapter 9

Sensor-based devices and their applications in smart healthcare systems

Sapna Singh Kshatri, Devanand Bhonsle, Sourabh Tiwari, Ravi Mishra, Tanu Rizvi and Rimjhim Pandey

Today, Internet of Things (IoT) technologies are emerging in various fields, among which, healthcare is one of the most important sectors. Emerging health technologies improve healthcare services. Although there has been significant progress in the field of healthcare, more research is still needed to make healthcare systems robust and highly secure. Today's new era of healthcare requires the detection and control of disease. Smart healthcare systems require monitoring, management, and decision-making. However, in developing countries, emerging healthcare technologies are burdened with regulatory compliance; as a consequence, many healthcare systems depend on paperwork to capture, process, and mange health data. Using such systems, it is difficult to access, process, analyze, and store the huge amounts of data related to the healthcare field. During pandemics such as COVID-19, the abovementioned difficulties are increased. To handle such difficult situations, it is necessary to deploy new technologies based on cognitive sensors in healthcare systems. These sensors are either implanted in the human body or placed in the patient's clothes, chair, or bed to monitor the various activities or behaviors of the human body. This chapter discusses the roles and capabilities of sensors, communication networks, and cloud storage in healthcare systems. These are useful in many applications in the field of medicine, such as remote monitoring, tracking, telemedicine, wellness monitoring, treatment reminders, etc. Cognitive sensors not only make the system smart and efficient but also reduce paperwork, errors, and the time required to retrieve information.

9.1 Introduction

Good health is a primary concern of our lives. A healthy body makes life perfect. However, life has become complicated for almost all age groups in society. Hence,

doi:10.1088/978-0-7503-5326-7ch9

from time to time, we all need to go to a clinic or hospital for a checkup. Many countries take care of the health of their citizens. For this reason, they have significantly developed their healthcare systems and are constantly making their healthcare systems smarter and more secure using IoT-based technologies [1–5]. In general, IoT-based healthcare systems are now referred to as the Internet of Medical Things (IoMT) [6–8]. These developments are required in order to avoid emergency situations. In December 2019, COVID-19 emerged and spread so rapidly that almost all countries in the world were affected [9, 10]. To handle such unfortunate situations in the future, all countries must develop their healthcare systems [11, 12]. The IoT is a modern technology which has great scope to improve this application area [13, 14]. It may enhance the field of healthcare system in such a way that patients receive better treatments and can be cured of lethal diseases. The best part of IoT-based healthcare systems is that they involve the use of wearable network-connected gadgets which are installed in a patient's bed, chair, or clothes. In some cases, sensors are permanently implanted in the human body; these are referred to as invasive sensors. In this way, medical personnel or family members may directly monitor patient health and in the case of any alarming situation, the required action may be taken to avoid poor outcomes. Another benefit of IoT-based healthcare systems is that the patient need not be admitted to hospital or confined to a bed, but rather the patient may walk, run, or use a wheelchair [15]. The patient may be monitored frequently and his/her health may be traced in the long term. The details obtained may be used for further examinations of the patient's health. These records also confirm the effects of the treatments given to the patient. Hence, we can say that the IoMT is one of the most important application areas for IoT systems [14].

Connecting wearable sensors, clinical diagnostic centers, and end users, the IoMT functions as an interoperable medium that provides information for healthcare applications [13]. This interoperable medium offers options for disease diagnostics and the prediction and monitoring of end-user health using vital physiological indicators detected by wearable sensors. IoT communication and data exchange platforms face latency and overloading issues in diverse environments. In the current scenario, in which the whole world is facing a global problem in the form of the pandemic known as COVID-19 [9, 10], it is essential to make healthcare systems more powerful in all respects so that lives can be saved worldwide at such a crucial time. Healthcare is one of the basic needs of every human being, regardless of one's economic situation, but in reality, medical facilities are more expensive than ever. Another difficulty is the use of conventional techniques and management systems in this field. These techniques are time-consuming, they require more paperwork, and they are difficult to handle. Thus, many researchers and developers have been working to make smart healthcare systems in the last few years [16, 17]. Cognitive smart healthcare systems [18] based on the IoMT are used for various applications. Some of the most common applications are health monitoring [19, 20], the treatment of disease, research into diagnostic and treatment techniques, and applications used by healthcare providers such as hospitals, clinics, etc. The following sections discuss the components of IoMT-based smart healthcare systems.

9.2 Sensors used in medicine

In general, a sensor is a device which measures or detects a particular physical property and collects records or responds accordingly. In other words, a sensor produces an output signal for the purpose of sensing a particular physical phenomenon. Sensors are used in many application areas [21]. Since this chapter is related to IoMT applications, only sensors used for medical purposes are discussed. Medical sensors can broadly be categorized into two types, viz. invasive sensors and non-invasive sensors.

9.2.1 Invasive sensors

Invasive sensors are sensors which are permanently implanted into the human body; hence, they are also called implantable sensors. An operation is required for the implantation process. Sensors are difficult to implant properly; hence, expert surgeons are required. Examples of the invasive sensors used for smart healthcare monitoring systems are artificial cardiac pacemakers, cochlear implants, retinal implant, deep brain stimulators, wireless capsule endoscopes, electronic pills, implantable defibrillators, etc. Figure 9.1 shows some of the most important invasive sensors used for implantation in the human body.

- **Artificial cardiac pacemaker** This is referred to as an artificial pacemaker or simply a pacemaker. It is a medical device which is used to generate electrical impulses using electrodes. It is attached to a chamber of the heart, either the lower ventricles or the upper atria. The main purpose of a pacemaker is to ensure that blood is pumped; hence, pacemakers are used to regulate the electrical conduction systems of the heart. They are used in the human body when the natural pacemaker of the heart is not fast enough to maintain an adequate heartbeat rate. Modern pacemakers can be programmed externally

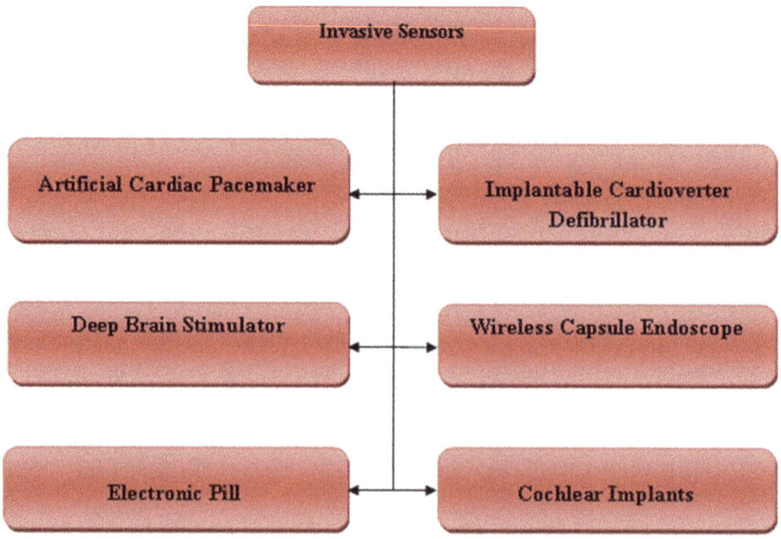

Figure 9.1. Invasive sensors.

and an optimal pacing mode can be set for each patient. They are invasive devices because they are implantable in the human body [22].

- **Cochlear implant**

  If a person has moderate or profound hearing loss, a cochlear implant can be used. Cochlear implants are surgically implanted neuroprostheses used to improve speech understanding in both noisy and quiet environments. They are suitable for both children and adults and require the use of a surgical procedure under general anesthesia. These implants have two main components, viz. the outside component and the inside component. The outside component is worn behind the ear and consists of microphones, a battery, digital signal processor chips, and a coil for single transmission, while the inside component includes a coil used as a signal receiver, some electronic parts, and an array of electrodes placed in the cochlea which stimulate the cochlear nerves. This system allows deaf persons to receive and process sound and speech signals. The processed signal is sent to the brain and can be understood by the deaf person. However, the quality of hearing is not same as that of natural hearing, and this approach is not suitable for everyone.
- **Deep brain stimulator (DBS)** This requires a neurosurgical procedure which is used to implant electrodes and electrical stimulation for the treatment of various disorders associated with dystonia, Parkinson's disease, essential tremors, and other neurological disorders.
- **Wireless capsule endoscope** This is used for endoscopic procedures. For this purpose, a small wireless camera is inserted into the digestive tract of the human body. This camera takes pictures of the interior of the body, which are used to diagnose problems inside the body. This is generally a safe procedure; however, bowel obstruction may happen if the capsule becomes stuck in a narrow passage. Sometimes, the patient may experience fever, abdominal pain, vomiting, nausea, bloating, etc. The length of the wireless capsule used for endoscopy is about 25 mm and its diameter is about 10 mm to 12 mm.
- **Electronic pill** This contains sensors or tiny cameras which are used to collect information as the pill passes through the gastrointestinal tract. It measures the pH values and temperature of the inner body, and is therefore used to check the acidity level of the human stomach. Some electronic pills are also used to measure muscle contractions, ease of passage, and other factors which may be useful to reveal problems inside the body.
- **Implantable cardioverter defibrillator (ICD)** This devices is placed in the chest to detect irregular heartbeats. It is a battery-operated device. It continuously monitors heartbeats and can deliver an electric shock if required. The electric shock helps to restore regular heartbeats. This is an invasive device which requires a surgical operation to be performed under the supervision of experts.

9.2.2 Non-invasive sensors

Non-invasive sensors need not to be implanted into the human body; rather, they are attached to the outer part of the body. Most of the non-invasive sensors which are used to monitor health issues are wearable; hence, in general, non-invasive sensors are also called wearable sensors [21]. Some examples of non-invasive/wearable sensors are elctrocardiographs [22], glucose sensors, electromyographs, electro-encephalographs, temperature sensors, pulse oximeters, blood pressure sensors, oxygen level indicators, etc. Figure 9.2 shows various types of sensor-based equipment used in healthcare monitoring systems [19, 20].

In the following, we discuss the abovementioned non-invasive sensor-based equipment used in healthcare monitoring systems.

- **Body temperature monitor**

  This is one of the most common monitors used when a patient is suffering from fever. It is used to measure the external body temperature. The two most common temperature scales are the Celsius scale, which ranges from 0 °C to 100 °C, and the Fahrenheit scale, which ranges from 32 °F to 212 °F. The Celsius scale is equally divided into 100 parts, while the Fahrenheit scale is divided into 180 equal parts. The body temperature normally varies with the person, the level of activity, age, and time of day. In general, the average normal body temperature is 37 °C or 98.6 °F. Thermometers, which are readily available, are used for body temperature measurements. Mercury/glass thermometers are required to touch the body of the patient, which may transmit germs. To avoid this, infrared thermometers can be used. They allow a person's temperature to be read from a certain distance and displayed within a second. Infrared thermometers are faster than conventional thermometers. Figure 9.3 shows a digital thermometer, which is the type of thermometer most commonly used to measure human body temperature.

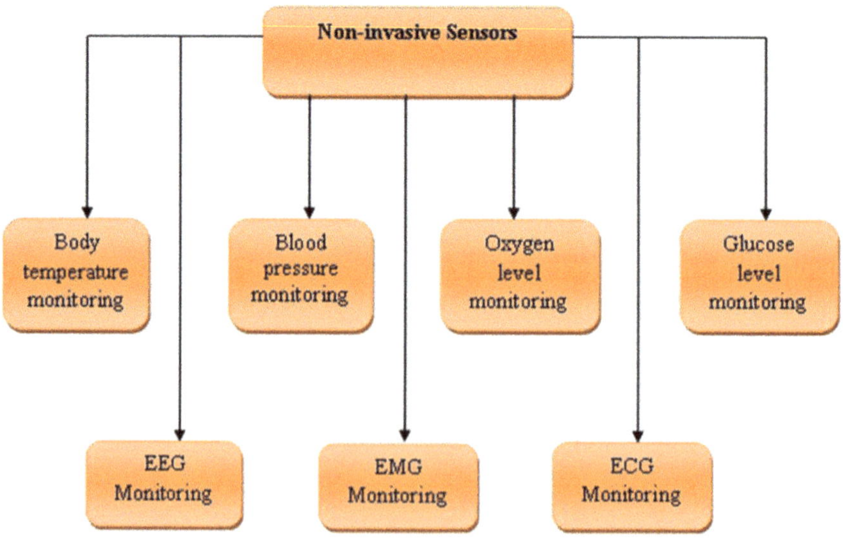

Figure 9.2. Non-invasive sensor-based equipment.

- **Blood pressure monitor**

  Blood pressure is one of the most important parameters used to monitor a person's health. The normal blood pressure range for an adult is 80 to 120. Conventional blood pressure measurements require the following steps:

  - Place the arm cuff just above the elbow
  - Inflate the arm cuff completely
  - Deflate the arm cuff slowly
  - Record the blood pressure reading using a stethoscope.

Today, many companies make blood pressure monitoring machines which can be used by non-medical persons. These machines do not need to be pumped manually and the readings are shown in the display. Figure 9.4 shows a blood pressure monitoring device.

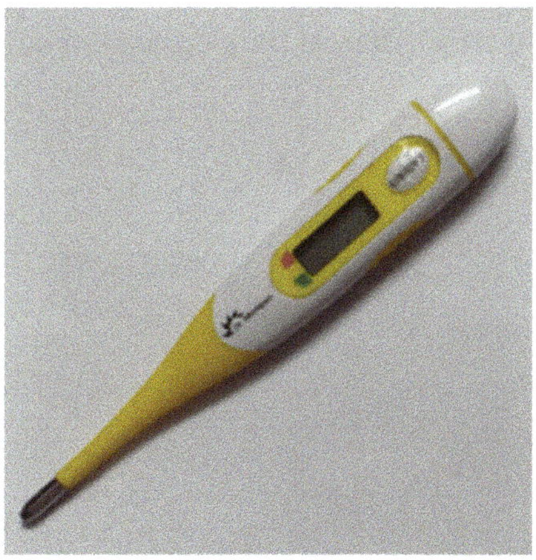

Figure 9.3. Digital thermometer.

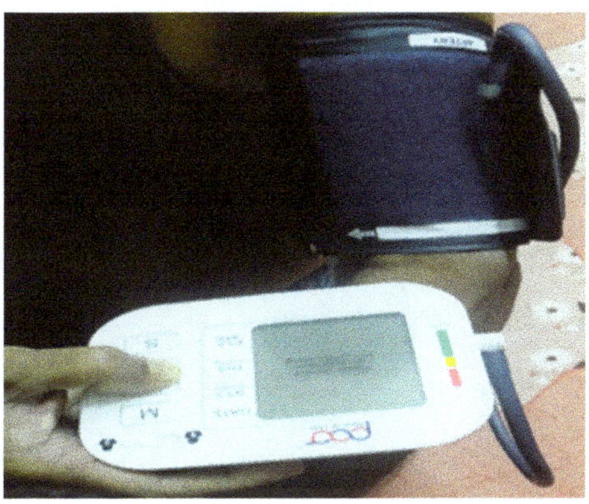

Figure 9.4. Blood pressure monitoring device.

- **Oxygen level monitor**

  This is a non-invasive test device. For oxygen level monitoring, a pulse oximeter is used, which is placed on the fingertip. This uses a light beam to estimate the blood oxygen saturation and the pulse rate. The saturation level indicates the amount of oxygen carried in the blood. Figure 9.5 shows a pulse oximeter probe which is applied to the finger of a person.

- **Glucose-level monitor**

  This is used to continuously monitor the glucose level in the blood. It can measure the glucose level throughout the day or night. It works through a small sensor, which measures the glucose found in blood cells. This monitor has a transmitter which sends the data to the receiver, which in turn displays the result. Figure 9.6 shows a glucose-level monitoring device.

- **Electrocardiography**

  This is the one of the most popular and common tests used to check the rhythm of the heart and its electrical activity. For this purpose, a sensor is attached to the skin which detects the electrical signals generated by the heart.

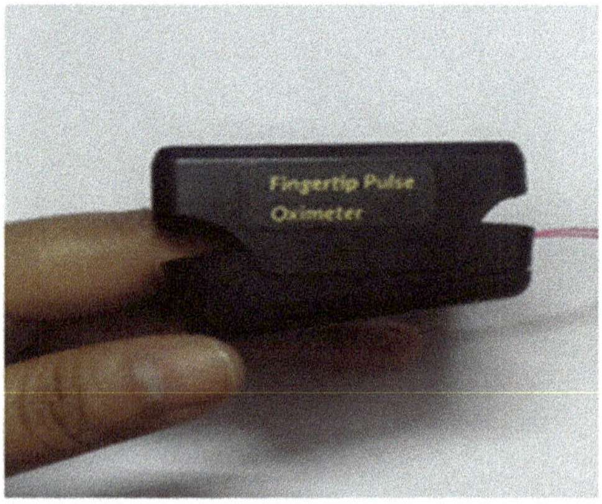

Figure 9.5. Pulse oximeter.

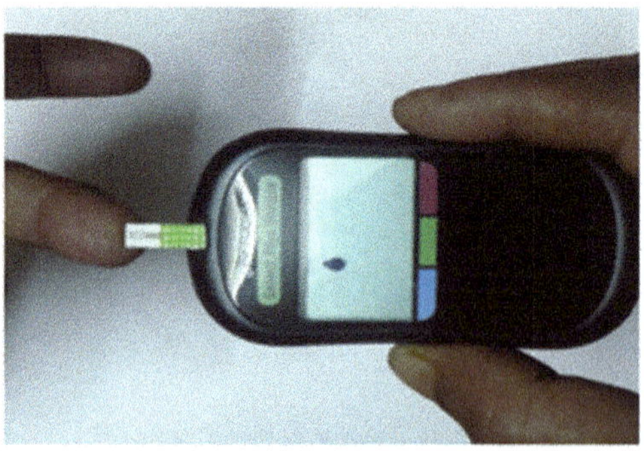

Figure 9.6. Glucose-level monitoring device.

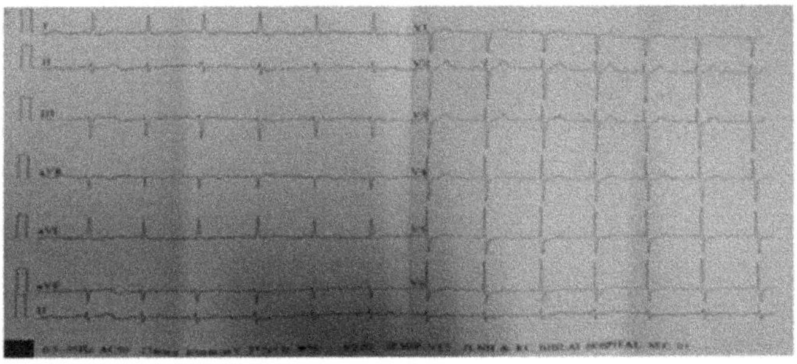

Figure 9.7. An electrocardiogram.

This signal is recorded by the machine and the results are further examined by the cardiologist to check whether they are unusual or normal. This test is only done by trained medical personnel. Figure 9.7 shows the electrocardiogram (ECG) of a female aged 62 with a suspected left anterior hemiblock.

- **Electroencephalography**
  This test is performed to measure the electrical activity in the human brain. It uses small metal disks attached to the scalp. These small metal disks are called electrodes. Brain cells communicate via electrical impulses. These brain cells are active all the time, even when a person is sleeping. All the brain cell activity is recorded in the form of wavy lines. Electroencephalography is used to monitor epilepsy, brain tumors, brain damage due to head injury, sleep disorders, brain strokes, herpes encephalitis, encephalopathy, and other brain disorders [22].

- **Electromyography**
  This is used to measure electrical activity due to the nerve's stimulation of the muscles, which allows neuromuscular abnormalities to be detected. The test requires electrodes which are tiny needles inserted into the muscles through the skin. The inserted electrodes pick up electrical activity which is displayed in the form of waves. These signals can be heard with the help of an audio amplifier. Muscle tissue does not normally produce electrical signals at rest, but when an electrode is inserted into it, an electrical signal is generated.
  In addition to the abovementioned techniques, some methods produce 2D or 3D images. Some of the most important and widely used imaging techniques in the medical field are computed tomography (CT) imaging [23–27], magnetic resonance imaging (MRI), ultrasound (US) imaging [25, 28, 29], dual-energy x-ray absorptiometry (DXA), etc. Figure 9.8 shows various medical imaging sensor techniques.

Table 9.1 presents a summary of the abovementioned imaging techniques.

These sensors have very wide uses in the medical field. For example, non-invasive techniques are used to examine the uterus or baby, to listen to fetal heart sounds, etc.

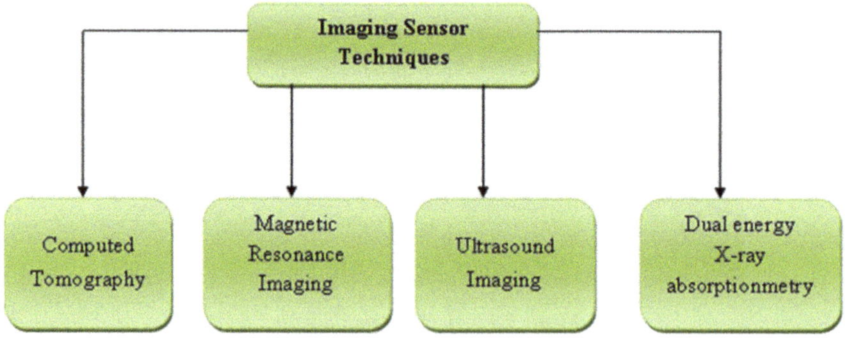

Figure 9.8. Imaging techniques.

9.3 Cognitive smart healthcare systems that use the Internet of Medical Things

Figure 9.9 shows a cognitive smart healthcare system that uses the IoMT; it has various parts, viz.:

- Cognitive sensors
- Low-power short-distance interface
- Hosting layer
- Wide are network (WAN) interface
- Cloud.

Cognitive sensors are transducers which can sense particular parameters of the human body. Section 9.2 discussed various types of sensor which are either wearable or implanted. Some of the most common imaging techniques have also been discussed. These sensors may be used to get multimodal signals, which are health-related data. The most common data are body temperature, heartbeat rate, blood pressure, glucose level, oxygen level, body movement, facial expression, respiration rate, electrocardiographic signals [22], electromyographic signals, electroencephalographic signals [31, 32], etc. All these parameters are useful in understanding patients' health conditions. Cognitive sensors can be embedded and installed in patients' surroundings. Cognitive sensors can communicate with other devices using the IoT [31–36]. Since these IoT-based sensors and devices are used for medical purposes, they may be referred to as the IoMT [37]. The data collected from the cognitive sensors are sent to the low-power, short-distance interface and then to the hosting layer. The main role of the hosting devices is to store the data acquired from the cognitive sensors [18]. The most common hosting devices are laptops, desktops, and smartphones. The stored data is further sent to the WAN interface which is employed by advanced communication networks. These networks are used to send the data to the cloud, which has

Table 9.1. Various medical imaging techniques.

Imaging Technique	Information	Advantages	Disadvantages
CT Scan	Density Shape	• It provides 3D images. • CT scanners are portable. • It is a fast technique. • It has better bone-tissue contrast than MRI	• It requires anesthesia. • It uses ionizing radiation. • CT scanners are expensive devices.
MRI	Density Shape	• It does not use ionizing radiation. • It produces 3D images. • It has better soft-tissue contrast than CT	• No metal is allowed due to the use of a magnetic field. • The image acquisition process is very slow. • It requires anesthesia. • MRI scanners are expensive devices.
Ultrasound	Density	• Anesthesia is not required. • Real-time observation is possible. • Ultrasound scanners are portable devices. • It is a fast technique. • Video recording is possible. • The price is reasonable. • There is no size limit	• It produces a 2D image • It has poor tissue contrast
DXA	Density Shape	• It is easy to handle. • It produces small amounts of radiation. • Its price is moderate. • It provides quick data analysis. • It is used for regional data analysis.	• It provides 2D information. • No direct data is provided for lean tissues.

three main units, viz. a cloud manager, a cloud engine and a deep learning server. All the health-related data is made available via the cloud and may be accessed by authorized persons. Since this data is very confidential, security [38] and privacy [39] are prime concerns of this system. This data may be useful for various types of application such as monitoring the health conditions of patients, the treatment of diseases following diagnosis, research into advances in treatment, etc.

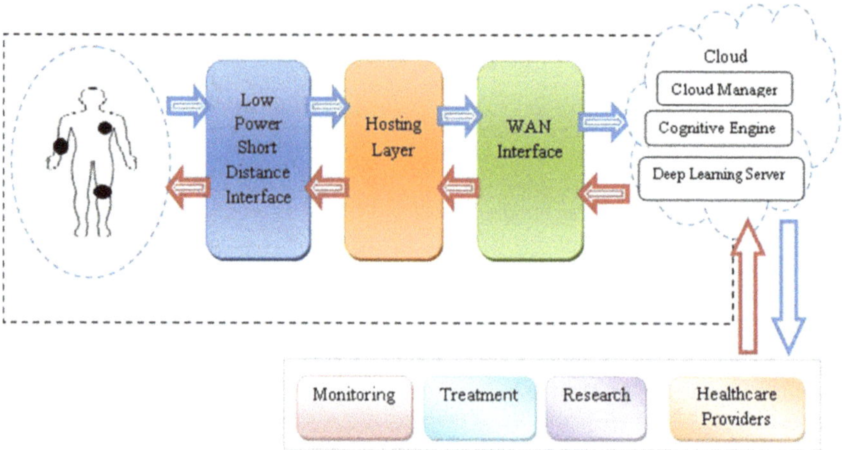

Figure 9.9. Cognitive smart healthcare system that uses the IoMT.

9.4 The Internet of Medical Things and its applications

Conventional healthcare systems are time-consuming and inconvenient for patients as well as their family members. Generally speaking, patients need to go to the hospital or the doctor's clinic for various tests and health monitoring, such as blood pressure, heartbeat rate, respiration rate etc. As soon as IoT systems were developed [40], healthcare systems were also improved and are now referred to as the IoHT. This has made healthcare systems faster and more convenient for patients as well as their family members. The following is a brief discussion of some applications of the IoT in healthcare [41–43]

- **Care:** IoT technology helps practitioners to use their knowledge to take care of the patients by taking the right decisions about their health conditions and avoiding complications if an emergency occurs. Using IoT systems, practitioners can get more accurate information, which may be helpful in resolving the situation, and the necessary actions may be taken accordingly.
- **Gadgets:** IoT systems use various types of wearable or implanted gadget. These gadgets may either be implanted into the body of the patient or attached to their clothes. In most cases, these gadgets are fixed to the patient's bed, wheelchair, etc. These gadgets consist of different types of cognitive sensor which are sensitive to minute changes and produce signals. These signals may be processed and used for further action, because they represent variations in the health condition of the patient.
- **Dissemination of medical information:** precise and recent medical information about the patient is among the most important data which can be used in the field of healthcare. IoT systems not only help patients in hospital but also in the home, public places, offices, etc.
- **Research:** in the real world, it is important to have large amounts of data about patients; however, at the moment, it is not available in such a way that all cases can be understood using the available information. The data

Figure 9.10. Applications of the IoMT.

provided by the IoT system should be available in the network and must be accessible by authorized persons. The data provided by IoMT systems is far better than the general analytical data produced by conventional equipment available in hospitals.

- **Emergency care:** in the healthcare system, it is desirable to get the knowledge about patients' health in such a way that any emergency cases may be detected as early as possible, so that the necessary actions may be taken and patients' lives can be saved. The IoMT provides all the information about patients' health, which is available in their profile. Since the IoMT is accessible from the network, however, the network should be secured in such a way that any unauthorized persons cannot access the system.
- **Telemonitoring:** this is a basic requirement in the current scenario, in which the whole world is suffering from the COVID-19 pandemic [9, 10]. As a response to this situation, most countries imposed lockdowns in various stages. As a result, patients suffered a lot, as doctors and medical personnel were not available for consultation. To avoid spreading the virus, many guidelines were realized and imposed on the public, among which, social distancing has played an important role in the pandemic. Hence, healthcare systems need to be upgraded in such a way that there is no direct physical interaction between patients and doctors or any other medical staff. In this scenario, telemonitoring plays an important role, as it allows the interactions

between patients and doctors to take place virtually rather than physically. This telemonitoring is also known as remote invigilation.

9.5 Conclusions

From the above discussion about the sensors used in the medical field, it can be seen that these sensors are very useful for monitoring health conditions and diagnosing any diseases or abnormalities in the human body. A great deal of research has been carried out in the field of sensors and their use in medicine. The IoT is the latest technology; it has allowed various applications to be automated. When the IoT is used in the field of medicine, it is referred as the IoMT. Cognitive sensors are intelligent sensors which monitor various body parameters; they are used for monitoring or diagnostic purposes. These sensors are either implanted in the body or are in contact with the patient. Implanted sensors are called invasive sensors, while wearable sensors are called non-invasive sensors. Using IoT technology, sensors can send data directly from the patient's body to the cloud using wireless technology. Patient data is stored in the cloud and can be retrieved by authorized medical personnel; it may be further used by them for treatment purposes. Such data is highly confidential and needs to be secured; hence, security and privacy are major concerns. Many developed countries have well-equipped, IoT-based smart healthcare systems, but most countries have yet to deploy smart healthcare systems. Healthcare is the prime concern for all countries and a fundamental right of every human being; hence, it should be made available to all on first principles.

References

[1] Mutlag A A, Ghani M K, Arunkumar N, Mohammed M A and Mohd O 2019 Enabling technologies for fog computing in healthcare IoT system sbFut *Gener. Comput. Syst.* **90** 62–78

[2] Min M, Wan X, Xiao L, ChenY B, Xia M, Wu D and Dai H 2019 Learning-based privacy-aware offloading for healthcare IoT with energy harvesting *IEEE Internet Things J.* **6** 4307–316

[3] Gope P and Hwang T 2016 BSN-Care: a secure IoT-based modern healthcare system using body sensor network *IEEE Sens. J.* **16** 1368–76

[4] Yeh K 2016 A secure IoT-based healthcare system with body sensor networks *IEEE Access* **4** 10288–99

[5] Mahmoud M M, Rodrigues J J, AhmedS H, Shah S C, Al-Muhtadi J, Korotaev V and Albuquerque V H 2018 Enabling technologies on cloud of things for smart healthcare *IEEE Access* **6** 31950–67

[6] Baker S E, Xiang W and Atkinson I M 2017 Internet of things for smart healthcare: technologies, challenges, and opportunities *IEEE Access* **5** 26521–44

[7] Marques G, Pitarma R, Garcia N M and Pombo N 2019 Internet of Things architectures, technologies, applications, challenges, and future directions for enhanced living environments and healthcare systems a review *Electronics* **8** 1081

[8] Elsts A, Fafoutis X, Woznowski P, Tonkin E, Oikonomou G, Piechocki R J and Craddock I 2018 Enabling healthcare in smart homes the SPHERE IoT network infrastructure *IEEE Commun. Mag.* **56** 164–70

[9] Rizvi T, Bhonsle D, Rahangdale R and Bagga J 2022 A Survey on Behavioral Change During the COVID-19 Outbreak in India *Computational Intelligence and Applications for Pandemics and Healthcare* (Hershey, PA: IGI Global) ch 9 pp 184–203
[10] Rizvi T, Bhonsle D and Uzma R 2022 Analysis and comparison of psychological constraints among various countries during COVID-19 *Computational Intelligence and Applications for Pandemics and Healthcare* (Hershey, PA: IGI Global) ch 12 pp 248–68
[11] Zhou W, Jia Y, Peng A, Zhang Y and Liu P 2019 The effect of IoT new features on security and privacy new threats, existing solutions, and challenges yet to be solved *IEEE Internet Things J.* **6** 1606–16
[12] Strangman G, Boas D A and Sutton J P 2002 Non-invasive neuroimaging using near-infrared light *Biol. Psychiatry* **52** 679–93
[13] Jaidka H, Sharma N and Singh R 2020 Evolution of IoT to IIoT applications and challenges *Proc. Int. Conf. on Innovative Computing & Communications (ICICC) 2020* (Rochester, NY: SSRN)
[14] Alam M M, Malik H, Khan M I, Pardy T, Kuusik A and Moullec Y L 2018 A survey on the roles of communication technologies in IoT-based personalized healthcare applications *IEEE Access* **6** 36611–31
[15] Jia X, Chen H and Qi F 2012 Technical models and key technologies of E-Health Monitoring *2012 IEEE 14th Int. Conf. on e-Health Networking, Applications and Services (Healthcom)* (Piscataway, NJ: IEEE) pp 23–6
[16] Goel A K, Rose A, Gaur J and Bhushan B 2019 Attacks, Countermeasures and Security Paradigms in IoT *2019 2nd Int. Conf. on Intelligent Computing, Instrumentation and Control Technologies (ICICICT)* (Piscataway, NJ: IEEE) pp 875–80
[17] Tuli S, Tuli S, Wander G, Wander P, Gill S S, Dustdar S, Sakellariou R and Rana O 2020 Next generation technologies for smart healthcare challenges, vision, model, trends and future directions *Internet Technol. Lett.* **3** e145
[18] Amin S U, Hossain M S, Muhammad G, Alhussein M and Rahman M A 2019 Cognitive smart healthcare for pathology detection and monitoring *IEEE Access* **7** 10745–53
[19] Talal M *et al* 2019 Smart home-based IoT for real time and secure remote health monitoring of triage and priority system using body sensors: multi-driven systematic review *J. Med. Syst.* **43** 42
[20] Ramya S, Karthikeyan P, Kaviya K, Ananthi B and Sachin M C 2019 IoT based smart health monitoring system a recent approach and analysis *Int. J. Adv. Res. Innov. Ideas Educ.* **5** 1403–10 http://ijariie.com/AdminUploadPdf/IoT_BASED_SMART_HEALTH_MONITORING_SYSTEM__A_RECENT_APPROACH_AND_ANALYSIS_ijariie9843.pdf
[21] Joshi S and Joshi S 2019 A Sensor based Secured Health Monitoring and Alert Technique using IoMT *2019 2nd International Conference on Intelligent Communication and Computational Techniques (ICCT)* (Piscataway, NJ: IEEE) pp 152–6
[22] Sharma V, Rathore S and Bhonsle D 2021 A novel approach for reducing attention deficit disorder in children using brainwave entrainment *Artificial & Computational Intelligence* **2** 18 http://acors.org/JOURNAL/Papers/Volume2/issue3/vol2_issue3_02.pdf
[23] Bhonsle D, Chandra V K and Sinha G R 2017 Noise removal from medical images using shrinkage based enhanced total variation technique *J. Adv. Res. Dynamical Control System* **13** 549–60

[24] Bhonsle D, Chandra V K and Sinha G R 2018 An optimized framework using adaptive wavelet thresholding and total variation technique for de-noising medical images *J. Adv. Res. Dynamical Control System* **10** 953–65

[25] Bhonsle D, Chandra V K and Sinha G R 2015 Gaussian and speckle noise removal from ultrasound images using bivariate shrinkage by dual tree complex wavelet transform *i-Manager's J. Image Processing* **2** 1–5

[26] Bhonsle D, ChandraV K and Sinha G R 2018 De-noising of CT images using combined bivariate shrinkage and enhanced total variation technique *i-Manager's J. Electronics Engineering* **8** 12

[27] Bhonsle D, Bagga J, Mishra S, Sahu C, Sahu V and Mishra A 2022 Reduction of Gaussian noise from Computed Tomography Images using Optimized Bilateral Filter by Enhanced Grasshopper Algorithm *2022 Second Int. Conf. on Advances in Electrical, Computing, Communication and Sustainable Technologies (ICAECT)* (Piscataway, NJ: IEEE) pp 1–9

[28] Bhonsle D, Chandra V K and Sinha G R 2018 Speckle noise removal from ultrasound images using combined bivariate shrinkage and enhanced total variation techniques *Int. J. Pure Appl. Math.* **118** 1109–31 https://acadpubl.eu/jsi/2018-118-18/issue18b.html

[29] Bhonsle D, Rizvi T, Mishra S, Sinha G R, Kumar A and Jain V K 2022 Reduction of Ultrasound Images using Combined Bilateral Filter & Median Modified Wiener Filter *2022 Second International Conference on Advances in Electrical, Computing, Communication and Sustainable Technologies (ICAECT)* (Piscataway, NJ: IEEE) pp 1–5

[30] Chaudhary S, Taran S, Bajaj V and Sengur A 2019 Convolutional neural network based approach towards motor imagery tasks EEG signals classification *IEEE Sensors J.* **19** 4494–500

[31] Islam S M R, Kwak D, Kabir M H, Hossain M and Kwak K-S 2015 The internet of things for healthcare: a comprehensive survey *IEEE Access* **3** 678–708

[32] Habibzadeh H, Dinesh K, Shishvan O R, Boggio-Dandry A, Sharma G and Soyata T 2020 A survey of healthcare internet of things (HIoT) a clinical perspective *IEEE Internet Things J.* **7** 53–71

[33] Machado F M, Koehler I M, Ferreira M, Sand and Sovierzoski M A 2016 An mHealth Remote Monitor System Approach Applied to MCC Using ECG Signal in an Android Application *New Advances in Information Systems and Technologies(Advances in Intelligent Systems and Computing* vol 445*)* (Berlin: Springer) pp 43–9

[34] Rodrigues J J, Segundo D B, Junqueira H A, Sabino M H, Prince R M, Al-Muhtadi J and Albuquerque V H 2018 Enabling technologies for the internet of health things *IEEE Access* **6** 13129–41

[35] Tiwari R, Sharma N, Kaushik I, Tiwari A and Bhushan B 2019 Evolution of IoT & Data Analytics using Deep Learning *2019 Int. Conf. on Computing, Communication, and Intelligent Systems (ICCCIS)* (Piscataway, NJ: IEEE) pp 418–423

[36] Alsamhi S H, Ma O, Ansari M and Almalki F A 2019 Survey on collaborative smart drones and internet of things for improving smartness of smart cities *IEEE Access* **7** 128125–52

[37] Rubí J N and Gondim P R 2019 IoMT platform for pervasive healthcare data aggregation, processing, and sharing based on OneM2M and Open EHR *Sensors* **19** 4283

[38] Arora A, Kaur A, Bhushan B and Saini H 2019 Security Concerns and Future Trends of Internet of Things *2019 2nd Int. Conf. on Intelligent Computing, Instrumentation and Control Technologies (ICICICT)* (Piscataway, NJ: IEEE) pp 891–896

[39] Zhou J, Cao Z, Dong X and Vasilakos A V 2017 Security and privacy for cloud-based IoT challenges *IEEE Commun. Mag.* **55** 26–33

[40] Elhoseny M, Ramírez-GonzálezG, Abu-Elnasr O M, Shawkat S A, Arunkumar N and Farouk A 2018 Secure medical data transmission model for iot-based healthcare systems *IEEE Access* **6** 20596–608

[41] Farahani B, Firouzi F, Chang V I, Badaroglu M, Constant N and Mankodiya K 2018 Towards fog-driven IoT eHealth: promises and challenges of IoT in medicine and healthcare *Fut. Gener. Comput. Syst* **78** 659–76

[42] Miotto R, Wang F, Wang S, Jiang X and Dudley J 2018 Deep learning for healthcare: review, opportunities and challenges *Brief. Bioinformatics* **19** 1236–46

[43] Siyal A A, Junejo A Z, Zawish M, Ahmed K, Khalil A and Soursou G 2019 Applications of block chain technology in medicine and healthcare challenges and future perspectives *Cryptography* **3** 3

IOP Publishing

Cognitive Sensors, Volume 1
Intelligent sensing, sensor data analysis and applications
G R Sinha and Varun Bajaj

Chapter 10

Electromagnetic fields and the effects of Internet of Things infrastructure on human health

Hitesh Singh, Vivek Kumar, Kumud Saxena, K V Arya and Ramjee Prasad

Technology developments in the field of wireless communication are creating concerns about the effects of wireless communication on human health. Many studies are being conducted to identify the impact of electromagnetic fields in living spaces on human health. The majority of electromagnetic fields (EMFs) are created by cell phone towers, cell phones, laptops, Wi-Fi, and Internet of Things (IoT) devices. This chapter concentrates on this issue and elucidates the different guidelines for EMF radiation limits that have been published worldwide. We also report an experimental study that has been conducted in order to ascertain the effects of EMFs in a college premises.

10.1 Introduction

Electromagnetic (EM) radiation is the flow of energy in space. This radiation, which is prevalent in both natural and artificial environments, is made up of waves of electric and magnetic energy traveling in tandem at the speed of light. Natural occurrences of EM radiation include ultraviolet light from the Sun or the lightning in thunderstorms. The artificial generation of EMFs is due to the cell phone towers, the broadcasting of television and radio signals, high-voltage electric wires, and various electric appliances present in homes and offices.

The term 'EMF' is given to a flow of photons in space. Each photon contains some energy, and the amount of energy present in the photon determines the type of radiation. The electromagnetic spectrum is the range of all types of electromagnetic radiation. Radio waves from radio stations or x-rays used in hospitals are all part of this spectrum. Depending on their frequencies and power levels, EMF emissions are categorized as either ionizing or non-ionizing. Electromagnetic radiation that produces ions is referred to as ionizing radiation; its waves contain enough energy

doi:10.1088/978-0-7503-5326-7ch10

to overcome the electron-binding energy of an atom or molecule. Some of examples of ionizing radiation are ultraviolet rays, x-rays, cosmic rays, gamma rays, etc. Non-ionizing radiation refers to a type of electromagnetic radiation that does not have enough energy per quantum to ionize an atom or molecule. Low-frequency radiation such as radio waves, microwaves, and infrared radiation are examples of non-ionizing radiation.

Other significant uses of electromagnetic radiation in daily life outside of telephony include: photosynthesis, which transforms solar energy from the Sun into chemical (food) energy in plants; x-rays, which are used in security scanners at airports and retail locations; microwaves, which are used in microwave ovens and radar; and radio waves, which are used in radio and television broadcasts. Frequency modulation (FM)/amplitude modulation (AM) radio, broadcast TV, the Global System for Mobile communication (GSM), code-division multiple access (CDMA), wireless local area networks (WLANs), Bluetooth, Zigbee, Wi-Fi, and mobile networks using Worldwide Interoperability for Microwave Access (WiMAX) technology, which use the very high frequency (VHF), ultra high frequency (UHF), long (L), and short (S) frequency bands, are the most prevalent exposure sources.

Electromagnetic radiation in the frequency range of 1 Hz to 1 THz (1000 GHz) is called non-ionizing, as it does not have enough energy to change the chemical bonds in the human body. The health effects of EMFs associated with non-ionizing radiation include the impacts of extreme levels of thermal stress on tissues. At frequencies above 1 THz, electromagnetic radiation is called ionizing and has sufficient potential to alter the chemical bonds in human tissues and cause serious genetic damage upon prolonged exposure. Some types of radiation have a negative effect on living organisms because they can ionize atoms/molecules. These can break chemical bonds and damage essential molecules. If this damage is minor, cells can heal; otherwise, apoptosis can occur as dead cells cannot be replaced quickly.

10.2 EMF exposure studies

Medical research takes many forms, including epidemiological, laboratory, and clinical research. Epidemiological studies collect real data and draw conclusions from the information collected. In medical research, epidemiological studies observe and compare people with a specific disease or exposure to see whether the risk differs between groups. In general, when epidemiological studies show consistent and strong associations with risk factors, researchers develop a plausible theory to estimate the disease incidence caused by such exposure, which is termed a biological mechanism.

Epidemiological studies alone are not sufficient to prove cause and effect, because the results are statistical conclusions rather than direct evidence. Beyond epidemiological studies, evaluating the actual health effects of exposure to magnetic fields requires laboratory studies of cells and animals, as well as clinical studies of human volunteers. Controlled laboratory studies are conducted to test hypotheses at migration levels and in laboratory animals. In medical laboratory research, researchers have complete control over the research conditions, which helps them to identify the actual biological mechanisms by which potential factors such as magnetic fields can trigger disease.

Clinical studies use biological mechanism theories and laboratory results to quantify the impacts to humans. In clinical trials, volunteers test a variety of treatments to accurately measure the actual impact. When studying the effects of EMFs, medical researchers use volunteer intensity-controlled exposures to look for measurable changes such as brain activity and hormone levels.

In order to identify the effect of EMF radiation on recombinant DNA (RE-DNA), a study was conducted by Del Re *et al* [1]. Samples were taken first in a controlled environment and then in an environment exposed to EMF radiation. Their results show that radiation has a significant impact on the transcription of RE-DNA and that the effects depend on cellular radiation and cell type. The authors of [2] conducted a review to identify the effect of radio waves on the human brain, and also discussed the physiological context of brain activities and their technical requirements for the analysis.

Schoolchildren are exposed to cell phones, which is also a health concern; hence, a study was conducted and an analysis presented by Ali *et al* [3] based on the number of hours that electronic devices were used by children. A study [4] concluded that long-term exposure to the EMF results in stress, anxiety, depression, and poor sleep. They also observed a relationship between increased stress level, depression, anxiety, and extremely low-frequency EMF (ELF EMF) exposure and clearly mentioned that the quality of sleep in technicians is low compared to that of other persons.

Another experiment by El-Naggar *et al* [5] was performed on rats in order to identify the effects of electromagnetic field exposure due to cell phones. They categorically concluded that morphological changes and behavioral changes occurred in rats as a result of mobile phone EMF exposure. The relationship between ELF EMF and metabolic currents was analyzed in [6], and the conclusions drawn using the experimental results suggested that metabolic current is the key indicator in identifying the effect of ELF EMF on living organisms. The effect of long-term exposure to ELF EMF radiation is to increase the absorption of radiopharmaceuticals by the pulmonary system, which may affect the lung-to-heart specific ratio. This in turn affects diagnostic process used by physicians [7].

Montzeri *et al* [8] established that radio signals from Wi-Fi affect the cell membranes, cellular proteins, and enzymes. They concluded from their research that due to the proven medical effects of EMF radiation on human health, more studies are required in order to determine the ranges and rates of these effects. Another experimental study was performed on 24 adults in the West Bank, Palestine. The purpose of this study was to identify the effects of EMF radiation on human beings. Exposure was measured using the EME SPY 140 device. The total exposure to cellular EMFs varied widely among participants, ranging from 0.2 V m^{-1} to 0.9 V m^{-1} [9]. Another important experiment performed by Ramirez-Vazquez [10] showed that the EMF exposure level fell below the recommendations of the International Commission on Non-Ionizing Radiation Protection (ICNIRP). They observed that the main sources of EMF exposure were Digital Enhanced Cordless Telecommunications (DECT), cell phones, and Wi-Fi.

The review performed in [11] showed that at present, there is a limited amount of scientific evidence for the harmful effects of EMF on humans and that a lot more study still needs to be done. They also described some regulatory measures used for EMF protection. Articles [12, 13] described the EMF exposure caused by telecommunication systems, and stated that equipment in urban environmental conditions does affect the health of the general public.

The effect of EMF radiation on Italian honeybees was studied by J Jungwirth [14]; the results of his experiment showed that the particles extracted from bees were magnetic in nature. Using a further experiment, he showed that exposure to 2.4 GHz waves for 30 days had no effect on average particle size.

A report published by New Zealand researchers Elwood and Wood [15] described the effect of radio-frequency EMF radiation on the health of living beings. Their report specifically mentioned that the enormous available scientific literature is mostly based on scientific studies and animal studies and is very variable in nature; hence, a systematic scientific study is still required. The effects of EMF radiation on insects, birds, and other living beings such as rats, cows, plants, and human beings were studied in [16, 17]; these studies observed that long-term EMF exposure may cause different diseases in living beings. It is necessary to study the biological effects of EMF exposure caused by various types of electronic equipment. A comprehensive review of this topic was presented in [18]. An experimental study was carried out in order to measure EMF exposure in Stockholm, Sweden in March and April 2017 [19]; the result indicated that all the measurements exceeded the target level of 30–60 $\mu W\ m^{-3}$. The effect of radio-frequency EMFs generated by cell phone technologies was systematically reviewed in [20]. Different strategies for protecting people from such harmful radiation have also been discussed. A simulation study was carried out by I Nasim [21] in order to identify the EMF radiation level of a 5G communication system over a short distance; it determined that that the radiation significantly exceeded the regulatory level. Various studies have been established to determine the effect of EMFs on male reproductive system [22]. This experimental study, which studied a radio frequency of 2.45 GHz emitted by Wi-Fi equipment, concluded that EMFs are hazardous to the male reproductive system.

Yassin *et al* [23] performed an experiment in the Gaza Strip in order to measure the effect of EMF radiation produced by cell phone towers and observed that the radiation level was less than the nationally and internationally recommended levels. A study was done by Hosseini *et al* [24] to measure the effects of Wi-Fi signals on the cognitive functions of the mind. The results showed that no significant impacts on reaction time, short-term memory, and reasoning ability were exhibited by the students who took part in the experiment. A proper scientific review based on current research findings related to the effects of EMFs on humans was performed by T Butler [25]; this also suggested ways to reduce the risks of EMF exposure for children and adults.

10.3 EMF guidelines

Different organizations have produced different guidelines for EMF exposure, some of which are discussed in this section. According to the guidelines for electromagnetic radiation exposure given in [26], the risk factors for human health can be divided into four groups, namely, low risk, slight risk, moderate risk, and high risk. Further details, including the ranges of the EMFs associated with abovementioned risk categories, are given in table 10.1.

Guidelines have been produced by the National Radiological Protection Board, UK for frequency ranges from 12 MHz to 300 GHz [27]. According to these guidelines, the exposure limit is divided into five different frequency groups. These frequencies are from 12 MHz–200 MHz, 200 MHz–400 MHz, 400 MHz–800 MHz, 0.8 GHz–1.55 GHz and 1.55 GHz–300 GHz. The exposure levels for the different frequency groups are given in terms of electric field, magnetic field, and power density in table 10.2.

The International Commission on Non-Ionizing Radiation Protection (ICNIRP) produces guidelines for frequency ranges from 10 MHz to 300 GHz [27]. These guidelines are the most widely used throughout the world. Most researchers and engineers use these guidelines for their calculations. These regulatory guidelines are mainly used for base transceiver stations (BTSs) or cell phone towers. These guidelines are divided into three different groups, corresponding to the ranges from 10 MHz to 400 MHz, 400 MHz to 2000 MHz, and 2 GHz to 300 GHz. Details of the guidelines for occupational exposure are given in table 10.3 and those for public exposure are given in table 10.4.

Table 10.1. Risk factors to health associated with EMF exposure.

Risk level	Electric field (mV m^{-1})	Power density (μW m^{-2})	Description
1	0.00–10.0	0.00–0.27	Low risk
2	10.0–100	0.1–26.5	Slight risk
3	100–650	26.5–1120	Moderate risk
4	More than 650	More than 1120	High risk

Table 10.2. NRPB guidelines for EMF exposure.

S. No.	Frequency range	Electric field (V m^{-1})	Magnetic field (A m^{-1})	Power density (W m^{-2})
1	12–200 MHz	50	0.13	6.6
2	200–400 MHz	250f	0.66f	165f2
3	400–800 MHz	100	0.26	26
4	0.8–1.55 GHz	125f	0.33f	41f2
5	1.55–300 GHz	194	0.52	100

Table 10.3. ICNIRP guidelines for occupational exposure.

S. No.	Frequency range	Electric field (V m^{-1})	Magnetic field (A m^{-1})	Power density (W m^{-2})
1	10–400 MHz	61	0.16	10
2	400–2000 MHz	3f1/2	0.008 f1/2	f/40
3	2–300 GHz	137	0.36	50

Table 10.4. ICNIRP guidelines for public exposure.

S. No.	Frequency range	Electric field (V m^{-1})	Magnetic field (A m^{-1})	Power density (W m^{-2})
1	10–400 MHz	28	0.073	2
2	400–2000 MHz	1.375f1/2	0.0037 f1/2	f/200
3	2–300 GHz	61	0.16	10

Table 10.5. EMF exposure norms for BTS towers in India.

S. No.	Frequency	ICNIRP norm (W m^{-2})	DoT Norm (W m^{-2})
1	900 MHz	4.5	0.45
2	1800 MHz	9	0.9
3	2100 MHz	10.5	1.05

In India, the Department of Telecommunications (DoT) is the regulatory authority responsible for monitoring the electromagnetic radiation emitted by BTSs. The DoT has issued a list of instructions that establishes the limits for EMF radiation and the testing procedures to be followed for BTS towers. The DoT randomly monitors the radiation emitted by BTSs and also performs tests following user complaints in order to check whether the regulations are being followed by the various service providers. The DoT has adopted the EMF exposure limits given in the ICNIRF guidelines. The EMF exposure limit was also amended by the DoT in November 2008. The regulations for Indian settings are given in table 10.5. Similarly, the EMF exposure limits followed in different parts of the world are listed in table 10.6.

10.4 The effects of EMF exposure on human health

A lot of research is being performed to study the effects of EMF exposure on human health. These electromagnetic hazards are present as a result of the rapid growth of wireless technologies. This section deals with the various research findings for the effects of EMF exposure on humans.

Table 10.6. EMF exposure limits for different parts of the world.

Country	Power density (W m^{-2})
USA, Canada, and Japan	12
EU recommendation 1998	9.2
Australia	9
Belgium	2.4
Italy, Israel	1.0
Auckland, New Zealand	0.5
Luxembourg	0.45
China	0.4
Russia, Bulgaria	0.2
Poland, Paris, Hungary	0.1
Italy	0.1
Switzerland	0.095
Germany	0.09
Austria	0.001

Some of the observations are based on the review published in [33]. A study by the International Agency for Research on Cancer (IARC) observed that radio-frequency EMF exposure has an adverse effect on human health. Another observation was that EMF exposure caused by cell phones has an adverse effect on children's brains because radiation reaches further into their brains. This is caused by the thin outer structure of their skull and skin compared to those of adults. Other experimental studies have shown that men who keep cell phones in their trouser pockets are more prone to low sperm counts and impaired sperm motility rates. According to a literature review [26], ELF EMF exposure has detrimental biological effects, depending on the current intensity, magnetic field, and exposure time. Cumulative epidemiologic data show a correlation between ELF EMF exposure and childhood cancer incidence, Alzheimer's disease (AD), and miscarriage. However, ELF EMF does not increase the risk of adult cancer. In addition, there is no conclusive evidence for cardiovascular death from ELF EMF exposure. There is no comprehensive mechanism that explains the biological effects of ELF EMF. Ultimately, more research is needed to clarify the mechanisms of these magnetic fields.

At a ELF of 935 MHz, Zielinski *et al* [27] investigated the effects of EMF on apoptosis, autophagy, oxidative stress, and electron metabolism in N9 microglia and SH-SY5Y neuroblastoma cells. Cells were given 2 h and 24 h of exposure to 4 W kg^{-1} or a placebo treatment. Regardless of exposure duration, EMF exposure to both cell types had no effect on apoptosis, viable cell count, or apoptosis-inducing factor (AIF). The autophagy marker protein ATG5 was found to be more abundant when cells were exposed to EMF for 24 h as opposed to 2 h; however, LC3B-I was unaffected by either the cell type under investigation or the exposure duration.

Transitory increases in glutathione (GSH) were only seen in SH-SY5Y cells, not in hydrogen peroxide or cytochrome c-oxidase. This suggests that transient EMFs at the specific absorption rate (SAR) are responsible.

Another study [28] examined how magnetic field (MF) exposure at frequencies of 50 Hz and voltages of 0.4 mT affected cellular frequencies. The findings demonstrated that sphingosine kinase 1 (SphK1) activity was markedly increased by MF exposure, and that MF-induced cell proliferation could be greatly reduced by blocking the SphK1–S1P–S1P (S1PR) receptor pathway. Additionally, S1P extracellular tests have shown that MF-induced S1P can indirectly operate on S1PR1/3, mediating scattering effects in a paracrine and/or autocrine way. Interestingly, the SphK1–S1P–S1PR1/3 cascade regulates MF-induced proliferation by activating the extracellular kinase (ERK) rather than the protein kinase B (Akt) pathway, despite the fact that MF activates the ERK and Akt signaling pathways. The findings of this study suggest that the SphK1–S1P–S1PR1/3 cascade plays a key role in MF-induced cell proliferation.

The effects of regular, daily exposure to a magnetic field at a frequency of 50 Hz were investigated in [29]. The EMDEX II dosimeter was used to record the magnetic field every 30 seconds during the day for one week in order to observe the effects of chronic exposure to a magnetic field at a frequency of 50 Hz for one to 20 years at home and at work in 15 males (aged 38.0 ±0.9 years). Individual exposures' average geometric values per week ranged from 0.1 to 2.6 T. Every hour between 20:00 and 08:00, blood samples were obtained. The chromogranin A (CgA) patterns of exposed participants and controls were compared for age. The results from 15 unexposed men were compared to those of the compounds under investigation. This study shows that serum CgA levels peaked in the control group with a gradual decrease in serum concentration and a morning trough. The CgA profile and serum concentration, an indicator of neuroendocrine tumors and stress, did not appear to decrease in patients with chronic exposure to the magnetic field for long periods of time (up to 20 years), even for maximum exposure levels ($>0.3\ \mu T$). However, this does not preclude the risk of potentially damaging ELF EMF for vulnerable populations, such as children and the elderly. Their risks may be higher at lower exposures, so, at a minimum, they should be registered for domestic exposure.

Age is not important here, because the effects of short-term overnight exposure to EMF on sleep are occasional for both young and elderly men [30]. However, this assumption cannot be conclusively tested until data from healthy young women can be compared with data from older women. However, the current results do not suggest that there is a negative effect on health. Kim and Nasim [31] investigated the impact of EMF on humans in 5G systems and compared it with measurements of the previous generations of cellular systems. Observing the minimum distance between the transmitter and human users kept EMF exposure below statutory safe levels, giving users assurance about the safe use of 5G communications.

To warn against unintended public health effects, data can be used to explain how 'devastating danger' exists and disrupts human wellbeing at wavelengths and access levels close to those provided by Wi-Fi networks [32]. Such danger can take the form of barrier, Ca++ flux, hormonal swings, neurological conditions, diminished brain wave function, depression, memory loss, slower response rate, tinnitus, fainting, skin problems, extreme exhaustion, constant pain and nausea, respiratory system failure, and so on. As the population grows, more people are exposed to electromagnetic waves.

The effects of EMF on humans (especially children) based on clinical studies was presented by J H Moon in [33], which discusses the current views of the World Health Organization (WHO) and the IARC on the causes, biological effects, and carcinogenicity of EMF exposure as well as the effects of EMF exposure in children. Scientific knowledge must be interpreted objectively, since well-controlled experiments on EMF in children are almost impossible. Precautions for children are recommended until the potential health effects of EMF have been identified.

A systematic and comprehensive evaluation was performed by Mannan *et al* [34] in order to study the distribution and spread of radiation produced by electrical products. This study investigated how radiation diffusion varies with building materials, along with measurements of the radiation exposure in buildings. In addition, the effects of radiation on human health were studied. In an environmental evaluation of five buildings, gypsum walls provided the best shielding (with a transmitted electric field strength of 18%), while wood and glass walls served as poor shielding materials (with electric field strengths of 96% and 97%, respectively). A comparison of the standard values set by international organizations shows that low field-strength values have been defined by some regulatory agencies, but these values are high compared to the protection limits in some countries, indicating that such low field-strength values may have a potential impact on human health in the built environment. Therefore, this study recommended that a suitable shielding material should be used to maintain the exposure distance and shorten the exposure time. The negative and positive biological effects of cell phone radiation on various human organs were analyzed by Singh *et al* [35]. Their study aimed to capture all the positive and negative health risks to the population caused by the excessive use of cell phones. It was concluded that prolonged exposure to EMF radiation from excessive cell phone use has many health effects, including brain cancer. It also creates some positive health benefits, including improving bone healing and reducing the toxicity of chemotherapy. Some employees did not experience a significant health impact as a result of cell phone use. Given all these facts, long-term research and analysis is needed. From the above discussion, it can be concluded that exposure to EMFs may alter neurobehavioral activity, cytokine levels, oxidative stress, and DNA integrity.

Binboga *et al* aimed to study the immediate effects of ELF EMF on heart rate (HR) and heart rate variability (HRV) [36]. Their sample included 34 healthy men between the ages of 18 and 27. A double-blind design was used in which participants were randomly allocated to the EMF group ($n = 17$) or the control group ($n = 17$). Repeated measurements were made up of three five-minute experimental periods. The chest region of every member of the EMF group was continuously exposed to a linearly polarized, 50 Hz, 28 T EMF during the EMF exposure. The intraoral imaging detector continually recorded HR and HRV data. Statistical analysis of the subjects showed that the HR was significantly slowed in the EMF and control groups. However, only the EMF group experienced an increase in the standard deviation (SDNN), the root mean square of successive differences (RMSSD), and the low-frequency (LF) and high-frequency (HF) power root mean square values of NN intervals, while the control group members maintained steady values. In addition, they compared the HRV values obtained throughout the simulation period as well as the EMFs of the two experimental groups. Numerical analysis revealed that the EMF group had considerably greater SDNN, RMSSD, LF, and HF values than those of the control group. The LF/HF ratio did not significantly differ across groups or among groups. These findings led to the conclusion that short-term chest ELF EMF exposure could increase parasympathetic dominance while the subject was at rest.

10.5 Experimental setup and results

An experiment was performed in order to understand a real EMF exposure scenario in an indoor environment under Indian conditions. The experimental site was the college building of the Noida Institute of Engineering and Technology (NIET), Greater Noida, India. EMF measurements were taken on each floor of the building and in the classrooms in order to determine the exact amount of radiation present due to Wi-Fi, cell phone towers, and other factors. The instrument used for the measurements was an RF three-axis field-strength meter working in the range from 50 MHz to 3.5 GHz. Its measurement range was from 38 mV m^{-1} to 20 V m^{-1}. It is shown in figure 10.1.

The experiment was performed in the D block of NIET, Greater Noida, India. Measurements were recorded on each floor of the building inside a classroom. A photograph showing the layout of the building used in the experiment is shown in figure 10.2. A floor plan of a classroom is shown in figure 10.3.

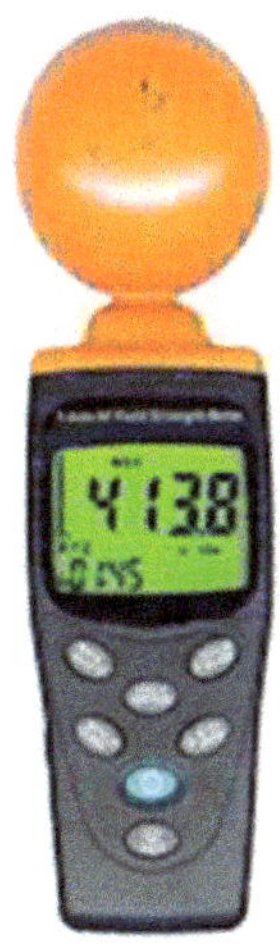

Figure 10.1. Instrument used to obtain EMF readings.

Figure 10.2. NIET D block.

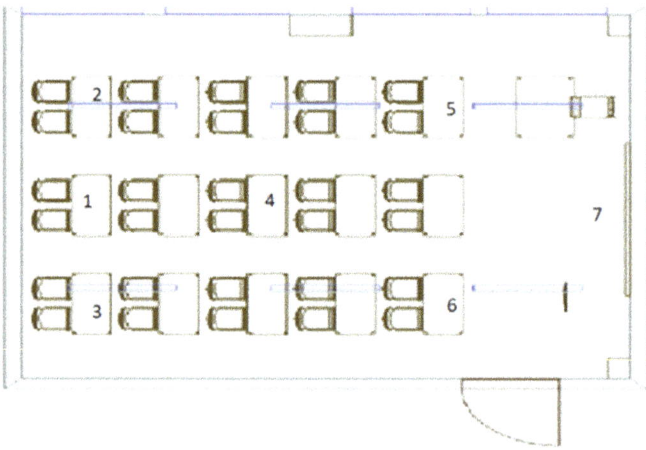

Figure 10.3. The layout of the classroom.

The methodology of the experiment was as follows: readings were taken using the instrument, the instrument was taken to the classrooms present on each floor of the building, and measurements were collected from one classroom per floor. The results are shown in table 10.7. In each classroom, readings were taken with students present in the class at a class strength of 52 students per room. The measurements were recorded at different positions (see figure 10.3) in the classroom. The results of the measurements are tabulated in table 10.8. It is clearly evident from the results depicted in tables 10.7 and 10.8 that the EMF radiation levels present in the building are higher than the recommended levels.

The results clearly show that there is a lower field strength of about 105.1 mV m^{-1} in the basement, compared to the figure of 900.7 mV m^{-1} for the third floor. In the

Table 10.7. Experimental results obtained for the measurements collected from the classrooms.

S. No.	Floor	Electric field mV m^{-1}
1	Basement	105.1
2	Ground floor	769.4
3	First floor	865.8
4	Second floor	7983
5	Third floor	900.7

Table 10.8. Experimental results obtained by measuring at different positions in the classroom.

Position No.	Electric field (mV m^{-1})	Magnetic field (A m^{-1})	Power density (μW m^{-2})	W cm^{-2}
1	701.6	1592.1	1282.6	0.022
2	522.4	1156.3	442.2	0.254
3	435.3	1673.8	1496.3	0.159
4	913.5	1527.6	1544.1	0.942
5	457.9	1692.5	1255.1	0.125
6	600	1670.3	1180.7	0.041
7	572.2	1734.2	1405.8	0.094

case of a classroom where students were present and using their mobile phones, the readings showed that an EMF of more 1000 μW m^{-2} was present, which was higher than the recommended value. It was also observed that all the readings exceeded both national and international regulations, as given in table 10.1 to table 10.6. Comparisons are shown on a floor-by-floor basis for the whole building in figure 10.4. Figure 10.5 compares the EMFs measured at various locations inside the room with the recommended levels.

10.6 Conclusions

The impact of the EMF radiation present in the urban environment is significant. The majority of the EMF radiation in this environment is due to cell phone towers, mobile handsets, Wi-Fi, laptops, and other IoT devices present in the smart city environmental conditions. EMF radiation has a significant impact on human health as observed by experimental study carried out at college premises. The results of the experiment showed that the levels of EMF radiation present in the classroom environments were significantly higher than those prescribed by national and international regulations. Therefore, there is a need to find a way to control the radiation.

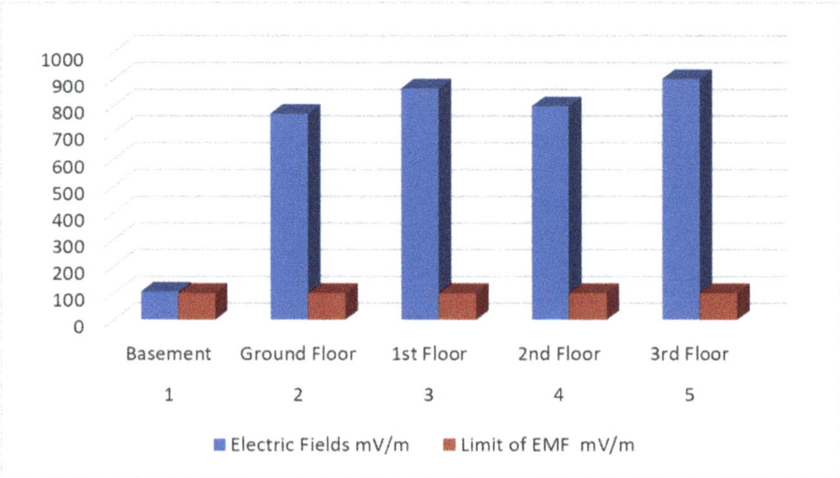

Figure 10.4. Comparisons of the measured EMFs and the regulatory limit for each floor of the building.

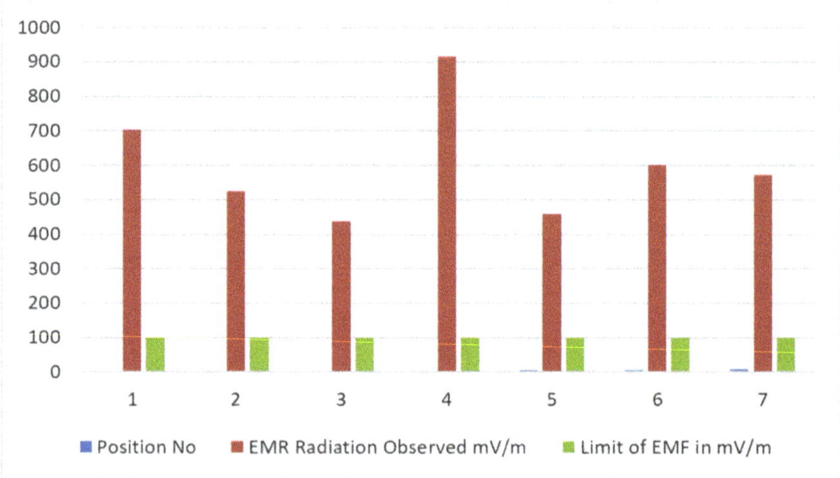

Figure 10.5. Comparisons of the measured EMFs and the regulatory limit for various positions inside the classroom.

References

[1] Del Re B, Bersani F and Giorgi G 2019 Effect of electromagnetic field exposure on the transcription of repetitive DNA elements in human cells *Electromagnetic Biology Medicine* **38** 262–70

[2] Danker-Hopfe H, Eggert T, Dorn H and Sauter C 2019 Effects of RF-EMF on the human resting-state EEG—the inconsistencies in the consistency. Part 1: non-exposure-related limitations of comparability between studies *Bioelectromagnetics* **40** 291–318

[3] Ali A H, Salman M D, Saleh R, Fayadh E N, abd Al-Hameed S, Falah H, Thamer R, Saad M and Ali A H 2019 The effect of the electronic devices on children. *J. Phys. Conf. Ser.* **1178** 012002
[4] Bagheri Hosseinabadi M, Khanjani N, Ebrahimi M H, Haji B and Abdolahfard M 2019 The effect of chronic exposure to extremely low-frequency electromagnetic fields on sleep quality, stress, depression and anxiety *Electromagnetic Biology Medicine* **38** 96–101
[5] EL-Naggar M I, EL-Sagheer A E and Ebaid A E S 2019 The possible protective effect of vitamin E on adult albino rat's testes exposed to electromagnetic field emitted from a conventional cellular phone *Egyptian J. Hospital Medicine* **74** 873–84
[6] Shi Z 2019 Geobacter sulfurreducens-inoculated bioelectrochemical system reveals the potential of metabolic current in defining the effect of extremely low-frequency electromagnetic field on living cells *Ecotoxicology Environmental Safety* **173** 8–14
[7] Jadidi M, Bokharaeian M, Esmaili M H, Hasanzadeh H and Ghorbani R 2019 Cell phone EMF affects rat pulmonary Tc-MIBI uptake and oxidative stress *Iranian J. Sci. Technol. Trans.* A **43** 1491–97
[8] Montzeri A, Pooladi M, Odoumizadeh M, Nazarian N and Karani S 2019 The effects of WiFi network (2.45 GHz) in rats with induced stroke associated with an increased risk of heart attack *Archives Adv. Biosciences* **10** 1–9
[9] Lahham A and Ayyad H 2019 Personal exposure to radiofrequency electromagnetic fields among palestinian adults *Health Phys.* **117** 396–402
[10] Ramirez-Vazquez R, Gonzalez-Rubio J, Arribas E and Najera A 2019 Characterisation of personal exposure to environmental radiofrequency electromagnetic fields in Albacete (Spain) and assessment of risk perception *Environ. Res.* **172** 109–16
[11] Muntjir M and Rahul M 2019 An empirical exploration to impact of Wi-Fi (wireless fidelity) on human health *Asian J. Technol. Management Res. (AJTMR)* **8** 1–8 http://ajtmr.com/papers/Vol8Issue2/Vol8Iss2_P1.pdf
[12] Massardier-Pilonchery A, Nerrière E, Croidieu S, Ndagijimana F, Gaudaire F, Martinsons C, Nicolas N and Hours M 2019 Assessment of personal occupational exposure to radiofrequency electromagnetic fields in libraries and media libraries, using calibrated on-body exposimeters *Int. J. Environ. Res. Public Health* **16** 2087
[13] Michałowska J, Mazurek P A, Gad R, Chudy A and Kozieł J 2019 Identification of the electromagnetic field strength in public spaces and during travel *2019 IEEE Applications of Electromagnetics in Modern Engineering and Medicine (PTZE)* (Piscataway, NJ: IEEE) pp 121–24
[14] Jungwirth J 2019 *The Effect of Electromagnetic Fields Produced by WiFi Routers on the Magnetite (Fe3O4) Particles of the Italian Honey Bee (Apis mellifera ssp. ligustica)* St. Cloud State University https://repository.stcloudstate.edu/biol_etds/42/
[15] Elwood M and Wood A W 2019 Health effects of radiofrequency electromagnetic energy *New Zealand Medical J.(Online)* **132** 64–72 https://journal.nzma.org.nz/journal-articles/health-effects-of-radiofrequency-electromagnetic-energy
[16] Christianto V, Boyd R N and Smarandache F 2019 Wireless technologies (4G, 5G) are very harmful to human health and environment: a preliminary review *BAOJ Cancer Res. Ther.* **5** 066
[17] Batool S, Bibi A, Frezza F and Mangini F 2019 Benefits and hazards of electromagnetic waves, telecommunication, physical and biomedical: a review *European Rev. Medical Pharmacological Sci.* **23** 3121–28

[18] Kim J H, Lee J K, Kim H G, Kim K B and Kim H R 2019 Possible effects of radiofrequency electromagnetic field exposure on central nerve system *Biomolecules & Therapeutics* **27** 265–75
[19] Carlberg M, Hedendahl L, Koppel T and Hardell L 2019 High ambient radiofrequency radiation in Stockholm city, Sweden *Oncology Letters* **17** 1777–83
[20] Boehmert C, Freudenstein F and Wiedemann P 2019 A systematic review of health risk communication about EMFs from wireless technologies *J. Risk Research* **23** 571–97
[21] Nasim I 2019 *Analysis of human EMF exposure in 5G cellular systems* Georgia Southern University https://galileo-georgiasouthern.primo.exlibrisgroup.com/permalink/01GALI_GASOUTH/1fi10pa/alma9916230293602950
[22] Jaffar F H F, Osman K, Ismail N H, Chin K Y and Ibrahim S F 2019 Adverse effects of Wi-Fi radiation on male reproductive system: a systematic review *Tohoku J. Experimental Medicine* **248** 169–79
[23] Yassin S, Musleh M and Abuzerr S 2019 Electromagnetic radiation exposure from nearby cellular base stations in the Gaza Strip, Palestine: a concern for public health *J. Biosciences Medicines* **7** 46–59
[24] Hosseini M A, Hosseini A, Jarideh S, Argasi H, Shekoohi-Shooli F, Zamani A, Taeb S and Haghani M 2019 Evaluating short-term exposure to Wi-Fi signals on students'reaction time, short-term memory and reasoning ability *Radiat. Prot. Dosim.* **187** 279–85
[25] Butler T 2019 On the Clear Evidence of the Risks to Children from Non-Ionizing Radio Frequency Radiation: The Case of Digital Technologies in the Home, Classroom and Society (Cork: University College Cork) p 33
[26] Karimi A, Moghaddam F G and Valipour M 2020 Insights in the biology of extremely low-frequency magnetic fields exposure on human health *Molecular Biology Reports* **47** 5621–33
[27] Zielinski J, Ducray A D, Moeller A M, Murbach M, Kuster N and Mevissen M 2020 Effects of pulse-modulated radiofrequency magnetic field (RF-EMF) exposure on apoptosis, autophagy, oxidative stress and electron chain transport function in human neuroblastoma and murine microglial cells *Toxicol. In Vitro* **68** 104963
[28] Chen L, Xia Y, Lu J, Xie Q, Ye A and Sun W 2020 A 50-Hz magnetic-field exposure promotes human amniotic cells proliferation via SphK–S1P–S1PR cascade mediated ERK signaling pathway *Ecotoxicology Environmental Safety* **194** 110407
[29] Touitou Y, Lambrozo J, Mauvieux B and Riedel M 2020 Evaluation in humans of ELF-EMF exposure on chromogranin a, a marker of neuroendocrine tumors and stress *Chronobiology Int.* **37** 60–7
[30] Eggert T, Dorn H, Sauter C, Schmid G and Danker-Hopfe H 2020 RF-EMF exposure effects on sleep – age doesn't matter in men! *Environ. Res.* **191** 110173
[31] Kim S and Nasim I 2020 Human electromagnetic field exposure in 5G at 28 GHz *IEEE Consumer Electronics Magazine* **9** 41–5
[32] Agrawal A and Razak R 2020 Wi-Fi effects on human health *J. Critical Reviews* **7** 758–63 http://www.jcreview.com/admin/Uploads/Files/61a59509c61167.62842791.pdf
[33] Moon J H 2020 Health effects of electromagnetic fields on children *Clinical and Experimental Pediatrics* **63** 422
[34] Mannan M, Weldu Y W and Al-Ghamdi S G 2020 Health impact of energy use in buildings: radiation propagation assessment in indoor environment *Energy Reports* **6** 915–20

[35] Singh A, Ashraf I, Jyoti A and Tomar R S 2020 Mobile phone radiations as an alarming tool for human health: a review *Indian J. Natural Sciences* **10** 18850–59

[36] Binboğa E, Tok S and Munzuroğlu M 2021 The short-term effect of occupational levels of 50 Hz electromagnetic field on human heart rate variability *Bioelectromagnetics* **42** 60–75

Chapter 11

Cognitive sensing in the brain–computer interface: a comprehensive study

Dhairya Shah, Haard Shah and Manan Shah

Cognitive sensing systems feature a perceptual processor that transforms raw sensor input into an informative perception of the environment; this, in turn, determines the next sensing action, simulating the perception–action cycle of human cognition. Cognitive sensing uses advanced problem-solving and machine learning (ML) to process sensor data, which results in timely insights produced by real-time processing. Brain–computer interface (BCI) systems provide a transmission medium that links the brain to various machines, allowing cerebral activity alone to operate external devices. Over the last two decades, the BCI has developed quickly in terms of research and development. Our aim in writing this chapter is to analyze the advances in the field of cognitive sensing in the BCI and provide the reader with a very insightful perspective on this topic to foster more research in this direction. We introduce cognitive sensing together with BCI and analyze the BCI-compatible signals produced by the human brain. Additionally, we explain how these signals serve as the backbone of the current clinical applications of this technology and what it holds for practical future operations; we also identify potential users of all types. The chapter is carefully structured and its sections are blended so that an overview of the topic along with its most important details, obscure aspects, and current limitations are highly comprehensible.

11.1 Introduction

Generation after generation, people have harbored dreams of developing technologies that can understand people's minds and communicate back to them. However, it has only recently become possible for humans to interact directly with the human brain, due to advancements in cognitive neuroscience and brain imaging technology [1]. The BCI interprets the brain's electrical activity at various points in the human brain, changing it into signals and actions that can operate a computer. Applications that

doi:10.1088/978-0-7503-5326-7ch11

utilize the BCI rely on the spinal cord and the brain. The functions of such applications include processing and integrating sensory inputs from peripheral nerves and returning impulses to actuators, such as muscles or glands, to trigger electronic or elective activity [2]. This capability is facilitated using sensors that can monitor certain cognitive states that correspond to physical brain functions.

The ability of BCI devices to directly integrate digital devices with the brain has increased their appeal [3]. The BCI has been imagined to be a practical method for enhancing/substituting neural recovery or an assistive device controlled directly by the brain. It is a promising means to restore basic communication skills and a measure of autonomy to those who are limited by health conditions [4]. People's curiosity has long been sparked by the idea of combining brains and technology, and this has recently come to fruition as a result of breakthroughs in engineering and neurology that have facilitated the ability to repair and possibly improve human physical and mental capabilities [5].

BCI research has emerged as one of today's most intriguing scientific fields, in part because of its potential to ameliorate the lives of those with acute motor damage, such as those with amyotrophic lateral sclerosis (ALS) [6]. The BCI can be used to create applications such as brain-controlled chairs, limbs, and speech systems [7]. The concept of the BCI has developed from a straightforward idea in the early days of digital technology to the incredibly complex signal detection, recording, and analysis systems of today [8]. Recently, research has focused on a different category of possible users: the general public. Commercially, such a broad potential audience might be a very valuable investment. Gamers are the best potential consumers, because they are quick to acquire new technologies if they might be useful and they make up a sizable portion of the population. The use of BCI control for real-world applications comes with several difficulties. These challenges can be grouped into a number of categories, for example, bandwidth, high error rates, autonomy, and cognitive load. The BCI also has many other problems, including the fact that it takes a lot of training to use and is slower and less precise than other modalities. Brain-signal emulation and offline data have been used to test all the current applications and interaction mechanisms; nevertheless, these are insufficient for conclusions to be drawn regarding the utility of different user interface (UI) models [9].

People, society, and healthcare professionals all need to keep up with the latest developments in the neurosciences and neurotechnology. One such developing technology in the field of neurosciences is the BCI. In a word, BCI technology enables direct brain-to-external-device communication to take place, bypassing the usual neuromuscular channels. In addition to the medical and healthcare industries, the BCI has applications in a variety of other spheres of human existence, including entertainment, gaming, education, self-control, marketing, and so forth.

The BCI was originally developed for biomedical purposes, which led to the creation of assistive devices for physically challenged people in order to restore their strength of movement and communication and to restore their lost motor skills. However, this field of study has expanded to include the development of BCI not only for healthy individuals through medical applications but also for the development of non-medical applications, such as generations and prototypes of hands-free products. Globally, BCI is currently in the experimental stage, although it has great potential and could soon be clinically useful. The use of these extremely cutting-edge technologies in the delivery of healthcare, notably in the field of neurology, will undoubtedly fundamentally alter the medical industry [10].

The tools used to simulate intelligent data processing will be included in the next generation of technologies designed to monitor the health of technical systems. The complexity of built technological systems, as well as the complexity of technologies in human society, necessitate the use of intelligent procedures. The phenomena and events that take place in these systems may be radically distinct from human experience in terms of scale, dynamics, and character. Event analysis in these systems is now concentrated on the conversion of a collection of patterns that we are familiar with into patterns that may be utilized to explain these events (which are frequently not directly interpretable).

If there are no strict time limits for this investigation, this approach is usable. However, if the source events' dynamics are more complex than those of our world, it is difficult to employ this method. The creation of technology systems capable of cognitive functions will enable the realization of control and management of technological events, offering a level of detail unreachable to human consciousness [11]. The field of structural health monitoring would be amenable to the use of ubiquitous computing principles, particularly with the advancement of cognitive technical systems [12].

Future computers may be able to feel and perceive emotions, which could expand their use beyond simply assisting people to include making judgments [13]. Based on physiological and behavioral factors, computers may be able to identify and elucidate primary affective conditions. Recent research has shown that the BCI has the potential to be used to study affective states, extending its potential applications to psychology [14, 15]. In order to distinguish both the good and negative emotions engendered by video stimulation, Huang *et al* proposed an electroencephalography (EEG)-based BCI [16].

In this chapter, we explore BCIs and their uses in neuroscience to raise awareness of this cutting-edge developing technology among those who may be concerned. Along the way, we will talk about recent developments in the BCI and how important cognition is to those developments.

11.2 Classification of advanced brain–computer interface technologies

Humans directing machines with their brains may sound like something out of a science fiction film, but BCIs are making this a reality. Before the BCI becomes commonplace, a better understanding of this upcoming technology can help to ensure that the appropriate rules are in place. If today's accomplishments in BCI technology already seem incredible, then it stands to reason that BCI advancements in the not-too-distant future could be truly momentous.

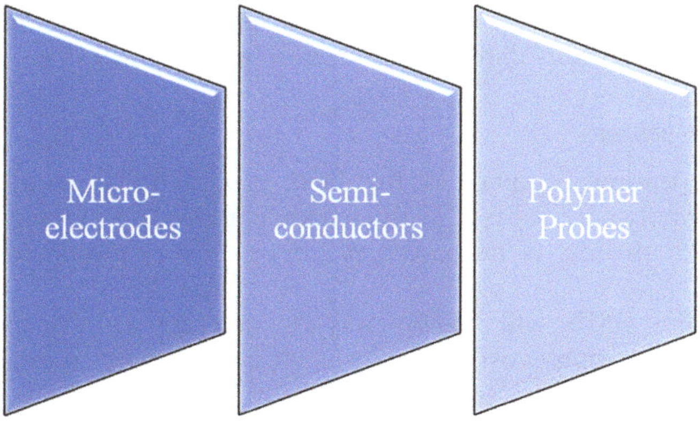

Figure 11.1. Promising technologies in the field of the BCI.

Noninvasive devices often use sensors that are positioned on or near the head to measure and record brain activity. Although these gadgets are simple to install and remove, their signals can be muddled and inaccurate [17, 18].

Surgery is required for invasive BCI. To target certain groups of neurons, electronic implants have to be placed in the brain, right beneath the skull [19]. The BCI implants currently being developed are small and thus require a massive number of neurons to be fired simultaneously. A research group at the University of California, Berkeley, for instance, has created implanted sensors that are about the size of a sand grain. These sensors are referred to as 'neural dust.'

Signals traveling between the brain and a gadget would most likely be much clearer and more precise if invasive techniques were used [20]. However, the operations necessary to implant them would raise health concerns, just as with any surgery (see figure 11.1).

11.2.1 Microelectrodes

The industry standard for recording action potentials is a microelectrode, although there is no therapeutically applicable microelectrode technology for extensive recordings [21, 22]. This would necessitate a system with high biocompatibility, safety, and lifelong material qualities. Additionally, for this device to eventually enable fully implanted wireless functioning, a viable surgical method and high-density, low-power electronics would be required.

11.2.2 Semiconductors

Arrays of electrodes composed of hard metals or semiconductors are the most common type of long-term brain recording equipment. Although rigid metal arrays can penetrate the brain more easily, mismatches in stiff probe size, Young's modulus, and bending stiffness can cause immunological reactions that limit the functionality and longevity of these devices [22, 23]. Additionally, because of the presence of vasculature and the set geometry of these arrays, there are limitations on the populations of neurons that may be accessible.

11.2.3 Polymer probes

Neuralink, the leading company in this field, proposed [21] an alternate robotic approach using polymer probes. The three main components of their technology are minute polymer probes, a neurosurgery robot, and sophisticated high-density electronics.

Future clinical BMIs will heavily rely on cerebral activity modulation, for instance to provide neuroprosthetic movement control with a sensation of touch or proprioception [23, 24].

Thin-film probes may exhibit improved biocompatibility, because their size and makeup are more similar to brain tissue material characteristics than those of the frequently employed probes made of silicon [25, 26]. Due to the versatility of probe placement, we may also develop unique arrays that concentrate on certain brain regions while avoiding vasculature, particularly in the case of subcortical structures. This characteristic is important for building a high-quality BMI, since it allows for customizable electrode deployment based on the demands of the task.

11.3 The role of cognition in brain–computer interfaces

Cognition incorporates mental functions including memory, perception, decision-making, reasoning, and language. Cognitive sensing systems feature a perceptual processor, which transforms raw sensor data into an informative perception of the environment, and an executive processor, which determines the next sensing action. These systems mirror the perception–action cycle of human cognition. The term 'cognitive sensing' refers to the use of intelligent sensors to recognize the environment's signals. Cognitive sensing processes sense data on a real-time basis and use ML and advanced analytics to produce timely insights. In order to support intelligent, reliable, and eco-friendly data collecting operations, cognitive sensing (CS) has been recommended for use in the future Internet of Things (IoT). CS denotes nodes' unique capacity to sense their surroundings intelligently via cognitive sensors. To provide IoT systems with the highly coveted intelligent data collection capabilities, CS makes use of data as well as other methodologies [27]. The use of cognitive computing in the IoT has attracted a lot of attention, since it allows intelligent smart objects to acquire a great deal of information from the real environment [28]. CS is an essential aspect of creating a connection between two devices by gathering brain signals. Device-to-device communication is a new paradigm in communication that enables pervasive connectivity between devices as well as the capacity for autonomous communication without the need for human intervention [29]. Cognitive computing is a subset of artificial intelligence (AI) [30–32]. An important future area of research in AI would be the combination of human brain cognition and intelligent computing [30]. In the area of research into human–computer synergy, a BCI is a successful outcome that allows for a direct brain-to-external-device connection that enhances, facilitates, and heals human cognitive sensing [1]. BCI technology has the potential to improve the learning environment in schools, which is now plagued by a number of challenges and shortcomings. A deep understanding of brain functions can be obtained from cognitive and affective BCIs, which may improve cognitive function and learning techniques. They can provide a more reliable empirical basis for teaching and learning strategies, such as modifying course material by considering student engagement and brain workload and even by assisting students to focus on a particular subject or topic. The human brain can be examined by using the BCI as a cognitive tool to evaluate, comprehend, and enhance the learning process. As demonstrated by cognitive and affective BCIs, identifying a student's emotional (such as feelings and moods) and cognitive (such as learning and memory) states can facilitate communication between students and teachers [33].

Numerous studies suggest that there is a connection between emotional states and the central nervous system's electrical activity. The electrical signals from the brain can be monitored for variations, locations, and functional correlations using EEG equipment [34, 35]. EEG signals have high temporal resolution and can be used to directly assess neural activity, which forms the fundamentals of cognitive sensing. These signals offer reliable information, since an emotional state cannot be duplicated by altering or replicating it. The challenge is to interpret this information and relate it to emotions.

Some researchers examining EEG-based functional relationships in the brain have observed a link between specific brain areas and emotional reactions. Asymmetric activity in the alpha band at the frontal site has been linked to emotion in studies that use single-electrode analyses. Ekman and Davidson discovered that enjoyment activated the left frontal regions of the brain [36]. Another study found that when individuals expressed fear, their left frontal activity decreased [37]. Greater power in the theta bands of the frontal midline has been linked to pleasant emotions, while the opposite has been linked to unpleasant sentiments [38].

11.4 General architecture

Figure 11.2 presents a general architecture of an EEG-based BCI structure intended to recognize cognition states.

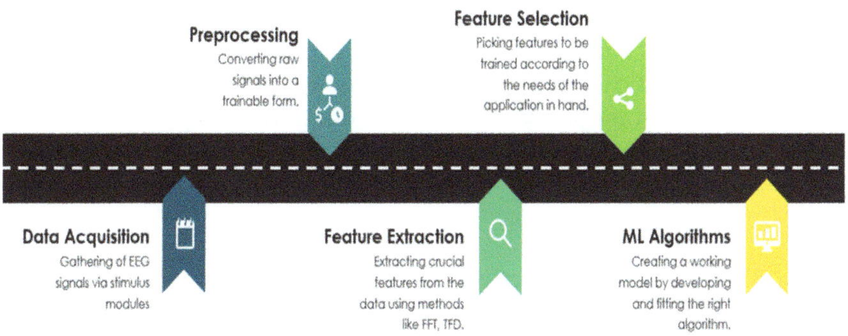

Figure 11.2. Components of a general application that uses cognitive sensing in the BCI.

11.4.1 Data acquisition

EEG helmets and headgear with noninvasive electrodes placed in parallel with the scalp can successfully gather EEG signals. Clinically, an EEG is described as an electrical signal that reflects brain behavior over time. Therefore, electrodes record, augment, and direct signals to a computer for processing and storage. There are several affordable EEG-based BCI products on the market at the moment [39]. However, many existing BCI models based on EEG are unpleasant in extended use. Thus, they still require improvement.

11.4.2 Preprocessing

Butterworth, Chebyshev, and inverse Chebyshev screens are favored for EEG signal filtering. All of these have distinct traits that must be researched. A Butterworth gauze has a uniform passband and stopband response, as well as a broad transformation zone. The Chebyshev filter is monotonic because it has a disturbance in the passband and a sharper crossover in the stopband. The last of the three has a smooth passband reaction, a modest transition, and a stopband ripple. A Butterworth zero-phase low-pass filter should be applied to avoid phase shifts. This screen passes the signal back and forth, which averts this problem [34, 40].

One more goal of this step is to remove disturbance that could be caused by shallow-frequency signals supplied through an exterior source, for instance power line intrusion [41]. Notch filters are utilized to prevent the transmission of a single frequency rather than a span of frequencies. They are intended to remove frequencies generated by electronic meshes, and normally vary between 50 and 60 Hz, depending on the frequency of the line power in the specific country.

11.4.3 Feature extraction

Signal noise that may occur when a signal is collected has a detrimental effect on the valuable features in the original signal. Muscular activity, for example, blinking of the eyes while the signal is being collected, and line power electromagnetic interference are the primary contributors of artifacts [42]. The feature collection strategy, which is supposed to distinguish a feature vector from a regular vector, is thus a critical phase in the procedure. A feature is a discrete or distinguishing calculation or structural component retrieved from a pattern segment [43]. Statistical attributes and grammatical descriptions are the two major categories of the traditional feature extraction modality. A feature extraction strategy's goal is to identify the most important features or information for a classification exercise [44–46].

11.4.4 Feature selection

The procedure of selecting features is critical, since it determines which features of the signal best represent the EEG characteristics to be categorized. The feature vector in BCI systems is typically of high dimensionality [47]. The number of input variables used for the classifier is reduced through feature selection (not to be confused with dimensional reduction). While both procedures reduce the properties of the data, the latter combines features to minimize their number.

11.4.5 Motor imagery algorithms

This step marks the final process of the architecture; it classifies the input data into the desired quantities and enables the researcher to dive deeper.

In EEG signal classification, there are a couple of selection procedures for the classifier that work well in specified situations [48]. The initial method determines the optimal classifier for a particular BCI device, whereas the second determines a classifier for a particular collection of attributes.

The ability to automatically classify EEG data is a major step toward making the use of EEG data more practical in various applications and less reliant on knowledgeable people. It is worth mentioning that, despite major progress in traditional BCI systems over the last two decades, research still faces significant challenges in the area of EEG categorization. Various biological and environmental disruptions to EEGs, their low signal-to-noise ratios (SNRs), and a reliance on human ability to extract critical features are among the challenges. Furthermore, much, if not all, of today's ML research is based on static data, making it incapable of identifying quickly changing brain signals [49, 50].

The recent availability of massive amounts of EEG data has resulted in the use of neural-network-based architectures, specifically, those used to extract pertinent information from signals that were previously impossible to obtain using standard approaches, and has demonstrated success in overcoming the aforementioned obstacles. Deep learning (DL) is simply a type of ML based on artificial neural networks that employs numerous layers of processing to extract progressively higher-level features from data [51].

For nearly all key EEG-based BCI systems, including the P300, SSVEP (steady-state visual evoked potential reflects the activation of cortical object representations), SCP (slow cortical potential), MI, and passive BCI, DL algorithms have been examined (for emotions and workload detection). This study made good use of various convolutional DL techniques which are described below.

A convolutional neural network (CNN) is a type of neural network design that specializes in dimensional data exploration. A CNN has at least one convoluted layer, which uses a convolution operator to map inputs to outputs [51, 52]. CNNs are adopted in BCI research to capture the distinct interdependences between the patterns associated with various brain impulses [50].

Deep belief networks (DBNs) are probabilistic generative models made up of a series of restricted Boltzmann machine (RBM) designs. In a DBN, the top two levels are connected by bidirectional connections, whereas the lower levels are attached via unidirectional connections. The RBM is made up of two layers: visible and hidden, with undirectional connecting lines between them [53]. Several studies have been conducted to investigate MI classification using DBNs [54–57]. In [56], a novel RBM-based DL strategy for EEG MI classification was suggested, in which a fast Fourier transform and wavelet package decomposition were used to train three

RBMs which were later augmented with an additional output layer. The authors of [55] suggested a DBN-based EEG MI data identification algorithm. The study's findings revealed that the identification rate of EEG MI data achieved by a DBN was higher than that of a typical support vector machine (SVM) model.

A recurrent neural network (RNN) model is a sophisticated DL classification algorithm designed exclusively for data containing elements ordered in sequences.

The authors of [58] suggested a unique cascaded RNN architecture based on long short-term memory (LSTM) blocks for the automatic recording of sleep phases using EEG, which achieved an average accuracy rate of ~87%. The authors of [59] also proposed an RNN based on an attention mechanism and bidirectional LSTM for the same purpose. This architecture outperformed the CNN model, however it required a longer training time. In [51], a unique DL framework called a spatiotemporal RNN (STRNN) was presented, in which dimensional and temporal details were combined for feature learning. The authors stated that the STRNN-based experimental findings were better than SOTA approaches for emotion identification.

An autoencoder (AE) is a DL approach for unsupervised learning that uses data coding and decoding. The input samples are frequently mapped to a lower dimensional feature space by a valuable feature representation during the encoding step [60]. Several significant AE architectures are used in the domain of EEG signal processing, including variational autoencoders (VAEs), stacked autoencoders (SAEs) and generative adversarial networks (GANs) [61–63].

11.5 Comparative analysis

BCIs have recently become very popular in applications involving human–machine interactions [64]. BCI systems carry out their functions by examining electrophysiological reaction (EPR) and EEG data related to human motion, expression, and mental state. EEG signals are frequently used to measure a variety of brain activity. ERPs appear at a particular time after a particular internal or external event has occurred [1]. They happen when a person is exposed to a mental or sensory experience or when a stimulus that usually occurs is missing. ERPs are used in order to ascertain how the subject's brain responds to various stimuli. Various BCI-related tasks can benefit from ERPs [66]. The classification of mental states, mental tasks, human emotions, and MI in BCIs is made easier by the properties of EEG and ERP signals.

A type of self-contained EEG, the MI-EEG, does not depend on any outside stimuli. The MI-oriented BCI technique instructs the subject to picture himself/herself moving various body parts in order to set off neuronal activity in the specific brain regions associated with the movements [67]. Event-related synchronization and desynchronization (ERS/ERD) patterns typically occur simultaneously in EEGs during the MI procedure. The ERD/ERS patterns produced by imagined or projected movements resemble those produced by actual movements, although they may differ depending on the scenario. Every subject therefore requires extensive instruction.

A crucial component of the BCI system is the decoding of MI, which converts the subject's intentions into orders that other devices may carry out. To make matters more difficult, the poor SNR of EEG recordings makes it difficult to accurately decode recorded EEGs of neural activity. Traditional methods for extracting distinguishing features mainly focus on the energy characteristics of the EEG, largely ignoring the need to further analyze temporal data in order to improve the efficacy of MI interpretation. Different types of mental action in mental-state-oriented BCIs generate various EEG signals and trigger different areas of the cortex [68]. These signals help to categorize mental tasks for the implementation of a BCI system. Diverse emotions are recognized by analyzing EEG signals in emotion-recognition-focused BCIs [69, 70]. However, BCI systems have significant difficulties with the effective feature extraction and precise classification of ERP and EEG signals connected with mental states, mental tasks, MI, and emotions. EEG-enabled BCI processes have low SNRs, poor spatial resolution, and are inherently dynamic. These limitations become much more severe as a result of users' changing mental states, eye movements, muscle motions, and varying electrode impedances. Analyses of brain dynamics and the classification of diverse signals associated with BCI are further complicated by all these factors. In these situations, ML algorithms offer very promising and effective methods for carrying out BCI-related activities. Even choosing the appropriate stage of BCI medicine for stroke rehabilitation benefits from the use of ML techniques [71].

ML is an advantageous and supportive technique for signal feature extraction and classification in BCI-related tasks due to its effective feature learning and complicated pattern recognition properties [72, 73]. Table 11.1 contrasts various feature extraction, selection, and classification techniques based on the year of publication, performance indicator, degree of accuracy, categories of BCI task, merit, and future direction [74].

11.5.1 Related work

Research into the application of DL frameworks to biological data has garnered a lot of attention during the last ten years. As a result of observing this progress and its use in daily life, we have examined related research methodologies.

Prior to the introduction of DL, regression and multiclass SVM models were introduced. The probabilistic SVM multiclass model described in [82] had an accuracy of 87.2% vs 85.4% for the regular SVM model. Confidence aggregation accuracy after careful adjustment and estimation was 91.2%. There are many significant uses for mental-state monitoring in many different domains, and [83] describes classification techniques that work well for BCI applications in a concise manner.

Powerful models such as *k*-nearest neighbors (KNN), whose classification accuracy was 83.3% when the dimensions were correlated using a linear regression classifier, were suggested by [84] for four EEG bands with nonlinear properties. This study also included the crucial EEG analysis stage for motor imagery [85]. The left and right hemispheres of a hybrid model comprising CNNs and LSTM were

Table 11.1. Comparative analysis of EEG signal categorization and MI classification.

Task	Classification Algorithm	Accuracy	Demerits/Scope	Reference
EEG signal categorization	Linear regression	95%.	Mathematically complex	[75]
	SVM	90.5%	Fewer people participated in the survey	[76]
	CNN	80.4%	A smaller amount of data, used data augmentation	[77]
	Adaboost extreme learning machine	79.5%	It is important to evaluate how well the proposed method performs in each category of mental task for the creation of a BCI.	[78]
MI classification	SVM	95%	Characteristics such as cortical potential should also be integrated	[79]
	Weighted naive Bayes	85.2%	Large time complexity	[80]
	Kernel extreme learning machine	96.5%	Experienced some data loss as a result of the energy function.	[81]

accurate to 99.12% and 97.66%, respectively. The authors of [86] estimated mental effort using ML and DL models. The accuracies for a CNN+LSTM network were 57% and 58%, while 61.08% was the best accuracy for the LSTM classifier. The accuracy of a KNN classifier was 57.3%. Recent advances in DL systems have made it possible to conduct research [87] that transforms EEG time series into multispectral images that contain spatial information. A recurrent hybrid network accurately predicted four levels of cognitive memory with a 92.5% reduction in classification errors compared to other methods. In [88], typical ML models such as SVM and logistic regression were contrasted with deep models. Improved neural networks and LSTM had accuracy rates of 78.9% and 71.3%, respectively. A DL model was developed by Das Chakladar *et al* in [89], which was based on the traditional processing of EEG signals. A system that uses deep neural networks and the gray wolf optimizer yielded accuracy scores of 86.33 for No Task and 82.54 for Multitask.

To widen the scope of application of cognitive workload evaluation, Gao *et al* [90] developed a spatiotemporal ESTCNN (EEG-based spatial temporal CNN) model for the assessment of driver fatigue in the context of trip safety. A core block with the advantage of temporal dependency extraction and a dense block integrated with that core block to enhance spatiotemporal information were utilized to combat the undervalued important information produced by electrode correlation. Wu *et al* provided a deep sparse contractive autoencoder (DCSAEN) model in [91] that assessed the mental workload of pilots. In all three trials, this model achieved a maximum accuracy of 83%. In order to predict driver fatigue, Budak *et al* [92] constructed an LSTM model that examined each attribute while also accounting for EEG signal processing. Lim *et al* [93] have made 48 participant readings and a raw EEG dataset available. A support vector regression model with a 66.7% accuracy rate was also presented by the authors of [94].

11.6 Exciting research in the brain–computer interface field

As the authors of [95–97] introduced subject matter about the progress and prospects of BCI systems, they stated that 'severe neuromuscular illnesses such as amyotrophic lateral sclerosis (ALS), brainstem stroke, cerebral palsy, muscular dystrophy, multiple sclerosis, and Guillain–Barre syndrome afflict millions of individuals worldwide.'

BCI systems have only recently begun to make major assistive communication technologies available to those who do not have other efficient means of communicating in their home contexts. As BCI R&D progresses and new individuals learn about its possibilities, BCIs may soon offer more specialized help to a bigger and more diverse user base (table 11.2).

The authors of [98] provided a scoping assessment of recent research in the area in order to understand the complexities of BCI usage. They aimed to illustrate the variety of approaches and issues from many perspectives and analyze the communal and human impact of BCI technology by looking at studies involving BCIs that use social research methods. In order to locate pragmatic research on BCIs, six databases from the social sciences, psychology, and medicine were thoroughly examined. According to the findings, whereas technical BCI issues like usability and feasibility are being researched in great detail, relatively little in-depth study has been done on the users' self-perception and self-experience. Additionally, the perspective of the caregiver is generally not focused on or examined, which is a factor that needs to be worked upon.

The authors of [99] proposed and evaluated a unique method to connect noninvasive BCI with a robotic operating system (ROS) through which a telepresence robot could be driven. They studied innovative ROS-based algorithms for navigation and obstacle escape to make the system safer and dependable and to make it easier to operate the robot via BCI. In this way, the robot made use of two maps of the environment—one for localization and the other for exploration—and used both to provide the BCI user with additional visual feedback which could be used to manipulate the robot's position. This was a preliminary attempt to bring together BCI and ROS, which can help advance the field of controlling BCI-driven robotic systems in the future.

As the authors of [100] discussed, the profound understanding of brain systems that cognitive and emotional BCIs can provide may enhance learning techniques, thus allowing cognitive abilities to be improved. Such an understanding can provide a stronger practical basis for teaching–learning pedagogies, such as adapting learning content depending on brain workload and gauging the level of student

Table 11.2. BCI research in modern areas.

Subject	Objective	Methodology	Result	Reference
Healthcare	Restoring mobility in people suffering from motor diseases	EEG and magnetoencephalography (MEG) were used to record all electrical and magnetic activities of the brain.	Motor ability was visibly improved.	[115]
Robotic arm	Detecting upper-limb imagery and controlling the robot arm	BMI module acquired EEG signals and used OPAL motion sensors to record movement.	Modules in the proposed system enabled the robot arm to execute high-dimensional tasks.	[116]
Gaming	To ultimately employ brain signals for control in virtual reality (VR) environments	The sample EEG signals were preprocessed and calculated over a set of ideal classifiers to determine the effectiveness of the classifier.	Findings were obtained by giving a BCI user multimedia feedback.	[117]
Augmented reality	Integrating an operational prototype of a mobile robot to control it in augmented reality (AR) using a BCI	Optical see-through head-mounted displays (OST-HMDs) were used to capture 5D data related to the visual layout of the BCI menu, namely: orientation, frame of reference, anchorage, size, and explicitness.	Combined the BCI configuration with the AR through EEG data, AR headset, and steady-state visually evoked potential brain pattern.	[118]
Haptic sensation	Creating an environment and a technology capable of acquiring brain signals and producing natural touch and texture sensations	EEG signals were acquired via a BrainVision recorder, a control device for tactile sensation was developed, and data analysis was performed at the end.	The subjects were able to use touch imagery for haptic analysis of varied tactile data.	[119]

engagement, thereby helping students to concentrate on a certain subject or topic [100]. Most BCI devices were created to help physically challenged individuals to regain their capacity to move and to compensate for lost or diminished motor abilities [101, 102]. However, BCI uses are not limited to the field of medicine [103]. Finally, the findings demonstrated that there is the potential to use BCI in education, as evidenced by numerous measures of the elements that influence effective learning.

As the authors of [1] discussed, the BCI is a successful outcome in the field of human–computer synergy research. The BCI involves direct brain-to-external-device communication that enhances, supports, and improves human cognitive functions. This paper discussed the progress of BCI systems and also the functioning of the human brain, through which we acquire data and signals. The architecture of BCI consists of three major sections, viz. signal acquisition, processing, and an elector device, which are discussed thoroughly. Additional BCI applications are made possible by new interaction paradigms, which also give rise to new research areas such as neural imaging for computational user experiences. It can be concluded that BCIs are becoming a more feasible, efficacious substitute for assistive technology.

As the authors of [27] have described, the challenges of domains such as data collection and IoT have motivated researchers, and intensive research is being carried out in those areas. CS is the utilization of smart sensors to obtain inputs from the surroundings in an intelligent manner. Additionally, the use of CS has been suggested for the future IoT in order to promote intelligent, reliable, and energy-coherent data-gathering procedures. Furthermore, this paper describes some cutting-edge strategies, potentials, and difficulties of AI methods for the chosen solutions. The survey shown in table 11.2 helps to clarify the AI methods that will be used for CS in the forthcoming IoT as well as the potential prospects for related research.

11.7 Challenges and future scope

It is unquestionably anticipated that BCI tech will be introduced into the market very soon; however, several significant drawbacks and challenges are present in each part of the BCI framework, and the BCI ecosystem needs to address these problems in order to advance the field's progression.

For example, in the case of EEG BCIs, the brain signals used to control the effectors are heavily influenced by an individual's capacity to form high-grade mental representations of movement, which dictates how readily identifiable the pertinent motor signals are [104, 105]. Visual motor imaging, which requires watching oneself perform a movement, generates less substantial signals recorded from the scalp than kinesthetic motor representation, which entails accurately experiencing the sense of movement [106]. Kinesthetic motor imagery generates more discernible brain signals. However, around half of all individuals find kinesthetic motor imagery challenging to complete [107].

It is difficult to champion the clinical applications of BCIs when the specific mechanisms underlying the functional advantages are unknown. To make credible

Table 11.3. The challenges of various aspects of BCI.

BCI paradigm	Challenge
EEG modality	MI, SSVEP, and P300 continuously face signal processing problems due to the nonlinear and dynamic nature of EEG signals.
Headset	Requires gel-based electrodes which are uncomfortable for the user. Dry electrodes produce more noise [108]. High ownership cost
Lack of an ideal data exploration strategy	Development of an algorithm for a concrete application with improved time complexity and accuracy for effective artifact attenuation.
Experimental environment	Current BCI research is extensively performed in controlled lab conditions. Multiple sensory impulses encountered in the outside world, such as sounds, motions, and odors, may impact the condition of EEG signals away from the lab. Thus, it is crucial to consider the environmental aspects in the design of the system.

forecasts of whether a patient will improve due to BCI training, clinicians must first understand which aspects of neural function are being addressed. The broad selection of BCI techniques available hampers attempts to understand mechanisms better, because various strategies are predicted to focus on various regions of the cerebral circuitry to yield the desired advantages. The challenges of BCI are summarized in table 11.3.

A big concern is making BCI tasks more appealing to users by providing a motivational response in a way that the user considers beneficial [109, 110]. Both hardware and software should be simple for patients and caregivers to use, which may improve enthusiasm for the technology [111]. In order to allow the user to easily identify the transition through stages as performance increases, tasks should avoid becoming fixed and repetitive [112]. The CRED-nf framework, which is used for reporting findings and making suggestions for future design studies, was developed in an effort to encourage scientific rigor and reproducibility in neurofeedback research [113]. Through the provision of some measurement standardization and the assurance of adequate experimental controls, it is intended that these coordinated efforts will increase our understanding of BCI mechanisms [114].

11.8 Conclusions

The BCI enables direct communication between the brain and other tools and technologies. Experiments with brain-signal activities fall under the purview of the BCI domain. BCI-based products are now accessible to researchers in a variety of fields thanks to extensive documentation, cheaper amplifiers, higher temporal exactness, and superior signal analysis methodologies. This chapter conducted a full review analysis of CS-based BCI, specifically to evaluate its methodological advantages and shortcomings along with the fundamental offerings necessary in this field. It presented a conceptual and architectural analysis of BCI. It was designed for individuals interested in learning about the current state of BCI systems and techniques. The fundamental concepts of BCI techniques were thoroughly studied. This chapter included a description of the fundamental BCI architectures, feature extraction, selection, categorization, assessment procedures, and strategies. It also explained the latest challenges and future scope of this technology.

The joint efforts of researchers and technology businesses are required to commercialize this vast realm for the benefit of ordinary people.

References

[1] Ramadan R A, Refat S, Elshahed M A and Ali R A 2015 Basics of brain computer interface *Brain–Computer Interfaces (Intelligent Systems Reference Library* vol 74*)* ed A Hassanien and A Azar (Cham: Springer)

[2] Sonam Y S 2015 A review paper on brain computer interface *Int. J. Eng. Res. Technol. IJERT) Ncetems* **3** 2015 IJERTCONV3IS10102 https://www.ijert.org/a-review-paper-on-brain-computer-interface

[3] Sadiq M T, Yu X, Yuan Z, Aziz M Z, Siuly S and Ding W 2021 Toward the development of versatile brain–computer interfaces *IEEE Transactions on Artificial Intelligence* **2** 314–28

[4] Hoffmann U, Vesin J-M and Ebrahimi T 2007 Recent advances in brain–computer interfaces *IEEE 9th Workshop on Multimedia Signal Processing* p 17–17

[5] Mridha M F, Das S C, Kabir M M, Lima A A, Islam M R and Watanobe Y 2021 Brain–computer interface: advancement and challenges *Sensors* **21** 5746

[6] Sellers E W, Vaughan T M and Wolpaw J R 2010 A brain–computer interface for long-term independent home use *Amyotrophic Lateral Sclerosis* **11** 449–55

[7] Chamola V, Vineet A, Nayyar A and Hossain E 2020 Brain–computer interface-based humanoid control: a review *Sensors* **20** 3620

[8] Moore M M 2003 Real-world applications for brain–computer interface technology *IEEE Transactions on Neural Systems and Rehabilitation Engineering* **11** 162–65

[9] Bos D P *et al* 2010 Human–computer interaction for BCI games: usability and user experience *2010 Int. Conf. on Cyberworlds* pp 277–81

[10] Mudgal S, Sharma S, Chaturvedi J and Sharma A 2020 Brain computer interface advancement in neurosciences: applications and issues *Interdisciplinary Neurosurgery.* **20** 100694

[11] Serov A 2017 Cognitive sensor technology for structural health monitoring *Procedia Structural Integrity* **5** 1160–67

[12] Weiser M 1994 Ubiquitous computing *CSC94: 22nd Annual ACM Computer Science Conf.* p 418

[13] Picard R W 2000 *Affective Computing* (Cambridge, MA: MIT Press)

[14] Piho L and Tjahjadi T 2018 A mutual information based adaptive windowing of informative EEG for emotion recognition *IEEE Trans. Affect. Comput.* **11** 722–35

[15] Song T, Zheng W, Song P and Cui Z 2018 EEG emotion recognition using dynamical graph convolutional neural networks *IEEE Trans. Affect. Comput.* **11** 532–41

[16] Huang H, Xie Q, Pan J, He Y, Wen Z and Yu R *et al* 2019 An EEGbased brain computer interface for emotion rECOGnition and its application in patients with disorder of consciousness *IEEE Trans. Affect. Comput.* **12** 832-42

[17] Buzsáki G, Anastassiou C A and Koch C 2012 The origin of extracellular fields and currents – EEG, ECoG, LFP and spikes *Nat. Rev. Neurosci.* **13** 407–20 [Medline: 22595786]

[18] Pesaran B, Vinck M, Einevoll G T, Sirota A, Fries P and Siegel M *et al* 2018 Investigating large-scale brain dynamics using field potential recordings: analysis and interpretation *Nat. Neurosci.* **21** 903–19 [Medline: 29942039]

[19] Kaiju T, Doi K, Yokota M, Watanabe K, Inoue M and Ando H *et al* 2017 High spatiotemporal resolution ECoG recording of somatosensory evoked potentials with flexible micro-electrode arrays *Front Neural Circuits* **11** 20 [Medline: 28442997]

[20] Yuste R 2015 From the neuron doctrine to neural networks *Nat. Rev. Neurosci.* **16** 487–97 [Medline: 26152865]

[21] Musk E 2019 Neuralink an integrated brain–machine interface platform with thousands of channels *J. Med. Internet Res.* **21** e16194

[22] Hong G and Lieber C M 2019 Novel electrode technologies for neural recordings *Nat. Rev. Neurosci.* **20** 330–45 [Medline: 30833706]

[23] O'Doherty J E, Lebedev M A, Ifft P J, Zhuang K Z, Shokur S and Bleuler H *et al* 2011 Active tactile exploration using a brain–machine–brain interface *Nature* **479** 228–31

[24] Sharlene N F 2019 Restored tactile sensation improves neuroprosthetic arm control? *Science* **372** 831–6

[25] Chung J E, Joo H R, Fan J L, Liu D F, Barnett A H and Chen S *et al* 2019 High-density, long-lasting, and multi-region electrophysiological recordings using polymer electrode arrays *Neuron* **101** 21–31 [Medline: 30502044]

[26] Timothy L H 2019 The 'sewing machine' for minimally invasive neural recording https://doi.org/10.1101/578542

[27] Osifeko M O, Hancke G P and Abu-Mahfouz A M 2020 Artificial intelligence techniques for cognitive sensing in future IoT: state-of-the-art, potentials, and challenges *J. Sens. Actuator Netw.* **9** 21

[28] Zaheer K, Othman M, Rehmani M H and Perumal T 2018 A survey of decision-theoretic models for cognitive internet of things (CioT) *IEEE Access* **6** 22489–512

[29] Aijaz A and Aghvami A H 2015 Cognitive machine-to-machine communications for internet-of-things: a protocol stack perspective *IEEE Internet of Things J.* **2** 103–12

[30] Wang G 2018 DGCC: a case for integration of brain cognition and intelligence computation, *2018 IEEE Int. Conf. on Data Mining Workshops (ICDMW)* pp 478–9

[31] Yao Y 2011 Artificial intelligence perspectives on granular computing *Granular Computing and Intelligent Systems* (Berlin: Springer) ch 2 pp 17–34

[32] Skowron A, Jankowski A and Dutta S 2016 Interactive granular computing *Granul Computing* **1** 95–113

[33] Jamil N, Belkacem A N, Ouhbi S and Guger C 2021 Cognitive and affective brain–computer interfaces for improving learning strategies and enhancing student capabilities: a systematic literature review *IEEE Access* **9** 134122–47

[34] Torres P E P, Torres E A, Hernández-Álvarez M and Yoo S G 2020 EEG-based BCI emotion recognition: a survey *Sensors (Basel)* **20** 5083

[35] Salzman C D and Fusi S 2010 Emotion, cognition, and mental state representation in amygdala and prefrontal cortex *Annu. Rev. Neurosci.* **33** 173–202

[36] Ekman P and Davidson R J 1993 Voluntary smiling changes regional brain activity *Psychol. Sci.* **4** 342–5

[37] Bhatti A M, Majid M, Anwar S M and Khan B 2016 Human emotion recognition and analysis in response to audio music using brain signals *Comput. Human Behav.* **65** 267–75

[38] Lee Y Y and Hsieh S 2014 Classifying different emotional states by means of eegbased functional connectivity patterns *PLoS One* **9** e95415

[39] Peterson V, Galván C, Hernández H and Spies R 2020 A feasibility study of a complete low-cost consumer-grade brain–computer interface system *Heliyon.* **6** e03425
[40] Bustamante P A, Lopez Celani N M, Perez M E and Quintero Montoya O L 2015 Recognition and regionalization of emotions in the arousal-valence plane *2015 37th Annual Int. Conf. of the IEEE Engineering in Medicine and Biology Society (EMBC)* pp 6042–5
[41] Sanei S and Chambers J A 2013 *EEG Signal Processing* (Hoboken, NJ: Wiley)
[42] Kousarrizi M R N, Ghanbari A A, Teshnehlab M, Aliyari M and Gharaviri A 2009 Feature extraction and classification of EEG signals using wavelet transform, SVM and artificial neural networks for brain computer interfaces, *Proc. Int. Joint Conf. on Bioinformatics, Systems Biology and Intelligent Computing (IJCBS'09)* (Piscataway, NJ: IEEE) pp 352–5
[43] Cvetkovic D, Übeyli E D and Cosic I 2008 Wavelet transform feature extraction from human PPG, ECG, and EEG signal responses to ELF PEMF exposures: a pilot study, *Digital Signal Process.* **18** 861–74
[44] Kordylewski H, Graupe D and Liu K 2001 A novel large-memory neural network as an aid in medical diagnosis applications, *IEEE Trans. Inf. Technol. Biomed.* **5** 202–9
[45] Güler I and Übeyli E D 2005 Adaptive neuro-fuzzy inference system for classification of EEG signals using wavelet coefficients, *J. Neurosci. Methods* **148** 113–21
[46] Al-Fahoum A and Al-Fraihat A 2014 Methods of EEG signal features extraction using linear analysis in frequency and time-frequency domains *ISRN Neuroscience.* **2014** 730218
[47] Lotte F, Congedo M, Lécuyer A, Lamarche F, Arnaldi B, Anatole L, Lotte F, Congedo M, Anatole L and Abdulhay E *et al* 2018 A review of classification algorithms for EEG-based brain–computer interfaces to cite this version: a review of classification algorithms for EEG-based brain–computer interfaces *Hum. Brain Mapp.* **38** 270–8
[48] Li X, Fan H, Wang H and Wang L 2019 Common spatial patterns combined with phase synchronization information for classification of EEG signals *Biomed. Signal Process. Control* **52** 248–56
[49] Rashid M, Sulaiman N, Abdul Majeed A P P, Musa R M, Ab Nasir A F, Bari B S and Khatun S 2020 Current status, challenges, and possible solutions of EEG-based brain–computer interface: a comprehensive review *Front Neurorobot* **14** 25
[50] Lotte F, Bougrain L, Cichocki A, Clerc M, Congedo M and Rakotomamonjy A *et al* 2018 A review of classification algorithms for EEG-based brain–computer interfaces: a 10-year update *J. Neural Eng.* **15** 031005
[51] Zhang X, Yao L, Wang X, Monaghan J, Mcalpine D and Zhang Y 2019 A Survey on Deep Learning based Brain Computer Interface: Recent Advances and New Frontiers *J. Neural Eng.* **18** 031002
[52] Fan J, Ma C and Zhong Y 2019 A Selective Overview of Deep Learning arXiv:1904.05526
[53] Abbas Q, Ibrahim M E A and Jaffar M A 2019 A comprehensive review of recent advances on deep vision systems *Artif. Intell. Rev.* **52** 39–76
[54] An X, Kuang D, Guo X, Zhao Y and He L 2014 A deep learning method for classification of EEG data based on motor imagery *Intelligent Computing in Bioinformatics. ICIC 2014* (*Lecture Notes in Computer Science* vol 8590) ed D S Huang, K Han and M Gromiha (Cham: Springer) pp 203–10
[55] Tang X, Zhou J, Zhang N and Liu L 2015 Recognition of motor imagery EEG based on deep belief network *Inf. Control* **44** 717–21

[56] Lu N, Li T, Ren X and Miao H 2017 A deep learning scheme for motor imagery classification based on restricted boltzmann machines *IEEE Trans. Neural Syst. Rehabil. Eng.* **25** 566–76

[57] Ortega J, Ortiz A, Martín-Smith P, Gan J Q and González-Peñalver J 2017 Advances in computational intelligence *IWANN 2017* (*Lecture Notes in Computer Science* vol 10305) ed I Rojas, G Joya and A Catala (Cham: Springer) pp 28–39

[58] Michielli N, Acharya U R and Molinari F 2019 Cascaded LSTM recurrent neural network for automated sleep stage classification using single-channel EEG signals *Comput. Biol. Med.* **106** 71–81

[59] Wang Y and Wu D 2019 Deep learning for sleep stage classification *Proc. 2018 Chinese Automation Congress, CAC 2018 (Xi'an)* pp 3833–8

[60] Alom M Z, Taha T M, Yakopcic C, Westberg S, Sidike P and Nasrin M S *et al* 2019 A state-of-the-art survey on deep learning theory and architectures *Electronics* **8** 292

[61] Rundo F, Rinella S, Massimino S, Coco M, Fallica G and Parenti R *et al* 2019 An innovative deep learning algorithm for drowsiness detection from EEG signal *Computation* **7** 13

[62] Idowu O P, Fang P, Li X, Xia Z, Xiong J and Li G 2018 Towards control of EEG-based robotic arm using deep learning via stacked sparse autoencoder *2018 IEEE Int. Conf. on Robotics and Biomimetics (ROBIO)* (Kuala Lumpur: IEEE) pp 1053–7

[63] Nair A V, Kumar K M and Mathew J 2018 An Improved Approach for EEG Signal Classification using Autoencoder *2018 8th Int. Symp. on Embedded Computing and System Design (ISED)* (Piscataway, NJ: IEEE)

[64] Fouad M M, Amin K M, El-Bendary N and Hassanien A E 2015 Brain computer interface: a review *Brain-Computer Interfaces* (*Intelligent Systems Reference Library* vol 74) (Berlin: Springer) pp 3–30

[66] Combaz A, Chumerin N, Manyakov N V, Robben A, Suykens J A and Van Hulle M M 2012 Towards the detection of error-related potentials and its integration in the context of a P300 speller brain–computer interface *Neurocomputing* **80** 73–82

[67] Padfield N, Zabalza J, Zhao H, Masero V and Ren J 2019 EEG-based brain–computer interfaces using motor-imagery: techniques and challenges *Sensors* **19** 1423

[68] Blankertz B, Tangermann M, Vidaurre C, Dickhaus T, Sannelli C, Popescu F and Müller K R 2009 Detecting mental states by machine learning techniques: the Berlin brain–computer interface *Brain–Computer Interfaces* (Berlin: Springer) pp 113–35

[69] Torres E P, Torres E A, Hernández-Álvarez M and Yoo S G 2020 EEG-based BCI emotion recognition: a survey *Sensors* **20** 5083

[70] Jiang X, Gu X, Xu K, Ren H and Chen W 2019 Independent decision path fusion for bimodal asynchronous brain–computer interface to discriminate multiclass mental states *IEEE Access* **7** 165303–17

[71] Mohanty R, Sinha A M, Remsik A B, Dodd K C, Young B M, Jacobson T and Prabhakaran V 2018 Machine learning classification to identify the stage of brain–computer interface therapy for stroke rehabilitation using functional connectivity *Front. Neurosci.* **12** 353

[72] Bablani A, Edla D R, Tripathi D and Cheruku R 2019 Survey on brain–computer interface: an emerging computational intelligence paradigm *ACM Comput. Surv.* **52** 20

[73] Chakladar D D and Chakraborty S 2019 Feature extraction and classification in brain–computer interfacing: future research issues and challenges *Natural Computing for Unsupervised Learning* (Cham: Springer) pp 101–31

[74] Rasheed S 2021 A review of the role of machine learning techniques towards brain–computer interface applications. *Mach. Learn. Knowl. Extr.* **3** 835–62
[75] Yi H 2022 Efficient machine learning algorithm for electroencephalogram modeling in brain–computer interfaces *Neural Comput. Appl.* **34** 9233–43
[76] Combaz A, Chumerin N, Manyakov N V, Robben A, Suykens J A K and Van Hulle M M 2012 Towards the detection of error-related potentials and its integration in the context of a P300 speller brain–computer interface *Neurocomputing* **80** 73–82
[77] Majidov I and Whangbo T 2019 Efficient classification of motor imagery electroencephalography signals using deep learning methods *Sensors* **19** 1736
[78] Gao L, Cheng W, Zhang J and Wang J 2016 EEG classification for motor imagery and resting state in BCI applications using multi-class adaboost extreme learning machine *Rev. Sci. Instrum.* **87** 085110
[79] Suk H I and Lee S W 2013 A novel bayesian framework for discriminative feature extraction in brain–computer interfaces *IEEE Trans. Pattern Anal. Mach. Intell.* **35** 286–99
[80] Miao M, Zeng H, Wang A, Zhao C and Liu F 2017 Discriminative spatial-frequency-temporal feature extraction and classification of motor imagery EEG: an sparse regression and weighted Naïve bayesian classifier-based approach *J. Neurosci. Methods* **278** 13–24
[81] Venkatachalam K, Devipriya A, Maniraj J, Sivaram M, Ambikapathy A and Amiri I S 2020 A novel method of motor imagery classification using eeg signal *Artif. Intell. Med.* **103** 101787
[82] Kai-Quan S, Xiao-Ping L, Chong-Ji O, Shi-Yun S and Einar P V W-S 2008 EEG-based mental fatigue measurement using multi-class support vector machines with confidence estimate *Clinical Neurophysiology* **119** 1524–33
[83] Klaus-Robert M, Michael T, Dornhege G, Matthias K, Curio G and Benjamin B 2008 Machine learning for real-time single-trial EEG-analysis: from brain–computer interfacing to mental state monitoring *J. Neurosci. Methods* **167** 82–90
[84] Behshad H, Mohammad H M and Rostami R 2013 Classifying depression patients and normal subjects using machine learning techniques and nonlinear features from EEG signal, *Comput. Methods Programs Biomed.* **109** 339–45
[85] Kumar S, Sharma A, Mamun K and Tsunoda T 2016 A deep learning approach for motor imagery EEG signal classification *2016 3rd Asia-Pacific World Congress on Computer Science and Engineering (APWC on CSE), Nadi* pp 34–9
[86] Pandey V, Choudhary D K, Verma V, Sharma G, Singh R and Chandra S 2020 Mental workload estimation using EEG *2020 5th Int. Conf. on Research in Computational Intelligence and Communication Networks (ICRCICN), Bangalore, India* (Piscataway, NJ: IEEE) pp 83–6
[87] Saha A, Minz V, Bonela S, Sreeja S R, Chowdhury R and Samanta D 2018 Classification of EEG signals for cognitive load estimation using deep learning architectures *Intelligent Human Computer Interaction. IHCI 2018* (*Lecture Notes in Computer Science* vol 11278) (Berlin: Springer)
[88] Nagabushanam P, Thomas George S and Radha S 2019 EEG signal classification using LSTM and improved neural network algorithm *Soft Computing* **24** 9981–10003
[89] Das Chakladar D, Dey S, Roy P and Dogra D 2020 EEG-based mental work load estimation using deep BLSTM-LSTM network and evolutionary algorithm *Biomed. Signal Process. Control* **60** 101989
[90] Gao Z, Wang X, Yang Y, Mu C, Cai Q, Dang W and Zuo S 2019 EEG-based spatio–temporal convolutional neural network for driver fatigue evaluation *IEEE Trans Neural Netw. Learn. Syst.* **30** 2755–63

[91] Wu E, Deng P, Qu X, Tang Z, Zhang W, Zhu L, Ren H, Zhou G and Sheng R 2020 Detecting fatigue status of pilots based on deep learning network using EEG signals *IEEE Transactions on Cognitive and Developmental Systems* **13** 575–85
[92] Budak U, Bajaj V, Akbulut Y, Atila O and Sengur A 2019 An effective hybrid model for EEG-based drowsiness detection *IEEE Sens. J.* **19** 7624–31
[93] Lim W, Sourina O and Wang L 2018 STEW: simultaneous task EEG workload data set *IEEE Trans. Neural Systems and Rehabilitation Engineering* **26** 2106–14
[94] Belsare S, Kale M, Ghayal P, Gogate A and Itkar S 2021 Performance comparison of different EEG analysis techniques based on deep learning approaches *2021 Int. Conf. on Emerging Smart Computing and Informatics (ESCI)* pp 490–93
[95] Allison B Z *et al* 2007 Brain–computer interface systems: progress and prospects *Expert Rev. Medical Devices* **44** 463–74
[96] Dobkin B H and Havton L A 2004 Basic advances and new avenues in therapy of spinal cord injury *Ann. Rev. Med.* **55** 255–82
[97] Tecce J J, Gips J, Olivieri C P and Pok L J 1998 Consiglio Mr Eye movement control of computer functions *Int. J. Psychophysiol.* **29** 319–25
[98] Kögel J, Schmid J R and Jox R J *et al* 2019 Using brain–computer interfaces: a scoping review of studies employing social research methods *BMC Med. Ethics* **20** 18
[99] Beraldo G, Antonello M, Cimolato A, Menegatti E and Tonin L 2018 Brain–computer interface meets ROS: a robotic approach to mentally drive telepresence robots *2018 IEEE Int. Conf. on Robotics and Automation (ICRA)* pp 4459–64
[100] Jamil N, Belkacem A N, Ouhbi S and Guger C 2021 Cognitive and affective brain–computer interfaces for improving learning strategies and enhancing student capabilities: a systematic literature review *IEEE Access* **9** 134122–47
[101] Caria A, Weber C, Brötz D, Ramos A, Ticini L F, Gharabaghi A, Braun C and Birbaumer N 2011 Chronic stroke recovery after combined BCI training and physiotherapy: a case report *Psychophysiology* **48** 578–82
[102] Chen C *et al* 2020 Neural activities classification of left and right finger gestures during motor execution and motor imagery *Brain–Comput. Interfaces* **7** 117–27
[103] Jamil N, Belkacem A N, Ouhbi S and Lakas A 2021 Noninvasive electroencephalography equipment for assistive, adaptive, and rehabilitative brain–computer interfaces: a systematic literature review *Sensors* **21** 4754
[104] Marchesotti S, Bassolino M, Serino A, Bleuler H and Blanke O 2016 Quantifying the role of motor imagery in brain–machine interfaces *Sci. Rep.* **6** 24076
[105] Chavarriaga R, Fried-Oken M, Kleih S, Lotte F and Scherer R 2017 Heading for new shores! overcoming pitfalls in BCI design *Brain Comput. Interfaces (Abingdon, England)* **4** 60–73
[106] Neuper C, Scherer R, Reiner M and Pfurtscheller G 2005 Imagery of motor actions: differential effects of kinesthetic and visual–motor mode of imagery in single-trial EEG *Brain Res. Cogn. Brain Res.* **25** 668–77
[107] Seiler B D, Newman-Norlund R D and Monsma E V 2017 Inter-individual neural differences in movement imagery abilities *Psychol. Sport Exercise* **30** 153–63
[108] Mihajlovi V and Peuscher J 2012 To what extent can dry and water-based EEG electrodes replace conductive gel ones? A steady state visual evoked potential brain–computer interface case study pp 14–26

[109] Kübler A, Holz E M, Riccio A, Zickler C, Kaufmann T and Kleih S C *et al* 2014 The user-centered design as novel perspective for evaluating the usability of BCI-controlled applications *PLoS One* **9** e112392
[110] Pillette L, Jeunet C, Mansencal B, Nkambou R, Nkaoua B and Lotte F 2017 PEANUT: personalised emotional agent for neurotechnology user-training, *7th Int. BCI Conf.* (Graz) https://hal.archives-ouvertes.fr/hal-01519480
[111] Käthner I, Halder S, Hintermüller C, Espinosa A, Guger C and Miralles F *et al* 2017 A multifunctional brain–computer interface intended for home use: an evaluation with healthy participants and potential end users with dry and gel-based electrodes *Front. Neurosci.* **11** 286
[112] Jeunet C, Jahanpour E and Lotte F 2016 Why standard brain–computer interface (BCI) training protocols should be changed: an experimental study *J. Neural Eng.* **13** 036024
[113] Ros T, Enriquez-Geppert S, Zotev V, Young K D, Wood G and Whitfield-Gabrieli S *et al* 2020 Consensus on the reporting and experimental design of clinical and cognitive-behavioural neurofeedback studies (CRED-nf checklist) *Brain* **143** 1674–85
[114] Simon C, Bolton D A E, Kennedy N C, Soekadar S R and Ruddy K L 2021 Jul 2 Challenges and opportunities for the future of brain–computer interface in neurorehabilitation *Front Neurosci.* **15** 699428
[115] Swaminathan R and Prasad S 2016 Brain computer interface used in Health Care Technologies Next Generation DNA Led Technologies (Singapore: Springer) pp 49–58
[116] Gantenbein J, Dittli J, Meyer J T, Gassert R and Lambercy O 2022 Intention detection strategies for robotic upper-limb orthoses: a scoping review considering usability, daily life application, and user evaluation *Front Neurorobot.* **21** 815693
[117] Coogan C G and He B 2018 Brain-computer interface control in a virtual reality environment and applications for the internet of things *IEEE Access* **6** 10840–9
[118] Si-Mohammed H, Petit J, Jeunet C, Argelaguet F, Spindler F, Evain A, Roussel N, Casiez G and Lécuyer A 2018 Towards BCI-based interfaces for augmented reality: feasibility, design and evaluation *IEEE Trans. Visualization and Computer Graphics* **26** 1608–21
[119] Kim M-K, Cho J-H and Shin H-B 2022 Recognition of tactile-related EEG signals generated by self-touch arXiv:2112.07123

www.ingramcontent.com/pod-product-compliance
Ingram Content Group UK Ltd.
Pitfield, Milton Keynes, MK11 3LW, UK
UKHW051859100926
8556IPUK00005B/691

9 780750 353243